# Special Thanks

I have to give very special thanks to my parents, whose support and encouragement even through tough times have made me who I am today. They not only gave me life but, more importantly, challenged me to get the most out of it!

A big thanks also goes to the designers and businesses that took time to shared their knowledge, their contributions have been amazing!

Author: Jonathan Cantin
Copy edited by: Noëlle Ji Sun Miller
Additional copy editing by: Albert Cantin



## Library and Archives Canada Cataloguing in Publication

Cantin, Jonathan, 1976-, author
CNCKing.com, rise of the CNC : ultimate CNC design course / Jonathan Cantin.

Includes bibliographical references and index.
ISBN 978-1-896369-51-8

1. Machine-tools--Numerical control. 2. Three-dimensional printing. 3. Computer-aided design. 4. Woodwork--Patterns.
I. Title.

TJ1189.C35 2013 321.9'023 C2013-905544-4

# Table of Contents

# Table of Contents



**Get more information about this book by scanning this code with your cell phone!**

QR2id.com

Serial No.
KGYV-JSFW-7H2Q

# Foreword by Benjamin J. Heckendorn

Ben and I share many striking similarities -- we are the same age, learned through trial and error until we mastered our skills in relative obscurity, absolutely love our CNC machines, and share our learnings with a worldwide audience both online and off. Where we deviate is in the work we produce, as he's an electronics guru who uses his skills to create amazing custom mods and projects that wouldn't be possible without CNC machines.

He's been written up in more magazines than I can list and has become a cultural phenomenon within the "hardware hacker" community, so I feel very honored that he agreed to write a short foreword to my book. He is introducing young and old to the wonders of the possibilities that are at your fingertips with CNC equipment!

Visit him online at BenHeck.com

---------

I first got started with CNC back when I was a graphic artist and did mechanical designs for signage. I understood vector graphics but was amazed to learn there was this magical world where those graphics could control machines and build the items in the real world! I immediately saw the potential for these machines and began building things outside of work that I could never have done by hand. Slowly but surely I gained a reputation, and now, 15 years later, I have my own shop and several machines of my own. I would not be where I am today were it not for computer numerical control.

The great thing about CNC/CAD CAM is that it provides a direct link from the creativity of our minds to physical objects -- a literal "if you can dream it, you can do it" situation. With it, we aren't limited by the resolution of our fingers, how steadily we can hold a tool, how well we can draw, or how fine of a detail we can see or measure. The CAD -- computer aided design -- lets us create our ideas with perfect accuracy in formats we can share, and the CAM -- computer aided manufacturing -- turns those designs into precise, real-world objects.

Safety is another benefit. Working with hand tools, especially at smaller, more accurate scales, increases the risks of error, slippage and injury. For example, holding a piece of wood steady while you cut it puts both of your hands in harm's way. You wouldn't think to put your hands near a 20,000 RPM router spindle on a CNC, but this is precisely what happens with hand tools. In this manner, CNC tools are like sending a robot to put out an oil well fire instead of walking up to it yourself with a bucket of water. Distance = safety.

Another great advantage is repeatability. Humans aren't very good at copying things precisely by hand and that's fine; we're very analog creatures. But repeatability is very important for production and interconnectivity of components. Again, CNC is a great solution, especially at smaller volumes. Large companies who need to make 3 million toothbrushes think nothing of getting some $30,000 injection molds made, but a small company could never justify that cost to create just one thing. For the same cost as a typical mold, you could invest in two or three CNC machines that will give you value and flexibility for years to come.

Yes, it will take longer to produce copies than with a mold or stamp. But they'll still be perfect copies, and you'll be able to make changes on the fly at no cost but your time and materials. Once your system is set up and running, and the parts are coming out nicely, it's almost as easy as the click of a mouse.

As an example, a friend of mine recently started a company to build small runs of pinball machines. We both own MakerBot Replicators, and they've been amazing. We have many parts (in pinball vernacular, "toys") that are made from ABS.

Rather than make expensive molds for each one, we simply run them on our 3D printers off the SD card, so the PC isn't even needed. I typically start prints when I get to work in the morning, and whenever I hear the "I'm finished!" tone, I pull the part off and start another one. This form of low quantity manufacturing would have been unaffordable to us a mere five years ago.

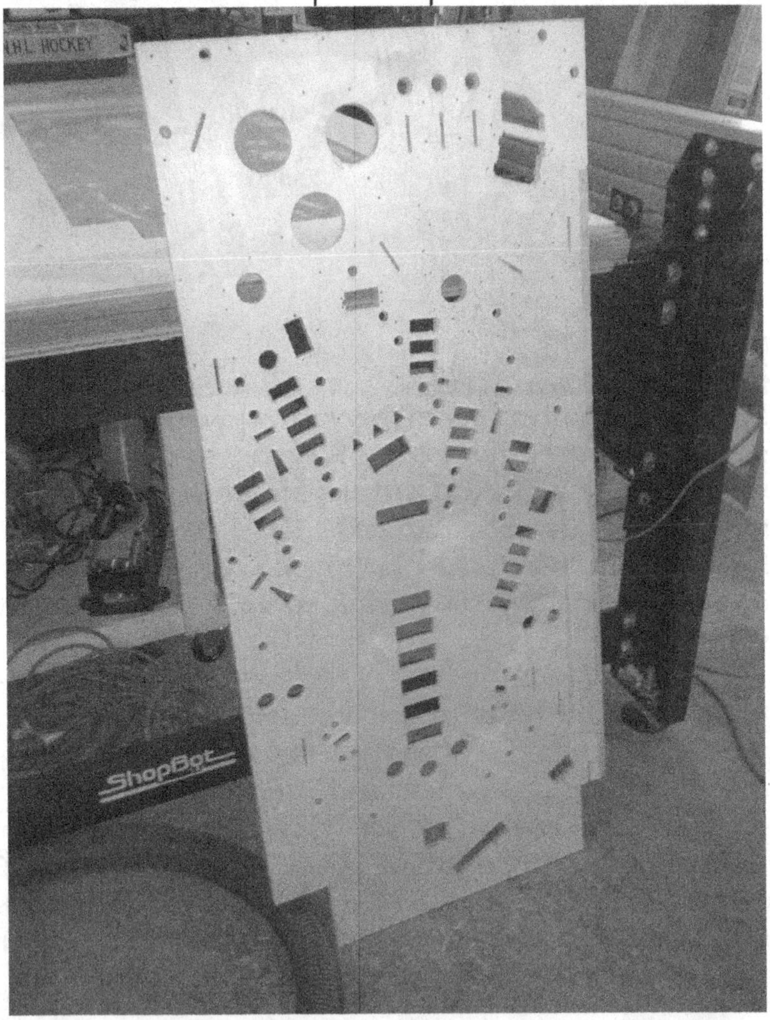


Laser cutters, often called laser engravers, are another great tool. I splurged a few years back on an American-made Epilog systems model and haven't regretted it for a minute.

As Jon mentions in this book, you can get lasers "off the boat" for a low price, but you're flying blind when it comes to setup and support. Being cheap can be very expensive, so I'd second the recommendation to get a laser with enough support so it's basically plug and play. Think of it this way -- are you buying a CNC for the "fun" of assembling it, or are you buying it to make other things?

My Epilog is so easy to use I've accidently printed my Gmails to it before.

Lasers are also one of the best tools for creating finished-looking, presentable parts. I like to build the structure of an item with my Shopbot mill or 3D printer, then "cap" it with laser cut engraving plastic or acrylic. If you're on the fence about getting one for a business, consider that it can do a lot of ancillary work that walk-in customers understand and want. Examples include plaques, trophies, small signage, etched glass, personalized pens and more. Architects can use them to create paper models of projects.

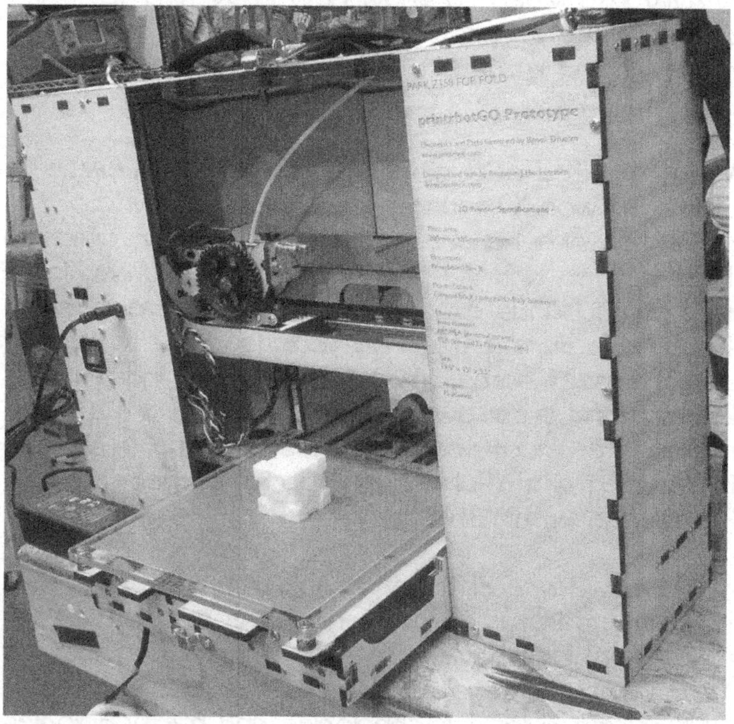


On the subject of paper, I cut a lot of paper patterns with my laser. It's a quick and cheap way to see if your design / template is correct before you cut the actual expensive material. I've even used it to prototype metal parts.

At the craft store I find a stock of paper the approximate thickness of the target metal, then I laser cut it and fold together to check that the parts work before sending the job off to the metal laser shop. Smaller CNCs can do practice runs for their big brothers.

I wish a book like this existed back when I first got started! Inside you'll find an amazing assortment of projects to make on CNC machines, information on how the machines differ, and the best processes for taking your designs from concept to reality. Once you're ready, you can order your parts from a number of online sources. Join a local hacker space / Techshop and use their equipment. Get plans online to construct your own CNC mill with a Dremel cutter. Or splurge on a 3D printer, which is safe, quiet and clean enough to use inside your home.

If you like building things, it's worth your time to learn all you can about CNC. It truly is the king of tools.

## Message from Jon

It has been two years since my last volume, and a lot has changed for me personally and professionally. I rebranded WoodMarvels.com into CNCKing.com to allow me to diversify across other CNC technologies and materials, and I got heavily into video to show how easy it is to design and build projects. Along the way I also got married and moved to Australia -- a lifelong dream fulfilled!

What started off as a site with only a few untested woodworking projects has grown to offer several hundred across CNC laser, plasma, table router and 3D printer and a customer base in the thousands around the world! I've been very fortunate during this five-year journey to be able to meet some amazing designers, industry leaders and shop owners who, like me, are passionate about the CNC field -- some of whom have been grateful enough to share their knowledge within this book!

I believe we are entering a golden period as consumers begin to awaken to the possibilities CNC machines have to offer. It's not as much a "new industrial revolution" as it is a "Rise of the CNC" using hybrid technologies and methodologies to make what was once difficult increasingly user-friendly.

These CNC technologies have a lot more in common than you may think. My hope is that after going through this book, the veil will be taken off and you'll realize the potential of this technology -- and the incredible opportunity that awaits a select few who learn to master these amazing tools!

Jonathan Cantin
Volume 1: Canada, 2: Korea, 3: China, 4: Australia!

Founder of CNCKing.com
The internet's largest source of tested and proven AI, EPS, CDR, DXF, STL, SLD, PDF files for CNC 3D Printers, Laser Cutters, Plasma Cutters and Table Routers!

Founder of CNCKingdom.com
Industrial Lead Generation

# CNCKing.com: How It All Began

On a very regular basis, people ask me about how all these things have come to be -- I'm an outlier's outlier when it comes to CNC, design and so forth. I'll be the first to admit that my life has had lots of swift turns; if anything, when I see an opportunity, I just can't resist taking it on. I hope this helps to give a little more context about my origins in the CNC world. I've made this in smaller type so that it doesn't take up too much space in this book, as this isn't a biography but an educational resource!

Growing up, I was never the best student when it came to academics. I would look outside and see a place of endless adventure, then go into a classroom and find it rather stifling. I never seemed to have the same values or goals as other people ... I always wanted MORE and DIFFERENT, so learning things only one way, following endless steps, was not of much interest to me.

<u>Awards and my perfect attendance record!</u>

In elementary school, there were awards to the graduating 8th grade class every year: the best student in math, best in the sciences, and so forth ... we all knew who would win these awards at graduation even before the new school year began, but there was one award that I knew I could get that nobody paid much attention to. The PERFECT ATTENDANCE AWARD! It was easy -- all I had to do was show up for class every day for a whole year. In other words, no sick days! I was driven to get this award ... it was something within my means! Suffice it to say, after a year, I was called up on the graduation ceremony podium and accepted my perfect attendance record with pride. I think I was the only student to get this award that year! All these students worked their butts off to be the best at a specific subject, and I got an award for just showing up!

<u>Longest serving newspaper delivery boy</u>

I learned that there is more than one way to succeed in life ... and going for things that most people think of as not important are equally as satisfying. I repeated this again when I was a paper boy. I grew my newspaper route from about 10-15 customers to over 100 over the course of 10 years ... I basically had a whole city block covered, and if it wasn't for my goal of working only 45 minutes a day (making $20 in the process at a young age,) it would have gotten bigger! I loved video games and growing/splicing plants -- my bedroom was a jungle -- I had my priorities! There was an award with The Standard for the paperboy who had the oldest route ... basically, the guy or gal who delivered papers for the longest period of time. Well, that was something, yet again, I could gun for -- and I did. I ended up delivering newspapers for a decade. My customers saw me grow from a small 8-year-old kid with a runny nose to an 18-year-old man with a mustache!

I got the recognition and only quit about a month before starting university. I saved a lot of time, collecting money from my customers only once every 2-3 weeks instead of every week. Some hated this but to me, why go to somebody's door every week for a little money when I could go every few weeks for more money using the same effort? Making over $20 for 45 minutes of work day after day was something my friends would both admire and joke to me about. I learned the value of money and saved as much as possible so that I could buy my own computer -- my parents lent me the money, and every week I'd give them most of my earnings to pay down my debt to them.

During this time, I was slowly seeing my grandfather die from emphesema -- he was a lifelong smoker. This influenced me greatly, and everytime I see a cigarette even to this day, I think of him. I never had any desire to start; I focused on being as healthy as I could from then on. My friends were getting drunk at parties and I'd be the volunteer designated driver.

Learning 3D

I'm not sure why, but I always wanted to get into animation. Doing 3D work was the coolest to me, but I couldn't draw if my life depended on it. My artistic skills are limited to stick figures ... and sharks that look like overgrown fish! I would tell my friends about 3D and back then, the response was "What's that?" With the money I made from my newspaper, I bought a $5000 Mac IIci.

It was FAST and capable of doing some rudimentary 3D stuff, mostly Strata 3D Studio Pro and Bryce, but I had a computer and taught myself day and night how to make stuff. I found yet another passion in life!

What would render in about a minute on my current machine would take days on this old computer, but I never gave up. It got to the point where I'd spend more time working on my 3D skills than doing homework at school -- as long as I got 70% in a class, I was happy. The amount of effort needed to get more than 70% didn't make sense to me, as my time was best used elsewhere. I decided to start a 3D animation studio. I figured that I was "all right" at 3D and nobody in my area could do this stuff, so I had the market! I can't draw, but I can build!

The beginnings of i3DS International Corporation

Between my second and third years of university, I applied for and got a government loan to pay for my studies, so I did what anybody of that age would have done -- I took the $5000 and bought a plane ticket to travel to Malaysia, Japan and Australia over the course of the summer break. When I returned, I had decided that I wanted to start an animation studio, but I had no money for the equipment or software required to have a go at it. I also left single and thought I had met the woman of my dreams during these travels, so things were looking great!

I soon realized trying to sell a service that costs thousands of dollars to local companies was going nowhere fast. Of course, I bought a $10,000 Dell workstation with DUAL Pentium processors! How did I get that kind of money? Well, I always wanted to fly, so I went to the airport to get information about getting my pilot's license -- and they told me about a guaranteed government program which would give me a $25,000 line of credit that I could borrow from to pay for the course. That summer I also worked at the airport, so I'd train before and/or after work while trying to shove as many passengers as possible into those same planes during the day.

Well, I thought it was perfect: I could buy an industrial 3D workstation AND learn to fly at the same time -- two lifelong goals in one shot! I went for it and got my private pilot's license in the process.

I had to call Dell and tell them I was an industrial manufacturer, as these computers were not available to the general public. My bedroom was a corporate suite! I just needed a way to pay it off, and 3D was going nowhere, but people were asking me about website design. Initially I kept saying "No, I do 3D, not web." After a few weeks, I turned around and said "I do 3D AND web" (internet and 3D solutions,) and a steady cash flow soon trickled in -- I had an office with staff, doing websites, custom programming and now, 3D animations for industrial companies around the world!

How was I able to expand globally like this? Well, my secret is that I decided to e-mail, call and send physical mail to EVERY INDUSTRIAL COMPANY IN THE WORLD. Crazy? Absolutely ... but it worked! International phone calls were expensive and there was no Skype, but e-mail was free! For instance, when Forbes would release their fastest growing company lists, they always included the name of the CEO ... what a perfect way to get into a company!

So, I'd send them a congratulatory letter and also introduce myself in the process. I got a lot of replies back, letters I've kept to this day from CEOs who manage billion dollar companies and have since retired ... inspiring me to push even harder to grow my business!

<u>You want to deal with decision makers!</u>

Why focus on CEOs? I never took a formal business course EVER, but almost every time I was waiting for a decision on a project, I was told that they were waiting for the CEO to make a decision. So I decided "Screw it, I'm dealing with the boss!" That was my sole focus unless he/she directed me to the marketing director or other high level executive. I like dealing with people with the power of DECISIONS. They tend to have a "Shit or get off the pot" attitude that I loved.

I still have this database of 30K+ leads with me to this day -- it took me five years to build it up one by one! My average was one paying customer for every 100 industrial CEOs I contacted ... it was profitable and fun work!

I became so well known in the industry that I was approached to build "The Industrial Sourcebook" website, which was a huge custom programming project that we built for a publishing company which essentially produced the phone book for all industrial companies in Canada! The company behind this site got bought out after we completed this project and moved internally, and the site died a slow death. My company was to develop a whole series of industrial websites, but this all disappeared over the course of the acquisition. It was a major blow as I was seeing money signs, but a few months later they all evaporated.

Success!

All my achievements allowed me to win an award for "High Tech Company of the Year 2000" -- and I was nominated for ACE Canada Entrepreneur of the Year (one of only three across all of Canada)!

I was probably the only Brock University Biology/Geography student to ever represent my university at an international business conference in Ottawa -- all expenses paid! As I spoke fluent English and French, I had no problem communicating with visiting dignitaries from Africa and Europe!

I actually was so much "into business" that I had to excuse myself from a very important business meeting with my universities' executives because I forgot I was graduating that day! They laughed ... and I saw them a few hours later at the ceremony. By the way, that year I also joined the Entrepreneurs' Club at my university ... they won the ACE CIBC best club across Canada. I can't say I learned much about entrepreneurship there, but I did learn about business -- notably, the power and influence you can yield if you modify other people's perception of you. I almost flunked my entrepreneurship class (which I took after my Japanese class was canceled) because the only part that interested me was meeting entrepreneurs when they would come for seminars and talk about their businesses. I couldn't care less about the "theory" ... I wanted the "practical" stuff!

Falling on hard times

For a while, business was running great ... projects were getting bigger, and so were the opportunities. I was in my early 20s and was interviewed for a provincial TV show, was in national newspapers, and had a few salespeople working for me -- and on afternoons I'd take a client up flying now that I was a real pilot! Life was great ... I was married, talking to business owners around the world, visiting industrial shops all over the USA and Canada, and looking for growth. I signed up to advertise in the Yellow Pages across an area that I calculated would cover 10-15% of Canada -- the bills were horrendous, and then the .com bust happened and I was knocked down severely!

Before this time, businesses were spending crazy money on marketing, money that was my lifeblood funding my dreams! After the stock market crashed, my large industrial clients all of a sudden has to put lots of projects on hold. I thought, no problem ... I have money and enough credit to survive. Things were going down and I made a critical mistake. I decided to ride it out ... acquiring huge debts in the process.

The stress was incredible. At the same time, after being married for a few years (felt like 6 months as things were flying so fast) to a woman I loved and had dated internationally for several years, it got to both of us and we divorced. I felt like an absolute failure. I had to fire my staff and get on the phone to try to get money from customers NOW, as all my lifelines were used. We also got broken into during this time and lost some equipment. It wasn't a happy time at all!

Hospital time

I was facing personal and business bankruptcy, a divorce and debts over my head -- several accountants suggested I call the bank and leave them the keys -- walking away from my commitments in the process. My appendix decided there were better opportunities outside of my body and decided to leave as well ... I spent three days drugged up in a hospital. After the operation, I decided that at least I wasn't like some of the other guys in the hospital with missing limbs -- my roommate was a diabetic who lost his legs, and his life story was interesting but not very inspiring.

A new life

I got out of the hospital with a new lease on life. I had a new girlfriend and began to pay down my debts -- I figured since I spent the money, the least I could do was pay it back -- and started up business again! I was in a bind -- a "regular job" wouldn't pay enough to allow me to pay down my debts and live a good life, so I had no choice but to bring things back up. Things were looking up ... then 9/11 hit and I was wiped out AGAIN! I got things back up a third time. My sister Dominique died from a drug overdose on Christmas, and I realized that I'm a victim of mortality just like her, but my number just hadn't come up yet -- I was young and had so many dreams to be fulfilled. So I sold everything I had, made some tough decisions, and although I was maxed out yet again due to the economy, decided to celebrate my 31st birthday on a tropical Australian beach.

I got a working holiday visa and a one-way ticket (no money for a return) and was off! I spent my birthday at Cape Tribulation, swimming in jellyfish and man-o'-war infested waters.

Teaching abroad, but first, more misery!

While I was in Cairns, I tried in vain to find a job -- the working holiday visa was great, but nobody wanted to hire a foreigner for six month stints.

I was generating some revenue on the side from a few clients and ended up working the midnight shift at Cairns Post.

I didn't see the sun very often; I became a hermit in paradise! I still had mounting debts that needed to be taken care of, so after 4 months, I found out about teaching English in Korea. They would pay the flight (great as I had no money,) sponsor my visa, and give me a salary working less than 20 hours a week. I thought ... great! Now I can automatically pay down my debts while traveling and building back my business.

As luck would have it, my passport was stolen in the mail while applying for a Korean E2 visa in 2008. I had to red-eye it to Sydney with less than a hundred dollars to last me a week! It was miserable ... I had no money for food, and everything was maxed out, again so I booked a room at a hostel that offered free breakfasts. I would eat for two hours straight in the morning until I wanted to throw up, and that would last me until the next morning! Things sorted themselves out with the consulates, and I managed to appear on Sydney television as well. The rail system broke and I saw a camera crew interviewing people, so I made sure to appear in the shot! Old habits die hard!

Business Opportunities

The internet was always interesting to me; it's what made me successful with 3D earlier in my career, so I decided to do some SEO testing, building DreamClue.com (I believe dreams mean nothing, so I could easily excel in this marketplace by making stuff up) and then I built a few more web properties to track my internet marketing efforts.

I was doing side projects during my teaching, from web to 3D, but the market never recovered. I wanted to get into woodworking! I was in a foreign country, not speaking a word of Korean, in an apartment the size of a western bathroom with a laptop and still not a dime to my name!

In the fall of 2008, WoodMarvels.com made perfect sense. I could create flat-packed projects and sell the files and kits cut on-demand locally where the customer was. The fact that I had next to no woodworking experience or CNC or product design experience didn't stop me. I never even knew lasers could cut this stuff, but I knew I hated measurements. I was going to base all my designs off relative wood thickness instead, so scaling was easy, and my goal was to design 100 projects.

After a few years, it had grown very nicely, generating sales around the world! I then built 3DMarvels.com to duplicate my wood design efforts in the 3D printing world. After a few years teaching back and forth in Korea, the Philippines and Canada, I decided it was time for me to learn Chinese. It was fun living in a new place and making new friends every 6-12 months, but that got old after a while.

The "teach in Korea" thing worked well, but now most jobs were requiring 40 hours a week at the same salary -- but not in China. I moved there, where I met my wife -- who ignored my advances for months after I told her I wanted to have a family with her the first time we met. I knew what I wanted and wasn't shy about letting her know -- she thought I was nuts, but the results speak for themselves. I then convinced her to write a book and produce videos with me to teach Chinese to foreigners -- that's how HowToChinese.com came about.

Although my salary in China was a quarter of that in Korea, the hours were averaging less than 15 a week, and my businesses were generating enough to live comfortably at this point.

<u>Moving to Australia</u>

I tried in vain for years to get into Australia. I didn't have any money, so the only way was through a sponsorship of some kind like I managed in Korea and China previously. I sent out e-mails to industrial companies across the country ... luckily, one was bright enough to take me on. A month after getting married, I moved back to Canada for what I thought would be a few months, so I told my wife that I'd shave my beard the next time we met.

Unless you've been in business or have been an entrepreneur, most people don't recognize any of those skillsets -- especially government bureaucrats who think certificates and diplomas are the only way to measure your abilities. My passion and drive were worthless to them unless certified!

The process took 8 long months but I had started making some good money from WoodMarvels.com and had a lot of designing to do, so I focused on that ... the time went fast! During this time I also acquired a ShopBot Desktop, so I'd spend hours designing in my room and then hours in my parents' garage cutting new projects (all the router projects you see in this book). I got camera equipment and started filming everything. I was happy to arrive: one of my dreams had been realized, and I was able see my wife again! I then rebranded to CNCKing.com to take advantage of new opportunities. We are now at the present day, June 20th, 2013.

Become like water

One of my idols is Bruce Lee ... I actually studied martial arts for 10 years and got my black belt right before starting university. His philosophy of changing your mind and body to suit local conditions is one that I've followed since I was a child.

I've never lived a conventional life like my friends. I've seen and experienced things they can only dream about, and most importantly, I'm a very content and happy person. It isn't easy, but I'd suggest trying things out regardless of what people tell you -- if somebody says no, it means you didn't try hard enough to prove your point.

I've not only managed to get my finances under control since then but also taught myself new skills for emerging industries!

I published six books during this "teaching" period of life, started a few businesses, and developed contacts that will provide me with the challenges for the second phase of my life.

Regardless of the troubles I put myself in, I never gave up the thought that what I got myself into I can get myself out of using my mind and body. It's important to treat both with upmost respect ... smoking, drinking, and taking drugs doesn't solve problems, it makes them worse. The body will only take so much abuse before it quits, and when it does, there is nothing you can do to get your health or life back.

My life's been quite an adventure so far... I can't wait to see what the next 100 years have in store for me (I'd like to be the oldest human one day). I know this ride hasn't been easy for my parents, but I'm sure everything will work out in the end. If you love what you do, good things come of it.

## Tools and Materials

If you have a laser cutter or CNC router, producing the work found in this volume will be relatively straightforward: scale the drawings using grid paper to match the thickness of the wood, and transfer these to your tool pathing or cutting software. Alternately, if you have a traditional woodworking shop, many of these models can still be constructed -- they will just need some modifications in order to be cut with less precise machinery. Regardless of how you wish to proceed, the following list of tools will come in handy:

1) Carpenter's wood glue
2) Extra-fine sandpaper
3) Jig saw or band saw
4) Clamps (lots and lots of them!)
5) Pencil
6) Pieces of wood
7) Rubber-tipped hammer
8) Laser cutter / CNC router

Please note that the information found within this book is based off my own personal experience and makes general characterizations between different CNC machine types, as with any machine or method, there are exceptions to every rule and technology is always changing.

The black square in the templates represents material thickness -- scale accordingly.

## CNC Machine Background

The focus of this book is CNC designs and designing, but I'm still very surprised by the fact that many people who use or write about these machines on a daily basis never actually realize that their 3D printer, plasma cutter, table router, or laser cutter is actually ... a CNC machine!

If you come from an industrial background, this is common knowledge; but if you come from a DIY or designing background, this fact may have eluded you.

The very fact that this misconception exists proves just how far these increasingly user-friendly machines have come from their engineering and programming routes! You no longer need to be fluent in G-code or a mix of electrical and mechanical engineering in order to operate them; all you need is a computer! As such, the reason for this book is to educate, motivate and inspire people around the world as to what is possible with these amazing machines! As Ted Hall, the founder of ShopBot Tools Inc. would say, "CNC machines are amazing, but they aren't magic."

Although I call these machines "command-n-control," the real meaning of CNC is computer numerical control ... an awful name to try to visualize, but it does describe exactly what a CNC machine is and does. You have a "computer" that sends zeroes and ones to a "control" which then does something -- usually very mundane -- repeatedly throughout its life until it is no longer needed, put up on a shelf, and then cannibalized for spare parts as it accumulates a layer of dust and debris before being recycled into a can of Pepsi. The life of a CNC machine is boring, but it doesn't need to be ... give it something exciting to do for a change! They are great fun once you get the hang of it!

Numerical control, as CNCs were called before the advent of computers, can trace its origins back to the second world war, when governments were pushing factories to build things faster, cheaper and better.

"We need that huge airplane in two weeks, not two months, for the front ... make it happen!" was usually followed by an invoice that didn't have 20 planes on it, but 100, to be produced within the same time frame!

Human limitations were well known at the time, and machines started playing a more prominent role on factory floors around the industrialized world -- they were stronger and less prone to emotional breakdowns. These machines, though clunky at first, were able to do the same task day in and day out with little error, no complaints and no coffee breaks.

The mechanicals involved with these machines were controlled using punch cards/tapes and other "early computers" that today's smart phones would easily overpower.

As these early CNC machines evolved along with engineering, material, programming and computer knowledge, they got smaller, easier to use and more closely integrated with one another. Only with the advent of cheap personal computers did CNC machines slowly make their way off shop floors and into people's homes; and, twenty or so years later, companies like CNCKing.com can exist thanks to another government-funded technology, the internet. The United States has been a major contributor to the creation of CNC machines from the very beginning! We owe American technological pioneers and inventors a great deal of gratitude!

Think about it! You probably wouldn't have CNC machines or the internet if it wasn't for fear of complete annihilation by world powers fighting each other till the bitter end -- for all the bad that comes with wars, at least some good does eventually seep out to benefit mankind. Mix machines, computers and entrepreneurs, and you have CNC!

**What is the relationship between CAD, CAM and CNC?**

These three acronyms get thrown around a lot, though you don't hear much about CAM in the designing community -- probably because it means computer aided manufacturing while CAD stands for computer aided design.

If you are a designer, you usually use a computer to build your design, but since the scale is usually very small, an entire manufacturing process isn't needed -- hence why you bought that little 3D printer in the first place! As for CAD programs out there, you can use whatever you wish, as long as you can export the file into something recognizable by your CNC machine. Generally speaking, 3D printers use STL while lasers, plasmas and table routers use CDR, AI or EPS as their primary "import into job production" software.

For instance, I've been using Autodesk's 3D Studio Max due to my origins in industrial 3D modeling and animation, but you can most certainly design using Corel Draw (huge with laser/table router small shops,) Solid Works (huge with the industrial crowd,) or Blender (open source 3D modeling and animation program).

There are a host of other free and paid programs out there, including some that run off phones! 2D software is vital, though. Moving to 3D software helps even more!

Whatever you end up getting to help you with the design process, realize that your design needs to be optimised for the CNC machine you will be using to form, cut or print your design. For instance, you can't print a 3D model using a laser cutter any more than you can create a closed internal negative space using a 5-axis CNC table router.

The tool you have on hand decides the design process and programs you use, not the other way around. Also, don't let your education (or lack thereof) get in the way of your inspiration -- I've had countless industrial design university graduates send me files that simply will never work (plunge lines for a laser cutter?) as I've had brilliant designs from hobbyists with no formal design experience, such as myself.

There is no way around actually USING a CNC machine; theory is nice, but don't expect things to go according to plan.

For instance, I can design some amazing extrusion specific designs in 3DS Max and even with all the knowledge I have as to what I can or cannot do, still not be able to print the model with my 3D Systems Cube. Why? Well, simple issues like flow errors, the model unsticking from the platform because I have too much material density at the base, and so forth. Consumer level 3D printers are still very much at the beginning of their growth curve, and like any other technology they have some teething to go through due to market pressures that don't exist as strongly within the cash-rich industrial sector.

These issues all come from firsthand experience, which you must develop a feel for. Sometimes I really scratch my head because what I made in my software SHOULD work, but there are so many issues that can cause a model NOT to work regardless of which CNC machine you designed it for. That's why I wrote this book!

There are four main reasons that led to the Rise of the CNC turning a once strictly industrial tool to the perfect home digital fabrication platform.

**Reason #1: CNC Programming language**

Punch cards and tapes, the only method used after the second world war to program "computers," was incredibly slow, inefficient and user unfriendly to say the least. There HAD to be a better way!

The title of this book describes that one moment in history where the American Air Force decided in 1956 that it was time to produce a "programming language" which forever changed the course of personalized industrial manufacturing. From then on, through the decades that followed, the manufacturing process became increasingly efficient and precise, which has led to the globalization we see today.

CNC machines play a critical role in our daily lives -- how such a MASSIVE advance stays so well hidden from most people's thoughts is incredible to me. Think about it: if it weren't for the American Air Force's critical decision to develop a programming language all those decades past and consolidate a number of patents, you'd be using carrier pigeons with punch cards to interact with your Facebook friends. Most likely the internet itself wouldn't have been possible without CNC machines either!

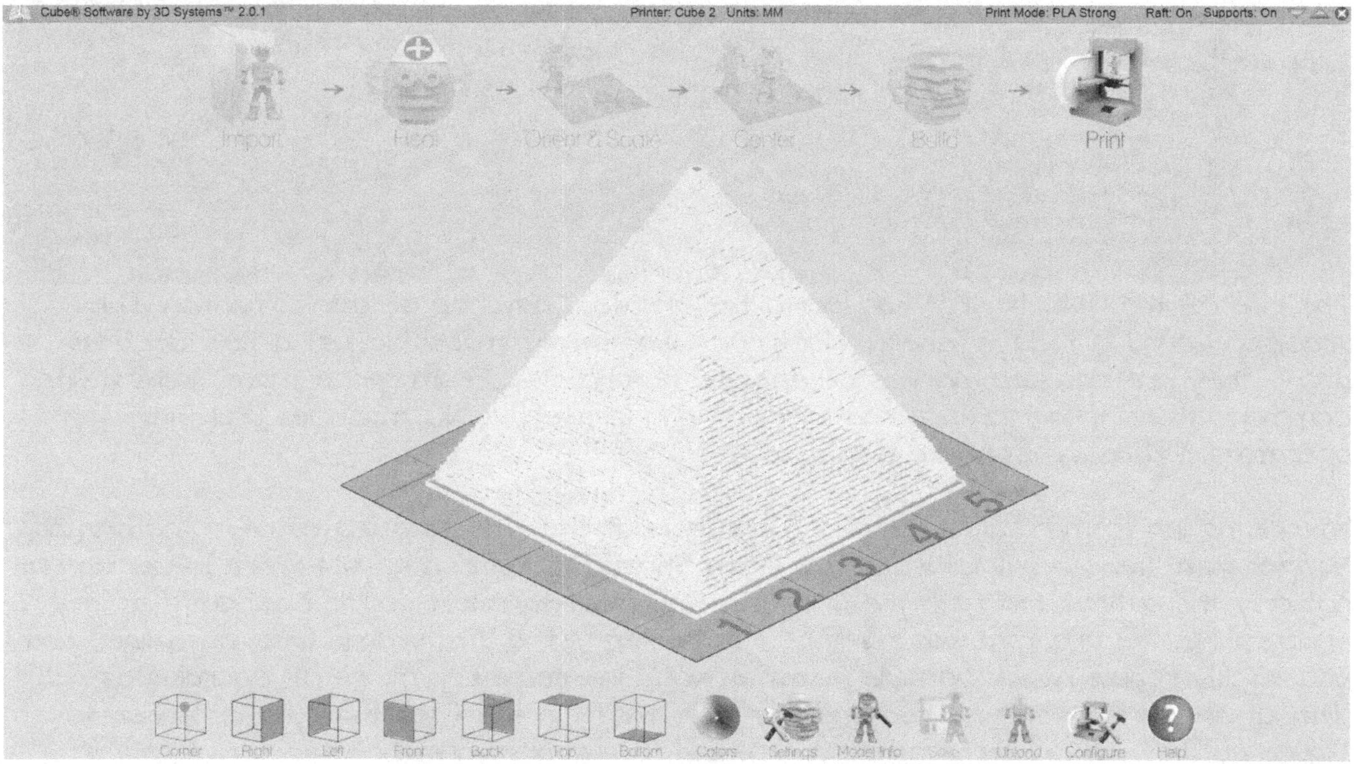

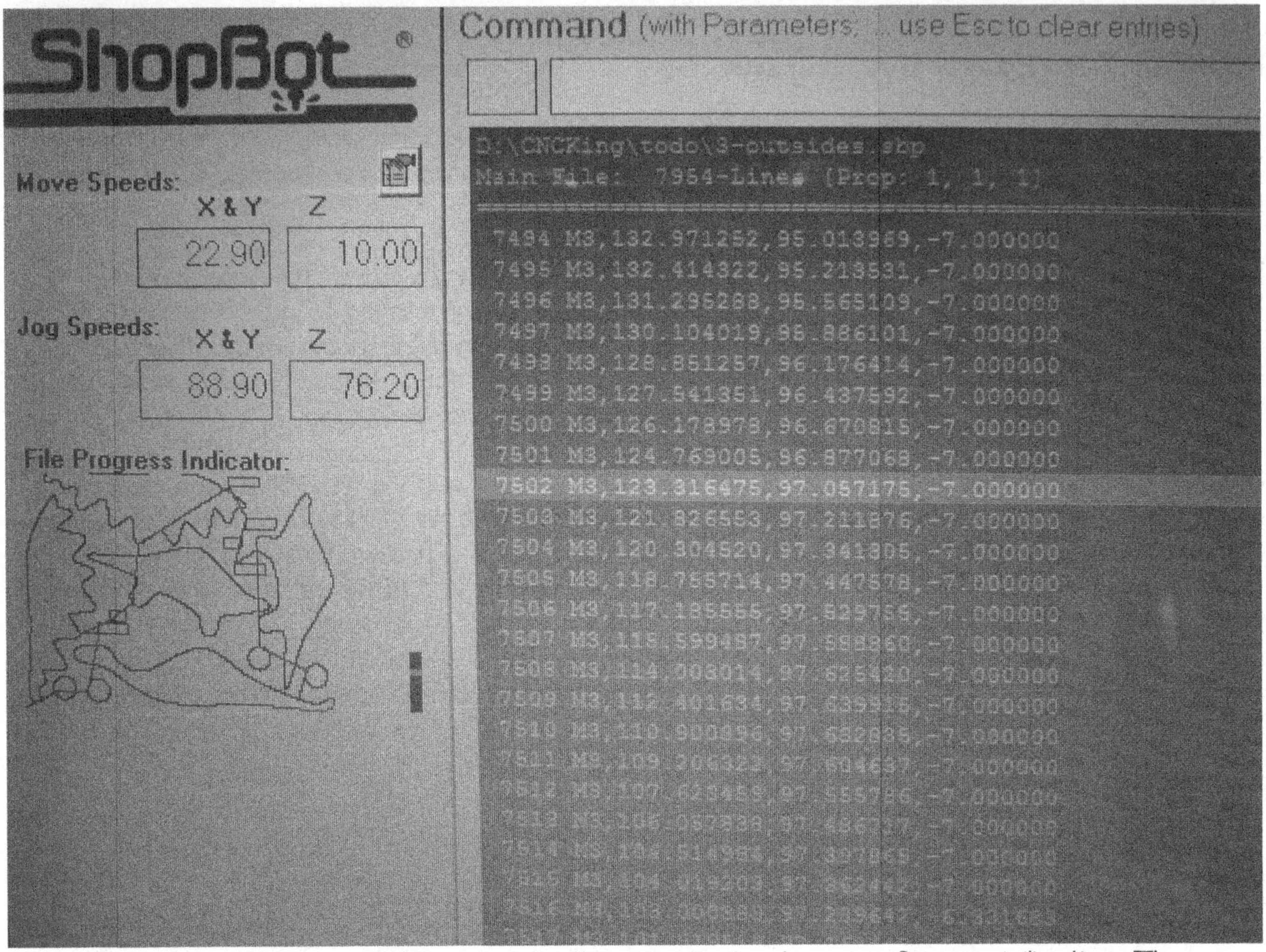


The rise of the CNC is what has led to the greatest advance of our civilization. The precision found in today's "cheap" electronics was simply impossible just a few decades ago -- they are absolute marvels of engineering prowess! From space travel and satellite communications to agriculture and everything in between, CNC machines and their operators and designers make it all possible!

Imagine having to manually, using punch cards, input the XYZ coordinates of something as "simple" as a stegosaur (picture above from ShopBot's job control software,) figuring out not only the optimal tool path the CNC machine must take but double-checking for any errors while the program was running -- or better yet, a 3D pyramid (previous page,) and you quickly realize how impossible a task it would be, requiring millions of coordinates and without a way to preview it realtime. For fun, throw in the fact that you also need to figure out how to create dynamic 3D rafting to support your model!

**Reason #2: CAD/CNC integrations**

Today's designers have it so good, we can visually interact with our models in 3D stereoscopic space! We no longer need to formulate Cartesian coordinates (X, Y, Z) for our creations and somehow hope that this will make something cool ... the money and time savings of CAD can't be overemphasized.

Although CAD started as a pet project a few years after the programming language was developed, it took almost a decade before the entire loop from initial design to finished product was closed by General Motors working with IBM in 1963. However, it wouldn't be until the 1970s that the CAD/CNC production cycle was adopted by the industrial sector.

Imagine how many man hours were needed simply drafting a model on paper, part after part, then combining them -- and if you had one measurement wrong, you had to redo EVERYTHING associated with that part. This was life before CAD.

The ability to preview tool paths alone is a massive improvement as it allows you to clearly see how things will be made once they are sent to the CNC machine.

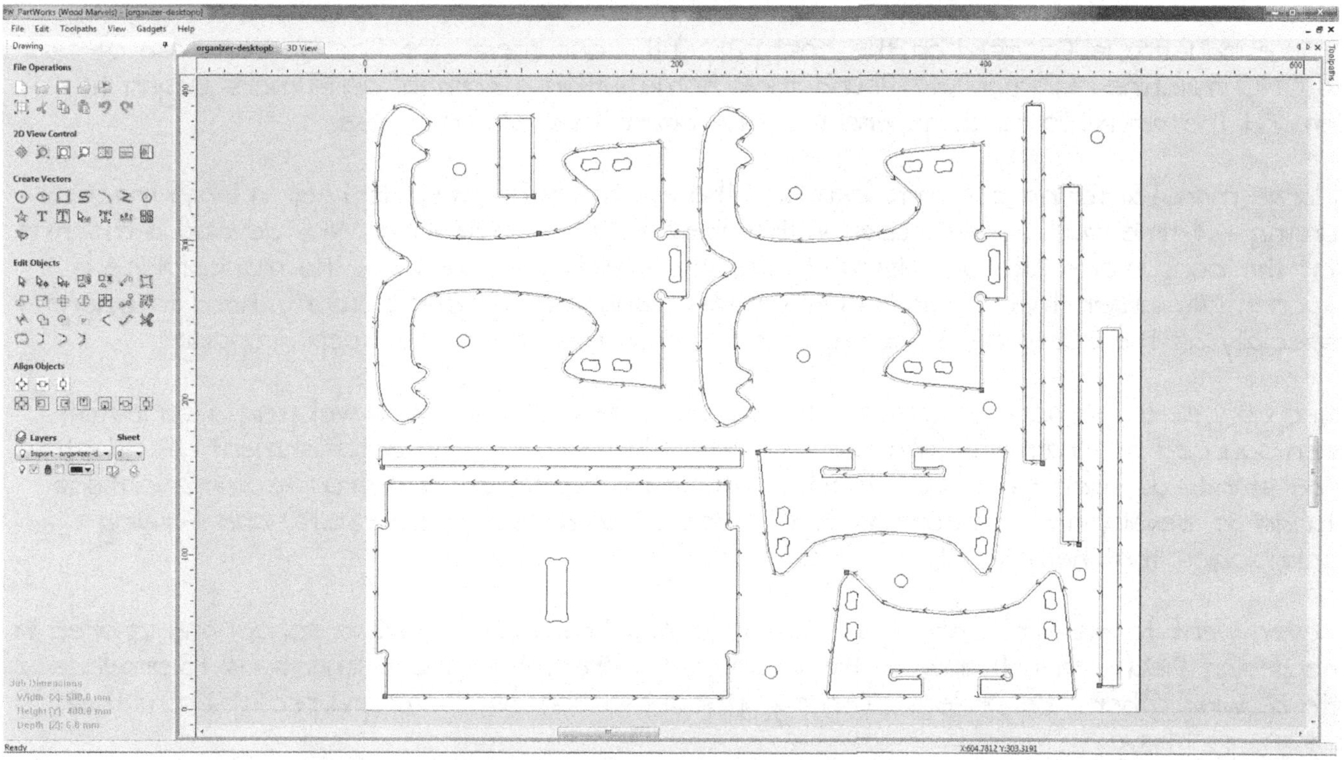

**Reason #3: Moore's Law**

Thanks to Moore's Law, computers and their associated hardware keep getting cheaper, smaller, yet more powerful in regular fashion. This affected industrial design massively because over time, things that were once processor-intensive (such as curves) became just as easy as straight lines. For example, a straight line has two X, Y, Z coordinates, one at the start and another at the end, while a curve can be made up of hundreds if not thousands of data entry points. Thanks to advances on the hardware and software front, stepper motors or servomechanisms also increased in precision and speed.

Not only have things gotten more efficient on the computer hardware end, but this extra power also allowed built-in redundancy within the CNC software itself, meaning calculations could now be double-checked even while the machine was operating, and fixed on the fly if needed! Want more than one arm doing something at once on a CNC machine? It isn't a problem, which leads to even faster machine operations as well. A machine can only run as fast as the coordinates being given to it and the forces generated by the creation process.

**Reason #4: Consumer Adoption**

Where industry is focused on the cost per unit, consumers are focused on ease of use. As CNC machines increasingly found their home within a garage, consumers began to demand the ability to make general machine operations easier to use.

A large manufacturing company can hire the expertise required to keep a large machine running full-time, but a hobby user with limited skills is constrained. Why develop a machine for the consumer? Although demands are different than industry, the marketplace is exponentially larger than the industrial sector ... and, like the car industry, hard to ignore -- especially as the costs of ownership due to mass manufacturing keep dropping.

Here as well, the American government helped hone CNC use by developing a series of open sourced programming languages, playing a large role in the establishment of industry-wide standards such as G-code, establishing safety, graphical and cross-communications standards, developing logic commands (if/then vs hard coordinate inputs) and funding technologies that helped with the proliferation of CNC machines.

I never went to drafting school and was rejected from 3D animation school and ignored in the design field -- but, thanks to the internet, I overcame all these limitations to produce a site allowing others, regardless of geographical location, to share and profit from their designs as I have.

Take a look on the following page at one of my setups that I have here in Australia for my ShopBot Desktop. This used to require a warehouse, but now I have all this manufacturing power in a corner! The camera is a Canon XF100, which is capable of broadcast quality filming; the computer has an Intel i7 Core and a CNC table router that does exactly what I tell it even if I'm wrong. If you look very carefully, you'll notice I have all this running off a normal power "home" socket with an extension cord! This power in this corner would blow GM's mind if I could have shown them this picture in the 1970s!

As for the costs, they are trivial compared to what any of these technologies would have cost decades back. There has been a massive change in the industrial sector, and this picture shows you just how much power a person can have in their home if they so desire. There is one more thing I want you to focus on -- that's me, the user. Me, myself and I am capable of following through with a project from concept to design to cutting and assembly. What would have taken several very specialized and highly educated people in the past to accomplish, I can do in the comfort of my office. I don't need a background in physics, mathematics and engineering to make my projects reality!

This bears repeating. Thanks to the advances made in job control and design software and miniaturization of technology, a person no longer has to attend an academic institution in order to enter the manufacturing and design field -- though it does help. There is some serious pushback from those who are highly specialized. Their skills are being increasingly commoditized, which allows those with a fraction of their skills to compete directly with their artistry, but the demise of low-level manufacturing processes is best left to industry experts to explain.

The software advances have been so tremendous that you can usually just trace a design and cut away without ever touching an algebra textbook. CNC machines are getting easier to design for and work with each successive year -- you can either learn now or be taught by your children later. In fact, 3D Systems just released Cubify Draw, allowing you to do just that -- to create 3D objects by drawing them on an iOS device. In a few years, you'll be able to model using only fingers movements on 3D holographic clay!

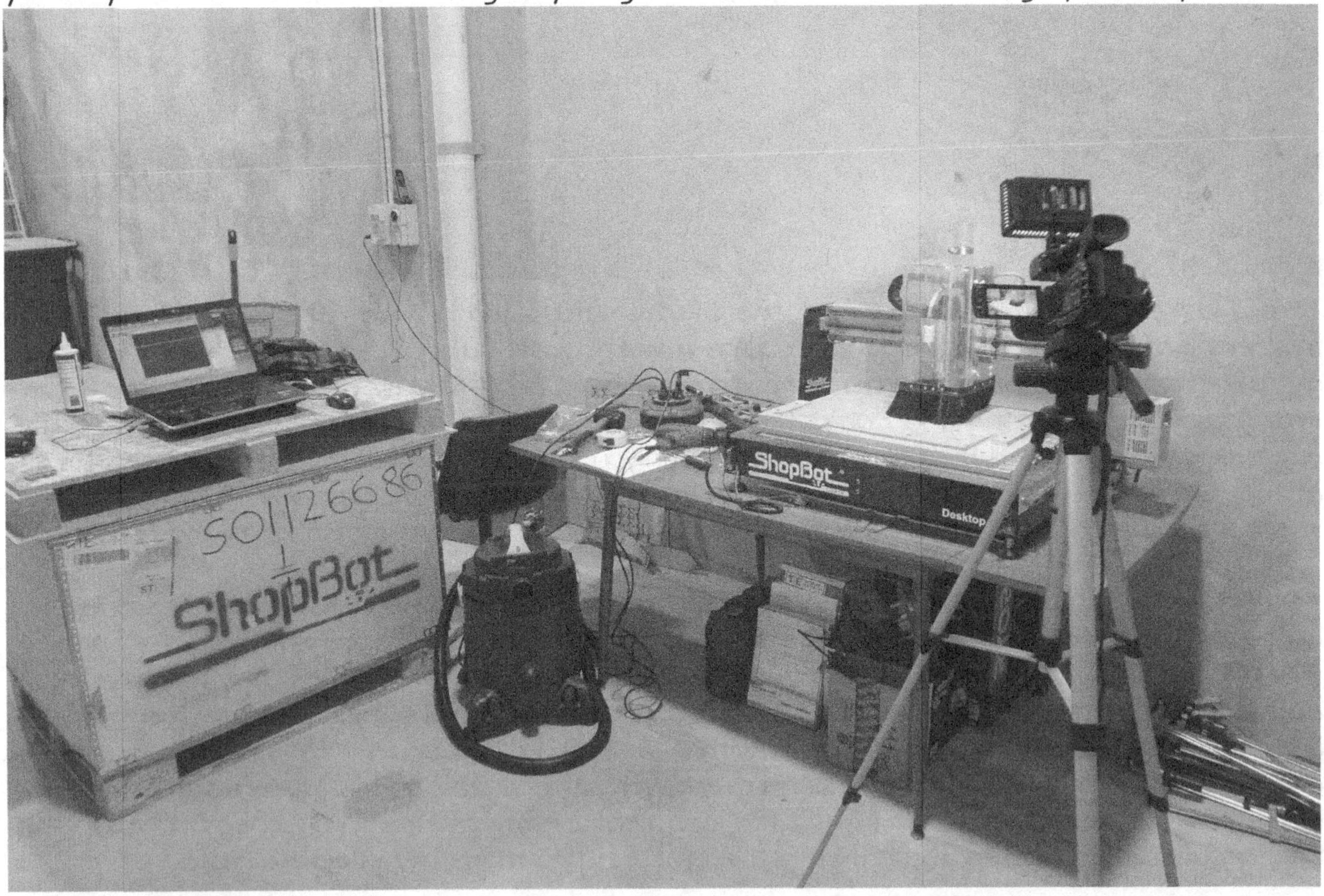

## CNC Laser vs Router vs 3DP

As my experience is extremely limited in the area of plasma cutting, I haven't included it in this comparison -- I hope to change that in the coming years.

There are three main tools out there that designers use to produce a real-world version of their designs: CNC table routers, CNC laser cutters and CNC 3D printers. Each has its advantages and disadvantages; and in general, a design must be optimized to use one of these technologies over the other. Compromises must be made with each CNC and machine type within that CNC category, though not as extreme..

In general, CNC table routers are the cheaper alternative compared to the other two, but it does come at a cost. Because you have a small bit spinning at several thousands of revolutions a minute, the material is ripped and torn away, which requires you to pay close attention to holding the material down (preventing it from flying away). This can be done several different ways including a vacuum table (suction from under the table though a MDF board,) screws (my preference,) squeezing it into place, jigs and/or nylon tabs. The solution that you will settle with is the one that best suits your application and intended design. The only part that "breaks" from usage on a CNC table router is the bit, which can be had for roughly $12 USD -- lasting between a few minutes to weeks depending on use (hours of operation, material type and thickness) and luck. Make sure you have lots of bits because they tend to break in batches when you most need them!

The second most used CNC technology by designers around the world is the CNC laser cutter. Although there are many advantages to using a beam of light over a spinning bit, the biggest are in the area of precision and shared lines. If you want to cut two holes with only a milimeter between them, it's just about impossible to do with a CNC table router because of material friction but easily do-able with a laser or mid-end 3D printer. In the area of shared lines, you cannot do this with a bit, while for a laser, this is an efficient method of cutting. This does come at a cost though, COST! High quality laser cutters in general are far more expensive than CNC table routers, and though maintenance may be less, a tube blowout can cost you several thousands of dollars when it does occur. Getting an extended warranty on the tube alone is a very worthwhile investment, as most of the cost of the machine is derived from this essential part.

From my experience, there are more shops around the world that have laser cutters than 3D printers at this point. How can that be? The engraving industry is MASSIVE. From Cermarking plaques to signage, very few technologies offer the versatility that a properly marketed laser can provide. Cutting 2D vectors or engraving text / images is easy compared to the very steep learning curve involved with modeling for a 3D printer.

The third most used CNC technology out there (and quickly gaining ground with consumers as well) is the 3D printer. Depending on the 3D printer technology you are using (there are more than several dozen types) quality commercial 3D printers offer designers ultimate freedom with their designs, with details measured in nanometers. But this comes at a very high cost in both material and the machine itself. Your build size (one solid unit) is more limited vs. laser or CNC table routers, but the major advantage over the other two technologies is that you can build a 3D model without requiring any assembly. Apart from the modeling aspect, 3D printing is very much "print and go." As Cathy Lewis told me repeatedly when we met, "... with 3D printers, complexity is free!"

I think a unique aspect of having conversations with designers from around the world is that you soon realize that they are as ambitious as they are creative. Where previously you needed an engineering degree to get these CNC machines to work properly, followed by a Masters in Programming to get it made, the simplification of both aspects has had a tremendous impact on this space. CNC machines have become USER-FRIENDLY!

For myself, I'm not a programmer, but I can quickly model a design and basically "print" it out using any of these CNC technologies. Not only that, but usually the software running the machine will tell me if I made a mistake somewhere in my design that should be more closely looked at! Things have evolved tremendously over the past few decades.

I often get asked by people interested in getting into the CNC field, "What would you get to start a shop?" My answer is always the same.

First thing I would do is get a CNC table router and/or consumer level 3D printer. A good quality machine for each is only a few thousand dollars and can be stored in a relatively small area with minimal maintenance. Before I'd go after any clients though, I'd spend a solid 6 months just designing by myself, using every waking hour to learn the software involved with both creating and running these machines. Having the machines is the easy part -- learning 3D modeling, optimizing paths and your files for efficiency (making more profits in the process) and wrapping your head around the endless possibilities you have with these two machines alone needs time. It took me two years to fully understand and comprehend what I know now, and there aren't any shortcuts. You must master not only the tools of the trade but also the software that runs it!

Where is the laser cutter? Cost is the main factor, but once you master a CNC table router, a laser cutter will be EASY. In the meantime, sub out your jobs to the local laser shop for a fraction of the cost and ask if you can hang around during the cutting.

# Top 10 General CNC Operational Tips

This book is full of tips, but I thought I'd make a specific section dealing with the top 10 mistakes or issues that you'll come across operating a CNC machine.

**Tip #1: Always err on the side of safety!**

These machines are powerful, and although many of their built-in safety features can be overridden, don't try to be a smart ass and think you can disable them for expediency. If you aren't sure about something or don't have a good feeling about it, just don't do it. It only takes one stupid move during a moment of inattention to seriously hurt yourself with these machines. Wood, steel, flesh ... makes no difference to them but it should to YOU!

**Tip #2: Measure material thickness before cutting**

If you are cutting a project that involves some type of interlocking or multiple layers of material stacked on onto the other, always measure the thickness of your material FIRST, then adjust the model SECOND based on these slight changes. You may think that a plan calling for 6 mm material is fine to be cut using 6.2 mm, but that extra 0.2 mm over five layers creates an extra 1 cm that you may not have accounted for, causing things to be too tight or not fit at all without some serious sanding.

**Tip #3: Keep your machine clean**

A dirty machine is a hazard waiting to happen, and it will cause excessive wear and tear and affect its future resale value. Like brushing your teeth, doing a little every day saves you massive bills later in surprise expenses. For instance, if you own a laser, check the lens often and remove any grime before it cracks it; if you have a router, check the gantry for chip buildup in the rails, if you have a 3D printer, check for fluid or material buildup around the platform and machine head.

**Tip #4: Keep your supplies topped up**

There is nothing worse than starting a project only to realize you didn't buy enough material or glue, or you don't have the right tools to do the job. Buy more than you need and find a place to store it. There are lots of things that can and will go wrong; having some extra on site will save you a lot of time and headaches.

Another common issue I used to come up against is clamps. Get a variety of them in different sizes and configurations -- they always come in handy.

**Tip #5: Don't mindlessly upgrade software or hardware**

If it's not broken, don't fix it! A few years back many CNC laser operators were left in the dark because they decided to upgrade their Corel to the newest version, which had a bug that prevented their laser from doing anything. If you have everything running smoothly, don't upgrade your primary system! Use a tertiary system for testing, and only after everything clears after a week or two, move it onto the primary.

**Tip #6: Don't customize your machine**

Your machine was made a specific way by the manufacturer for optimal use. Sure, there are some flaws, but tell the manufacturer about them and don't take it on yourself to fix them. Not only does this likely void your warranty, but you may end up with a broken and useless machine if you don't know the reasons behind the engineering decisions made at the CNC factory. Just because you can operate and design for a CNC machine doesn't make you an instant industrial or mechanical engineer.

**Tip #7: Don't focus on price, focus on ROI**

With everything in life, you need to overcome the sticker shock and figure out WHY machine A may cost double that of machine B, even if they do the same thing. This is never truer than in the CNC world ... not all machines are made the same! I've seen some laser systems out there that require the user, without an instruction manual, to place a glass laser tube into the machine, align it and do several test firings after hooking it up to a bucket of water. It may save them a lot of money, but this isn't the kind of stuff you want to be fooling around with. Actively question your sales representative to justify price differences and try to accrue the cost over the life of the machine. A $20,000 machine that lasts five years is worth far less than a $40,000 one that lasts ten years.

**Tip #8: Buy more than you need**

If your budget is restricted when looking at machine options in the marketplace, but the perfect machine for you is still over-budget, take a leap of faith and try to get that one rather than settling for a CNC machine that is just under your needs. People with CNC machines soon discover that what they thought was great when they purchased the machine may be too limiting a few months down the road.

Vendors are always open to additional methods of financing -- they only make money when steel goes out the door -- so be honest with them and try to figure something out that's mutually beneficial, even if this means having to lease a machine with an option to buy later down the track.

**Tip #9: Never stop learning!**

You didn't stop learning new things after you got out of school, nor should you stop after gettting a CNC machine. There is a tremendous amount of knowledge out there if you seek it out -- this book was a great first step! Each industry, whether laser engraving or cutting, 3D printing, table routing and so forth, has industry events, online forums and gatherings that you should try your best to attend. The worst thing that can happen is you make some new friends, but it also opens the door to a host of new possibilities, from strategic partnerships to new customers (which I'll delve into further).

**Tip #10: Try new things every once in a while**

The potential of your CNC machine is limitless. It will do exactly what you tell it -- so tell it to do some cool things every once in a while. If you bought a machine to do only one task, try to see what other areas of your business it may be able to help with, or even try expanding into new markets. A CNC machine that isn't moving is generating ZERO revenue; try to keep it as active as possible. You have in your shop a power of manufacturing magnitude that would give engineers wet dreams even just 20 years ago. Use it to its full potential -- you never know what doors will open up.

# CNC Laser Technological Overview

Before I knew that lasers capable of cutting my woodworking projects existed, I designed them all to be flat-packed for easy shipping -- borrowing that idea from IKEA. I thought any technology out there to cut my models << en masse >> but on-demand would most likely follow vectors from above to cut through material below. When I discovered laser cutting, my assumption was proven, but I had to redesign a few models. I was right on laser functionality but had mistakenly assumed wooden dowels were a consistent diameter around the world -- they are not! Rarely is a 6 mm dowel actually 6 mm!

The laser is more precise than the dowels, and this caused lots of guiding problems. Sanding down a solid wooden dowel is difficult at the best of times. After that redesigning effort, I decided that EVERYTHING I design would be cut from the same board without requiring any external hardware or trips to the lumber shop. This is how I've done things ever since, and I also suggest this advice to others!

Lasers are really quite fun to work with. They are faster than table routers (material depending) but they do have some technological flaws -- the biggest being precise z-axis vaporization, where they exceed in x, y but fail tremendously in z. I have seen some amazing engraving work that you'd swear was done with an extremely small table router bit, but these are by far the exceptions rather than the rule.

Another issue to keep aware of is that there is radiation production with a laser, which is why most models are enclosed or require special glasses to safely operate them. You can look at a bit turning all day, but do the same with a laser beam and you'll get a good tan and dose of UV radiation as a result -- wear your sunscreen! Where a CNC table router is loud, especially when you have an extraction system running, a laser cutter is incredibly quiet by comparison and the only "dust" produced is just that, vaporized material.

If you compare a 3D printer to a laser cutter, you have basically the same level of precision (even more so on higher-end machines) but across all three axes (x, y and z) vs. just two (x and y). You have the quietness of a laser without needing an extraction system, as it's an additive technology rather than a subtractive one.

Where a CNC table router exceeds the 3D printer and the laser to a lesser extent is material sizes -- industrial scale 3D printers are confined to a relatively small cubic area. You can remove material along a z axis with more precision than a laser using a CNC table, but you are still limited by the overhead gantry positioning, meaning you can't produce negative or inverted shapes -- the top must always be "smaller" than the bottom. As you can see, each technology has its costs and benefits!

I'm lucky enough to have played around on most of Trotec Laser's offerings at different wattages, cutting and engraving various materials. They are very nice high-end workhorses, and I've only heard positive things from customers buying plans on CNCKing.com and using Trotecs to cut and engrave the files with.

There is an incredible diversity of laser machines out there, from the low-end Chinese and Indian machines to their high-end European counterparts. Many people don't know what they should watch for when looking at purchasing a laser, so I hope these tips empower you to ask your sales representative the right questions.

**Laser Buying Tip #1: Remote Support and Diagnostics!**

Your machine, regardless of manufacturer, will just stop working sometimes. Is it a software issue (like a Corel conflict) or hardware issue like not being plugged in? As with most CNC machines, any kind of down-time always seems to happen when you are trying to get an urgent project in and out of the door. Waiting till Monday to contact a representative who can't get to your physical location for another week or two just doesn't cut it. Make sure that there is a system in place for remotely diagnosing what's wrong with your machine or computer by generating a file and sending it through, along with a phone number that's available nights and weekends. Sometimes representatives go on vacation as well, so be sure to have direct numbers to the head office. What's great about remote diagnostics is you can give the representative access to your actual production computer, and they can check the settings and fix them realtime!

**Laser Buying Tip #2: Passthrough Ability and Ergonomics**

Most lasers work within an enclosed box due to radiation concerns (not strong enough to make you glow in the dark but worse than laying in the Sun all day), but being able to override this and open it up either front to back or side to side is a massive advantage. There is always that one piece that's just slightly too large to fit within the bed or that customer who brings in an 8 ft ore that they are willing to pay top dollar to have engraved with their name. Being able to accept the job and having the flexibility to charge a premium on isn't something to be taken lightly.

On a related issue is ergonomics. You want the machine to be at a perfect working height that allows you to easily load and unload projects to and from the bed. An extreme example of this is the Trotec Speedy 400 flexx (next page) that allows you to completely open up the front to load just about anything at optimal height with passthrough ability in the rear! It's really an engineering marvel that you have to try at least once.

**Laser Buying Tip #3: More watts ... not less!**

Get as much wattage as you can afford ... and then some! The more power (watts) your laser has, the faster you'll be able to cut or engrave a job and send it out the door. This is vital for any business, and the relatively "cheap" expense of adding watts to your machine purchase at the beginning is one of the quickest ways to pay off your investment. An added benefit of having additional watts to your disposal is that you can cut through thicker material in one pass that may otherwise require multiple passes.

I have had many customers waste hours of their life trying to cut one of my models out of 4 mm thick plywood using a 30 watt laser-- don't be one of them! I'd suggest to anybody buying a laser to go no less than the bare minimum of 60 watts ($CO_2$).

**Laser Buying Tip #4: Amazing Job Control!**

You know what's worse than having a broken CNC? Useless job control software! It's incredibly frustrating to work with a machine that doesn't have two-way feedback. If I move the head to the left, I should have some measurements or better yet, a representation on the screen showing me realtime feedback for precise positioning.

This is one of those things that most people overlook, but it's as critical as choosing the right laser machine for your needs! Make sure that when a job is being made, you are able to see what's going on, and that when you are setting up a job, you can actively see where your project is relative to the platform your board is sitting on.

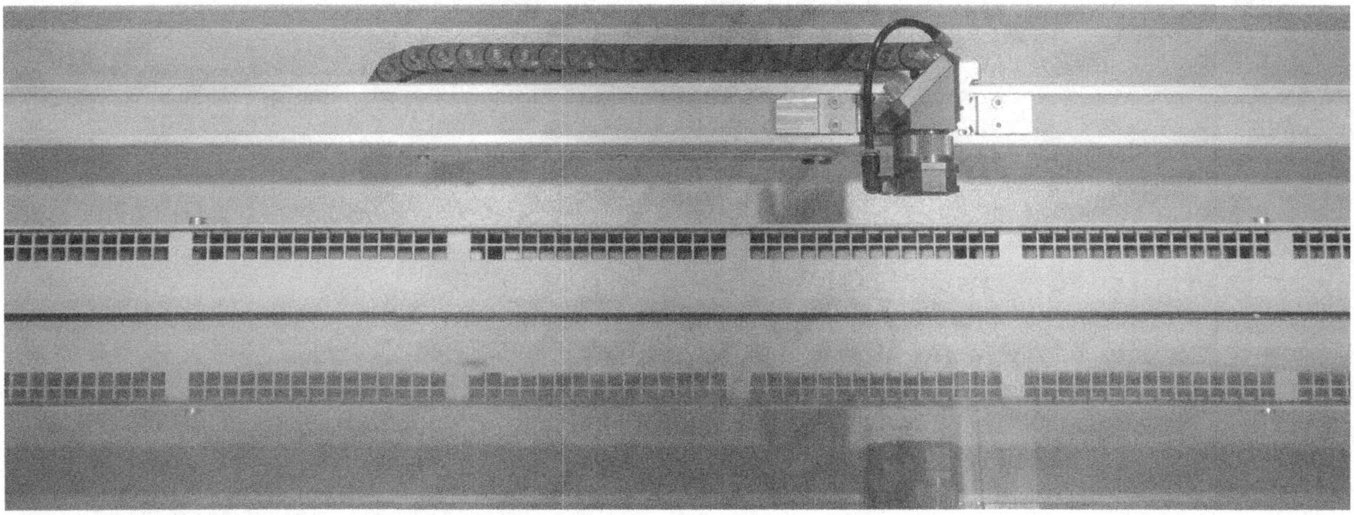

**Laser Buying Tip #5: Extraction System**

Extraction is a requirement when it comes to CNC laser machines ... whatever you vaporize needs to go somewhere, and that's preferably not all over its electronic components! You usually have two choices: extract the fumes directly outside or send them through a filtering system. Depending on your set-up, one will make the most sense to you. One aspect of lasers often overlooked, though, is how that extraction is performed. You want something that is powerful and adaptable to your needs. For instance, the above bed extraction system from a Trotec Speedy 300 $CO_2$ allows you, with tape, to cover holes at the back of the bed for smaller parts so you have maximum power where the cutting and engraving are happening. A vacuum table is also a great way to further augment your extraction abilities, something that should be seriously considered with any laser machine purchase.

**Laser Buying Tip #6: Consider $CO_2$ and Fiber**

If you are trying to get some jobs that require metal marking, you have two choices: use a $CO_2$ laser with Cermark (or equivalents) which allows you to permanently engrave the metal, or use a fiber laser. The $CO_2$ method sounds fun until you realize you have to apply a toxic paste to the metal, wait for it to dry, then engrave and wash it off vs. simply loading up the metal on the bed and using your fiber laser to do all the work. You will save a tremendous amount of time, hassle and health points by using a fiber laser, so if you see any metal marking in your future, you shoud seriously consider getting a machine capable of doing both, using the same machine preferably.

**Laser Buying Tip #7: Training and Set-up**

I've lost count of how many people bought my plans on CNCKing.com and sent me an e-mail complaining the plans don't work. There is nothing wrong with the plans; they just don't know how to use their laser! You wouldn't buy a car without taking it out for a test drive -- don't buy a laser unless you know how to run it! Make sure your representative spends a few hours with you, teaching you how to focus the laser, play with watts and speed settings, and most importantly, is easy to get ahold of if you run into any issues. A machine that's delivered to you by mail with a manual explaining to you how to set it up simply isn't worth the time, effort or money you've invested into joining the laser club.

**Laser Buying Tip #8: Laser Tube Warranty**

The biggest cost for any CNC laser machine comes from the laser tube itself. If you've been in the industry long enough, you'll hear of the horror stories that occur when the tubes "break" repeatedly, and the representative always find something in the shop to blame other than the laser tube. Don't be a victim of your laser tube ... many of the low-end machines come with a warranty that "innocently" neglects to include the tube. Make sure, in writing, that the tube IS included in the machine warranty; better yet, if you can get an extended warranty on the tube itself, you'd be very smart to take them up on the offer. In addition to the warranty, be sure to get an idea of the expected lifetime of the tube. A machine that's good for 10 years is worthless if the tube lasts 3.

**Buying Tip #9: High Quality Bed**

The bed of your CNC laser must be sturdy and perfectly flat. The laser is focused before the job is cut or engraved (manually or automatically,) and having a bed that's off kilter makes sure that your work will not look as good as it should with sections being out of focus or not engraved at all.

Your bed should be perfectly square with each line, horizontally and vertically, being perfectly straight from the zero x and y coordinate (usually top left corner). An easy way to test this is to power your laser and run it with a pointer (if it doesn't have an integrated one) the full length of the working bed on both the x and y axes; the pointer should follow the bed and bed inserts perfectly without any deviation. An added feature that's nice is having x and y axis rulers; it comes in handy when doing manual changes and to double-check material positioning.

## CNC Opportunities: Warehouse to Garage

Although I fell into the CNC industry as a complete outlier, producing industrial 3D animations for large global companies before the whole "3D craze" began almost 20 years ago, little did I know how much these massively heavy and complicated machines would shrink and increase in capability over these past two decades. If you think the evolution of computers has been massive through the years, just go to any industrial show and realize that many of the goliaths you see on the floor there can be replaced using entry level machines for a fraction of the cost. There is still lots of machismo in the industrial sector, which is why the lower-end industrial machines haven't been replaced by their prosumer variants, but that's changing rapidly.

In step with the evolution of the physical presence of CNC machines, the software running them has also greatly advanced to the point that you no longer need an industrial workstation to run a lot of what used to require a very hefty investment in both resources and time (remember my first 3D machine was an "industrial" workstation). The software has dropped in costs as the machines have gotten easier to use.

These three factors: increasing computer processing power, decreasing hardware requirements, and simplification of CNC machines themselves, are slowly seeping industrial opportunities in the consumer marketplace that few are aware of. Ask any industrial machine distributor what can be had nowadays compared to 20 years ago in the areas of capabilities, speed and most important, pricing, and you'll be blown away at the advancements made within the CNC field. Here's what's scary though -- most people are completely unaware of how these changes will affect their lives before it's too late.

When I was growing up and wanted something, I could either build it out of wood with a hammer and a limited selection of power tools that were off limits, OR go to the store and buy it. Today, if I want something, the first thought in my mind is "I wonder if I can build it?" How many power tools do I have today? I have a skill saw to cut board to size to fit within my CNC machines and a drill for screws. My jigsaw comes in handy for "edits."

My ShopBot Desktop CNC table router has effectively replaced any need for a drill press, bandsaw, scroll saw, and the list goes on. One machine replaces the capabilities of more than a dozen, and with a precision that's unmatched by any of the previous tools. Throw in a CNC laser and 3D printer, and I've basically become a one-man design and manufacturing company that can be run out of a shed!

I have several friends who, like me, own a variety of CNC machines, and they are making a KILLING in the marketplace because they can ask industrial level pricing using far less expensive equipment and run it all themselves.

These are people like you and me who decided to express their freedom and have their own companies capable of doing what used to require half a dozen people and a large warehouse. Consider for a moment that a few of my friends are making far more than $200 USD per hour on their machines, and it doesn't take too long to figure out how profitable running your own manufacturing plant out of your garage really can be! 200K in machines pays off fast! What's more -- they tell the machine to do the job and sit back and let it do all the work. Once a machine is given a job, there is nothing preventing you from giving another job to another machine -- quickly scaling your hourly rate!

You may have heard that China has become the manufacturing capital of the world. It's very true, but there is a tremendous amount of work that can't be done profitably offshore -- the production runs are too small, and there are tight deadlines for personalization plus shipping and customs duties. There is a massive market out there in just about any medium-sized city that's profitable and relatively easy to get into if you have the right equipment, knowledge and skill. Make a search online for 3D, laser or table router image product searches and you'll see a plethora of ideas to create your very own very profitable CNC production niche.

The way I look at it, either get into the CNC industry NOW or work for somebody in the CNC industry later for a fraction of your worth. Either way, the CNC revolution won't really start building up steam for another 5-10 years -- as big as lasers and routers are, 3D printers will be larger once they become more affordable and overcome their cubic limitations -- so there is still plenty of time to get into the market and establish your clientele before the market gets too saturated for all the low-hanging fruit out there.

Here are some top secret laser operational tips that will drastically improve the efficiency and productivity you get out of your laser.

**Top Secret #1: Bottom Right for Material Testing**

Your CNC laser will be at its strongest power at 0, 0 (x, y axis,) which is usually found at the top left corner depending on your power source and mirror setup because the beam has less distance to travel, hence less power degradation.

When you perform material testing, do it at the furthest point to guarantee that your settings will be strong enough. Otherwise you may find that your laser doesn't produce the optimal results throughout your engraving or cutting process on larger jobs. If you have an extremely small bed, this difference might be so small as to be unnoticeable, but on larger platforms spanning more than a meter, this can affect overall job quality even with a high quality tube. You also want to test two shapes, a circle due to its constant velocity and a square due to accceleration and deceleration lines with a square corner.

### Top Secret #2: Use Print Merge

If you have a billion name tags that need to be done -- save yourself days of work by using Corel's "print-merge" abilities (and a jig). First sort the names in an Excel document within their respective columns. After you've built your template within Corel, perform a "print-merge" with the Excel document, and you will have everything automatically populated in the right format and size! Be sure to run through each entry manually just to double-check, as sometimes people's names are longer than the space provided.

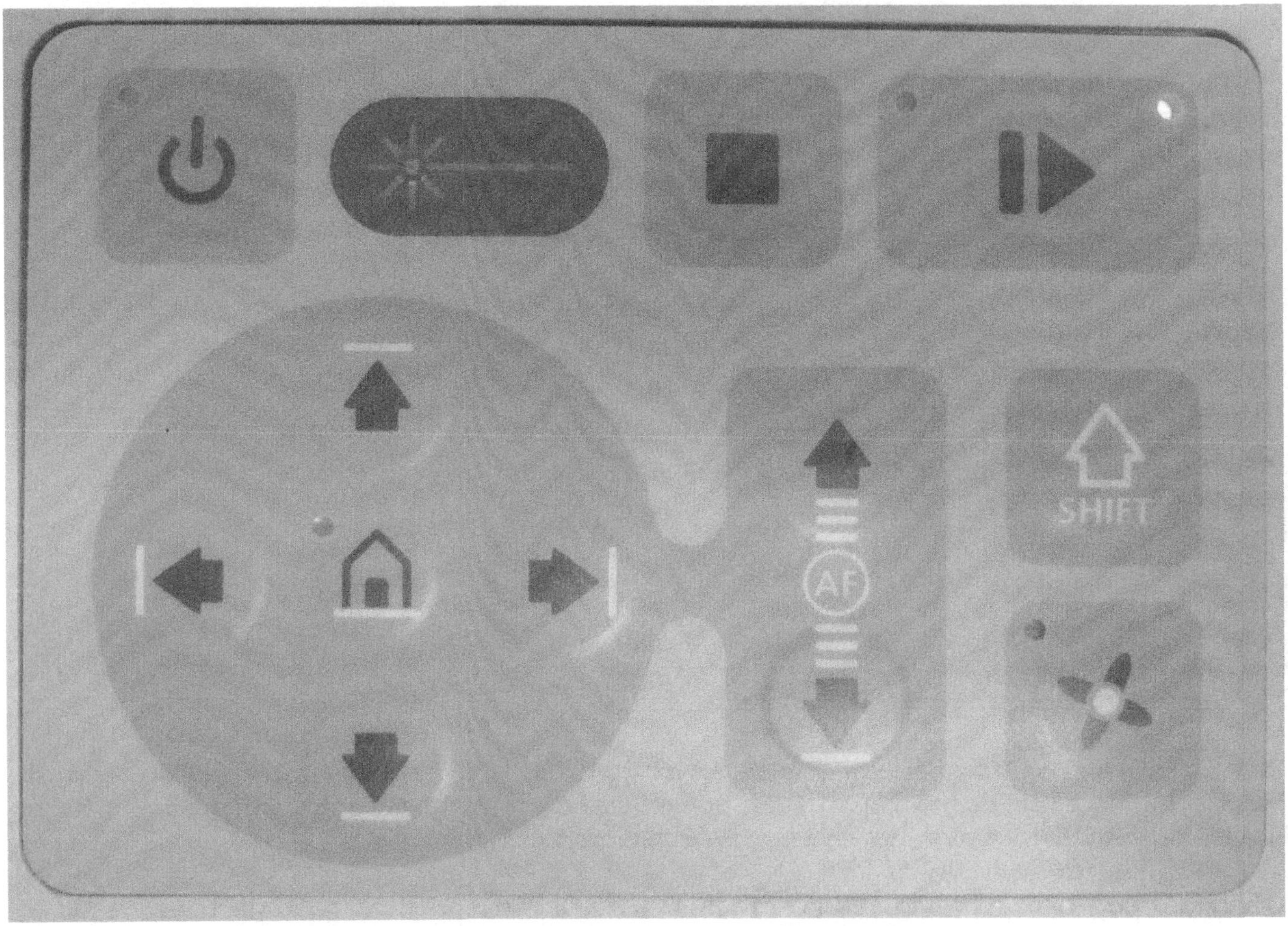


### Top Secret #3: Put your computer next to your machine controls

If your laser machine has an integrated control pad, save yourself lots of time by having your computer on the same side as the keypad. It will save you a lot of time going back and forth between the two for job placement and monitoring as manual focus is usually done on the keypad, not the laser software.

Unlike other CNC machines, you have a gradual increase of particles that build up under the cutting platform, requiring regular maintenance -- which in the above picture was very much overdue! Be sure to open up your machine on a regular basis and vacuum it!

# Interview with Simon Moore
# Director of Trotec Laser Australia

I've had the pleasure to meet-up with Simon quite a bit since moving to Australia; he has the enviable task of managing Trotec's entire Australian operations! As a result of how spread-out this country is, he spends quite a bit of time flying to and from Sydney helping with demonstrations, installing lasers and making sure their client's operations pay off their Trotec investment ASAP.

You can find out more by visiting TrotecLaser.com.au

**Can you describe how you first got into laser engraving and what your initial thoughts were on the technology?**

Back in the early 1990's, Evright.com was looking to further expand our business. We had offered traditional rotary computer engraving for 15 years, but started to look for new technology that would ensure a high quality result, with shorter lead times. We decided on a Trotec laser solution because it suited the existing customer base who were using our engraving services.

Initially the obvious attraction was the massive reduction in set up times and labor offered by a laser. At that time, customers could not readily supply vector files, so computer engraving had quite long lead times. The Trotec laser enabled Evright.com the ability to seamlessly enter new markets and take advantage of new opportunities. This has provided Evright.com with a sustainable competitive advantage that it has maintained to this day.

**Can you give me a little background about Trotec? What is it's unique value proposition in the CNC laser field?**

Customer support is at the heart of everything we do at Trotec Laser. Trotec Laser can truly lay claim to being the industry leader in flatbed lasers, with innovations like patented JobControl software and InPack-Technology. Trotec Laser has saved our users hundreds of thousands of dollars in production time. We exist to make our customers more profitable, that is our promise and what drives our innovations.

Trotec have now moved to CeramiCore Technology in our laser sources, after many years of rigorous research and development. This technology was previously only available to the military, but now it provides ground breaking improvements to the speed and quality

of Trotec laser systems. CeramiCore Technology has also reduced average running costs by 30%, as the new technology reduces the amount of heat generated and is overall even more energy efficient than our previous laser sources. The benefits of CeramiCore customers to our customers is huge.

**How have laser cutters and engravers changed over the past quarter century of you using them? What other advances do you see happening in this sector over the next 25 years?**

Laser technology has developed at a rapid rate, in line with technological developments that have occurred within the wider I.T industry sector. Customers today can benefit from these advances and take advantage of faster, more efficient systems with lower entry costs.

The additional developments in technology that are specific to Trotec Laser not only reduce entry costs, but can also offer the lowest overall cost of ownership of any laser on the market.

The greatest change for today's laser industry is to introduce the benefits of laser technology to a wider market and new industries. The return on invest in a laser is now so fast, it's very easy for small owner owners to justify the entry price of laser technology. Trotec lasers are also designed to be exceptional user friendly and easy to use - Trotec Laser has the only galvo system in the world that can operate from a standard printer driver.

The level of innovation in laser technology has never been greater, and will certainly continue as the use of cameras in laser systems increases. Flexx technology, which is the ability to combine Fibre and CO2 in one job, is incredibly popular and offers so many advantages that we see it becoming a standard rather than an option.

The combination of digital finishing and laser cutting will increase considerably, and Trotec Laser is proud to have already partnered with Ricoh in Europe to maximise the benefits that laser processing offers digital printers.

**If I were looking at buying a laser today, what are the key features I should be focusing on?**

The main driver of the purchase of a laser system for any business should be to increase the profitability of your business. Trotec Laser develops our systems with this as our primary consideration, with the total cost of ownership as the key indicator of a profitable laser system.

Three core aspects to consider when comparing laser systems are POWER, SPEED, RELIABILITY. Laser power is only useful if the motion system and electronics can deliver the power, many systems are simply unable to deliver the power when operating at high speeds. This means you have paid for power that cannot be truly used when you are trying to achieve quality engraving, marking and cutting results.

This is why Trotec Laser has developed CeramiCore Technology, because we are determined to achieve the finest detail engraving, at full power, at full speed and combined with the fastest acceleration of any laser system on the market.

The benefits of a laser source than can deliver power and speed are significant, when you consider that production times can be as much as half compared to other laser systems sharing similar laser power. That's when our customers enjoy the competitive advantages of a Trotec laser.

The operator cost can often be more than the laser system, so ease of use is paramount . The JobControl software that comes with a Trotec laser can often return the cost of the laser back to the business because of the amount of operator time it can save. When you combine this with the ability to control your laser system with the Laser Remote App for iPads and iPhones, you can really optimise your production and finish more jobs, sooner.

Every laser system is an investment in your business, and you want to ensure that investment returns over time. As part of our commitment to making our customers more profitable, we ensure the highest possible build quality in every laser we sell, so that you have a solid foundation from day one with a laser system that's been built to last.

Our relationship doesn't end after the installation - we offer a lifetime of free training, service and at designated training facilities Australia wide. We want to partner with our customers businesses for the long haul, and stand by our laser systems quality by offering up to ten years warranty under our Trotec Protection Plans.

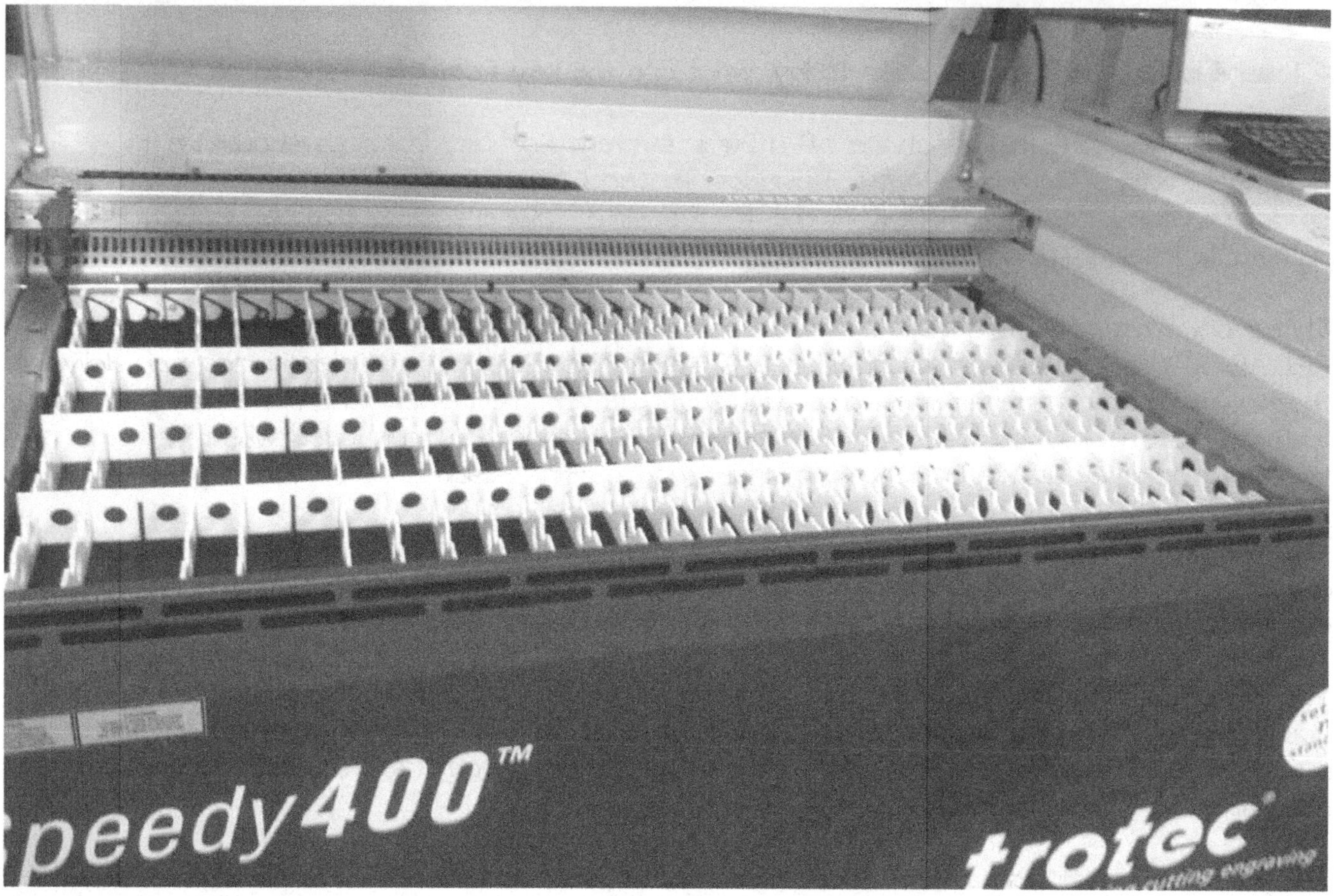


**You recently designed an acrylic cutting table for use with paper cutting - where did this idea come from?**

Laser technology is the perfect partner for the digital printing and finishing industry. Paper cutting is one of our fastest growing markets, because Trotec lasers can offer a high quality of cut without scorching. It's a demanding industry, which uses only the best technology to deliver the high quality results that customers demand.

The paper table was designed to provide zero flashback, whilst still providing a full vacuum.

Each Trotec laser system comes with jigs and tables designed with the customers' needs in mind - each Trotec laser is customised to suit a specific customers unique requirements.

There are a large range of options and attachments designed to make the customers laser as user friendly and versatile as possible.

**You went from managing a well known awards brand, Evright.com to managing a well known brand in the laser industry - how has this transition affected your perception about the industry?**

The transition itself was quite straightforward, as both brands have a customer centered approach and are focused on delivering the best results for their customers. The brands also share similarities in that both Evright.com and Trotec Laser invest very heavily in their market and development.

I have been proud to have been a part of both brands as it has given me the opportunity to give back to the industry, through sponsorships of industry initiatives such as the Engravers Register and the Trotec Laser WorldSkills Challenge in partnership with our industry association, VISA.

I have been surprised by the lack of support offered to machinery purchases in our industry, so I am always looking at ways we can make a difference and ensure that after sales service is a priority. Hence my desire to ensure that after sales service and training facilities are available to our customers Australia wide. It is vital that our customers are successful and profitable if Trotec Laser is to grow as a supplier of laser equipment. I believe that success of Trotec Laser is linked to the success of its customers. I want to ensure that every customer is fully supported every step of the way, so we are pleased to offer lifetime free support and training.

**What material do you enjoy working with most and why?**

Like any laser user, I love working with acrylic and engraving on the latest tech toys - we all have Trotec laser engraved laptop and iPhone.

I love acrylic because it is so compatible with the lasers and has seemingly endless applications from signs to promotional products.

I would expect the paper market is about to explode, with more printers and designers creating projects that would be impossible to do without a laser, they are so advanced that traditional knife cutting methods cannot deliver the same results.

trotec
A RAYJET!
C37
TROTEC LASER - WIN A RAYJET!
trotec
trotec
www.troteclaser.com.au
trotec
Speedy500
trotec
trotec

# A Word about CNC Model Complexity

One of the most common problems I see at my end of the stick talking to people buying files from CNCKing.com is jumping right into massive multi-board and 100+ piece projects on their first go. They just got their CNC table router or laser cutter and they just can't wait to make someting truly impressive. I try to tell the customer that they need to slowly get into it and not try to jump the boat, as they'll drown from the current.

Cutting a massive project is exhausting for the operator, but what's worse is looking at 100+ pieces lying in front of you and realizing that you have to put it all together using glue and clamps. Many people really aren't used to using clamps or seriously gluing things together at a massive scale -- yes, pieces may slide as they dry, and your hands get rough, and you will draw blood on more than a few occasions and get that wonderful sliver right under your fingernail! People start, then give up and never try it again.

Another common issue I see is designers looking at a few models and going nuts on their first one to design something truly impossible to make, that nobody else would want to make! Who in their right mind wants to cut and assemble a design with over 400 pieces? The cutting costs a fortune and the results could be just as easily achieved using a fraction of that piece count.

I'm astonished at how many students send me their stuff they spent a month on designing and argue with me that I don't know anything because I tell them their design is impossible to make. From my understanding of physics, you cannot have two people in exactly the same space at the same time, yet many designers see their computer capable of doing this with their design virtually and think this is also valid in the real world.

Generally, people confuse complexity with usefulness or esthetics -- they are wrong. If you can make a beautiful design using 10 pieces, why would you want to cut all those pieces up so that they now become 30 pieces with the same result?

I'm not sure where this idea comes from but I strive to use as few pieces as possible for all my designs. Sometimes I throw caution in the wind and make something with 200+ pieces, but I do this as a personal challenge and I KNOW IT WILL WORK. I also optimize the hell out of the design so those 200 pieces are the absolute minimum I can get away with.

If you want to get into complex models, learn the basics by buliding very simple ones first. There is a lot more to design than astronomical piece count and the complexity involved with having interlocking parts. Look at the award on the right -- it has lots of depth and complexity, but look closely and you'll see it's mostly made of large repeated parts.

1
E=MC2

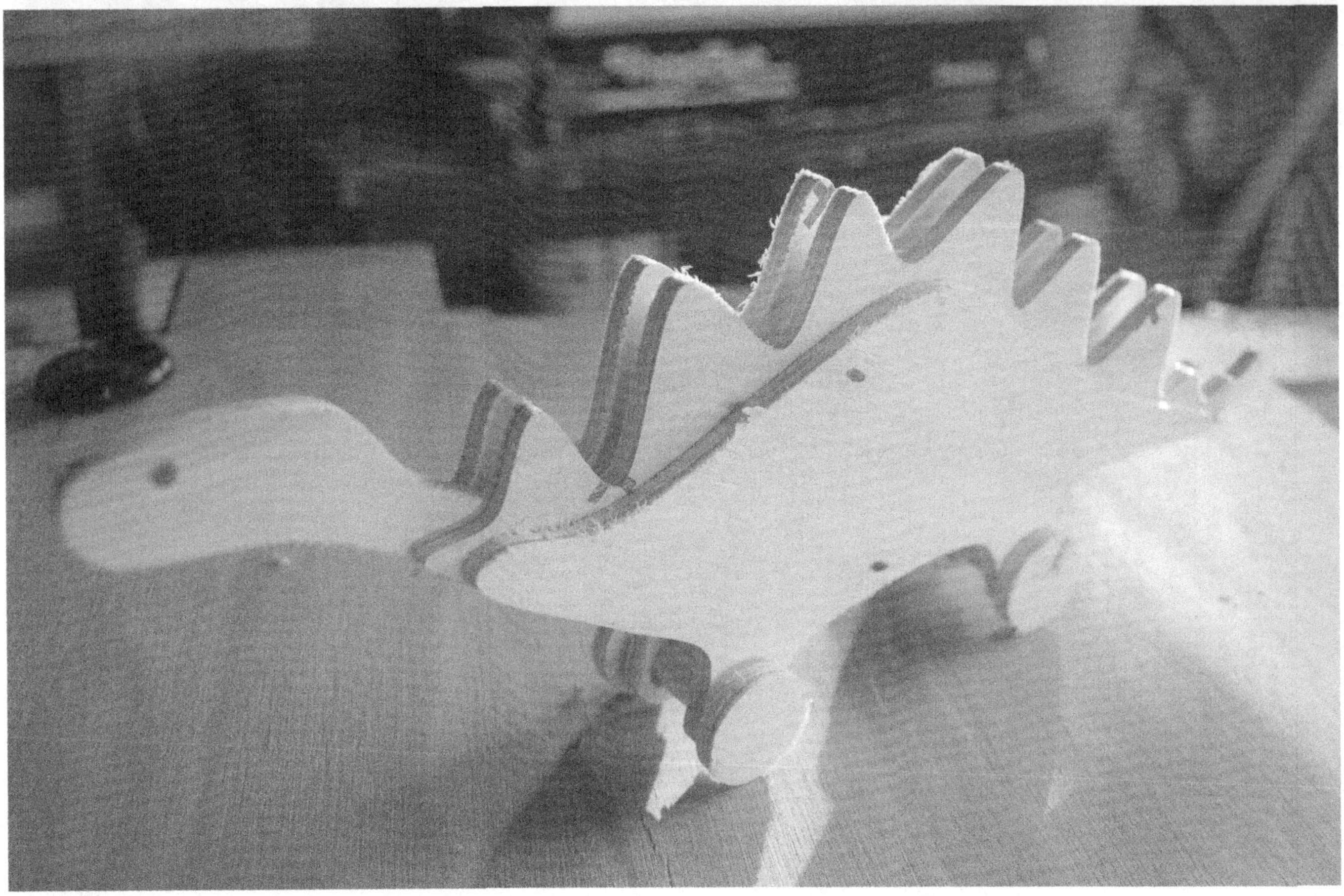

I know you may think it's a waste of time to make a Stegosaur (see table router section of this book for plans) as your first CNC table router or laser cutting project, but even after years of doing this stuff, that's the first one I cut and design whenever I get a chance. It teaches me about pockets and assembling with guides and is an enjoyable project to make even without instructions.

It has both straight edges and curves and is extremely easy to customize further -- want to make the body wider? Add more panels ... change the number of spikes on its back or make them sharp. How about adding a mouth or doubling up the wheels so that it doesn't tip over as easily? Add spokes to the wheels or extend the tail so that it can actually swing using a "living hinge."

It's really a wonderful project that I enjoy making time and again, and that's my suggestion to you. Start off small, customize and learn instead of jumping right into massive projects that will likely never see the light of day.

# Toy Design 101: Behind the Scenes

I have been interested in wooden toys for a very long time but fell into the design aspect of them only recently. I've never taken any design courses, so the experience I gained was from trial and error. YEARS of trial and error over thousands of hours!

What I enjoy most about building these items is how infinite the design options really are -- you can design just about anything, and as my skills improved and my exposure to other designers grew, so did my interest in the CNC design field. As a result, take everything I have in this first section of the book with a grain of salt, since what might work great for me might not for you; nor are these ideas static -- they evolve over time. Above all, don't fear making mistakes! I'm sure my mistakes to success ratios for these designs through the years was far higher than the average, but with persistance and undying perseverance, they began to diminish and the result speak for themselves.

**Establish your limitations early**

For vehicles, you want to start with the wheel base to establish total model dimensions and proportionality early, while for static objects, start with either the largest piece in the model or objects farthest away from the "center" of the model. For instance, the Ferris Wheel B began with the large wheel while the Stereo Bank began with the speakers on either side.

The reason for wanting to establish your limitations early is simple: you don't want to design something that can't be made! If your maximum board size is 500 mm square, there is no point in designing something where the dimensions of one part will be 600 mm!

Other limitations to your design can be wood availability (6 mm or 1/4 inch is generally easy to find globally), skill level of client, equipment availability (CNC plasma, CNC laser, CNC router, jigsaw, scrollsaw,) and even budget. Realize that these limitations shouldn't influence your creativity. Spending time figuring something out is half the fun!

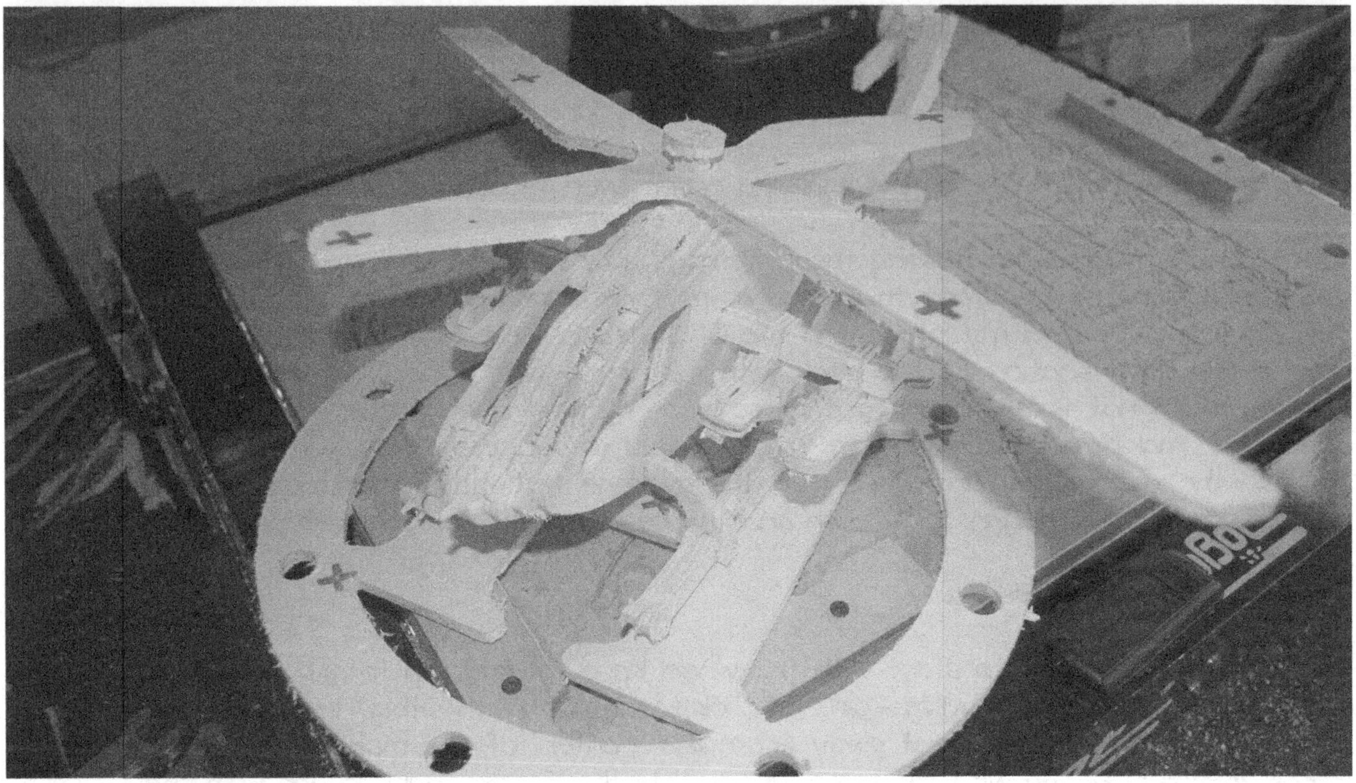

**Stop while you're ahead!**

Like a fine painting, there comes a time when you have to STOP and move onto the next project. I've been involved with several designers who will build a model, then rebuild it and then ... get another idea and rebuild it again! I think the most efficient way to model is to build something to the best of your ability at the time and if you get another great idea, build it into your NEXT model instead of your previous one.

The amount of knowledge you gain from building NEW stuff far outweights what you would have gained rebuilding past work. Of course, if your model has a design flaw, fix it, but keep in mind that you want to solve the problem, not rebuild the whole thing! Don't be a perfectionist, be a pragmatist ... you are a mortal, so don't waste your time! How do you know when to stop? When you spend more time thinking about improvements rather than implementing them. For instance, replacing a solid side panel with one that has additional engravings or cuttings is fine, but when you get to the point where you are considering splitting that side panel into three pieces, each with their own hinges for a back panel that won't even be seen -- you know you've gone overboard. Follow your gut!

Use IKEA's philosophy -- make what you see great and what you don't see functional.

**Give lots of clearance**

When I first started designing models with moving parts, I thought that they would eventually wear out into a smooth turn or spin, but when you think about it, the amount of wear and tear necessary to accomplish this using 6 mm (1/4 inch) wood is pretty massive. It's best to build your model with extra room built in; otherwise it will take a child years of rolling your toy car across the floor to accomplish something that would have only taken an extra fraction of a second on a CNC machine.

Regarding general piece clearances, I've always cut my laser models using only the amount of material vaporized as my clearance -- cutting directly on the line. Maybe a milimeter or a fraction of an inch is burned away, and that's fine. As for my ShopBot Desktop CNC table router, I need to give a little extra clearance, but it's so insignificant as to not even be a consideration when designing. If you ever really do get stuck, cutting another part down using some rough grit sandpaper more than does the job.

**Build your own models**

A body is nothing without a brain and a brain is nothing without a body. Think of designing in the same way, where both thoughts (virtual) and building (physical) are required to fully flesh out a design. When I first started WoodMarvels.com, I didn't have the space to design within my cramped Korean shoebox apartment, so all my work was done virtually.

Only later, after getting my ShopBot Desktop from ShopBot Tools Inc., did I fully realize how much I was missing by jumping this critical step. I did assemble earlier laser-cut models when some testing was required to verify the model or I wanted to confirm the functionality of a design, but nothing quite matches assigning tool paths, cutting, assembling and playing with the finished model YOURSELF!

If you are in a similar circumstance to mine, other options include making your model out of cardboard at a smaller scale, sending the design to other designers to have a look-see, or even going through a local design shop if your budget allows. Making simulations was my way around this limitation, but the work involved for this can be daunting if you aren't comfortable with 3D software. Master the software and hardware will follow.

Since my previous WoodMarvels.com volume, I've taken things a step further by filming the cutting, assembling and presenting of the models found on my site. This has increasingly refined my efforts and what I learn as I'm making the entire process behind my models public. This aspect of my business will expand greatly as I acquire the machines, partnerships and skills necessary to pull this off. Pressure is a great teacher, and I want to show both my triumphs and failures so WE can learn from them!

When you put yourself on the spot to cut and assemble your own models without any retakes, it really requires you to put your thinking cap on. This is yet another method to showcase a design in addition to the 3D assembly animation, numerous renders and books such as this. As the body is nothing without the mind, the CNC machine is nothing without the software running it.

**Take a Break!**

Designing is great fun, but if you do too much of it, your ideas start to merge and you end up going down the same comfortable path over and over again. New ideas come when you are essentially doing NOTHING ... go to the cottage for a week or even a month, and when you come back to the drawing board, all that built-up motivation will gush out of you -- ideas you never even imagined will seem like common sense. I've done this several times, and it leads me down fantastic paths!

Another common thing I do is not focus entirely on one project at any given time. Taking a break by multi-designing is also a wonderful way to give your creative muscles a rejuvenation as long as the model or CNC machine you are designing for is completely different. Converting files from one CNC type to another is also a fun exercise.

**Enjoy your headaches**

I'd say the best part of designing is the mental gymnastics you force onto yourself. It's incredibly exhilarating to start with a blank screen and a few hours later, or maybe even a week of designing on/off, have a finished model that others are willing to not only enjoy but purchase the plans for! CNCKing.com has been a passion of me bringing together skills that I've taught myself throughout my life, the love of 3D modeling and animation, the love of woodworking and the love of working with my hands. It isn't always smooth sailing though and I don't try to hide my faults -- I share my screw-ups!

Sometimes parts don't work the way you intend or there is that one critical piece that just happens to be 2 mm too long or wide. You have a design that's optimized for 6 mm, but after cutting all the parts, you realize you bought 7 mm by accident and nothing works. There are endless hours standing next to a machine cutting away on your design or worse, having a bit break right when the project was just about over. To me design is headache management -- you try to manage the things that can go wrong as best you can, but they WILL go wrong, and it's your job to keep on pushing through!

Why do I say enjoy your headaches? Because they mean you are achieving progress. If things always work or go according to plan, it's too easy, and you aren't pushing yourself! I do find it relaxing though to design one difficult model and then after, a far simpler one to recharge my mental batteries. Now that I've gotten into other CNC machines, I can do a laser model, then a 3D one and then power-up the router. Don't let stress of the past affect you. Move on!

**Keep at it and NEVER GIVE UP!**

As with anything in life, the more you do it and the more time you devote to an endeavor, the more you'll improve. Don't give up! Even if your designs suck and none of them works, keep going -- eventually the design part of your brain will light up and you'll be doing amazing things! If you show kids your design and they throw up, make something better!

Don't get into designing or CNC'ing seeking absolute perfection -- it doesn't exist. The human mind will never be able to match the precision of these amazing machines. Instead, focus on getting incrementally better each time. It's quite incredible, if you think about how many things can actually go wrong in software, hardware, environment, power and so forth, that anything of quality is ever produced -- grow mentally in your abilities and the error rate will go down, but don't set a specific timeline or pressure yourself. Eventually, you will be comfortable and be able to produce great stuff!

# Castle Bank

<u>Outline the model</u>

When I'm designing a structure, I find it best to build the outer parts first to establish what looks most in proportion within my size requirements. I knew I wanted a grate door on both sides with an arch.

<u>Add some visual loot signals</u>

As I knew this was going to be a bank, I needed an easy way for the customer to put money INTO the bank in addition to a visual cue as to how much was already inside. Adding a big hole in the top and small window slits on the sides (so money didn't fall out) did the trick.

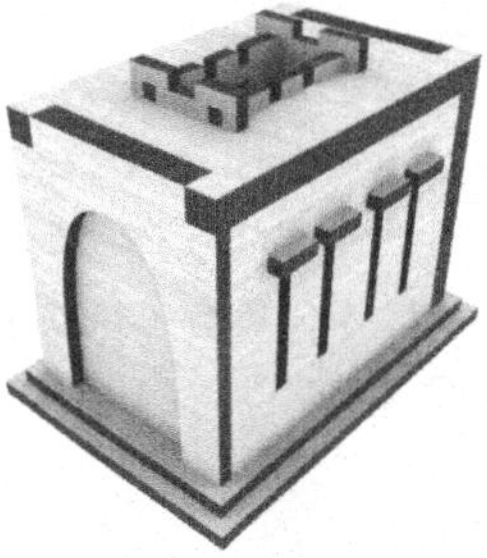

<u>Create a presence</u>

A castle is anything but inhibited -- it's a strong, imposing structure, and I had to build that into the model as well. This helps give it presence and also establishes that yes, it's medieval in nature.

<u>Add a side fence</u>

Adding small details like a fence along the side helps add a lot more depth to the model. I've found adding a "stepped" growth to certain models works very well in creating scale and mass instead of straight large pieces.

<u>All finished!</u>

I had to come up with a clever way of removing the money from the bank other than just turning it upside down. That's where the thought of swords on both sides hanging from the window slits would work best. The swords fit into a groove that doesn't allow you to slide the central piece out.

# CH47 Chinook Helicopter

Establish cockpit dimensions

As the cockpit is the most massive part of the helicopter, I started with establishing these dimensions and prorportions. I knew the largest piece would be the rotors above, but spacing here also established their maximum length.

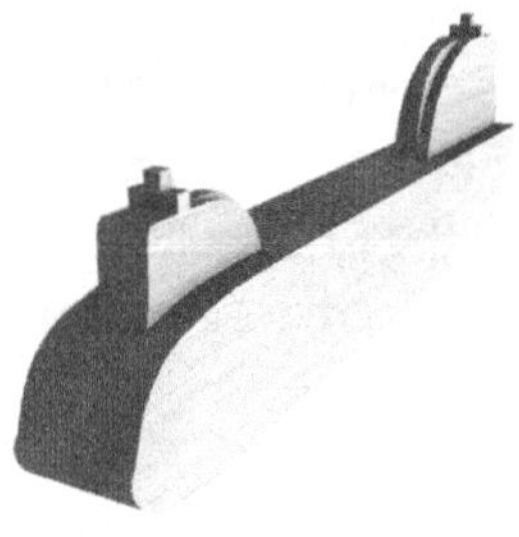

Add windows

The unique round windows along the sides and flat window in the front are obvious aspects of this model that had to be transferred over. Another major detail is the wide lower panels on this helicopter that follow the length of the cabin.

Add rotors

Rotors with end-caps so they don't fly away were locked in place, and wheels were added at the bottom of the model. I also added a stepped look so the model didn't look so flat along the bottom, along with guides that go all the way through the model for easy assembly.

Add an engine

There are two massive power sources on the back rotor pillar, another unique aspect of this helicopter which had to be added. I also further refined the wheel placement on the bottom of the model.

Make refinements

I noticed that the bottom panels didn't go all the way to the front cockpit, so I pushed them back a bit and lessened the length of the internal guides to make them less apparent. I also made the last side panel on both sides removeable so the internal cabin could be easily accessed.

## Easter Island Head

<u>Establish working dimensions</u>
As the Easter Island Heads found on Easter Island are so well known, I wanted to make sure I had the proportions right with a strong base so the finished product wouldn't fall over or seem out of place on a shelf. Instant recognizability was also a must!

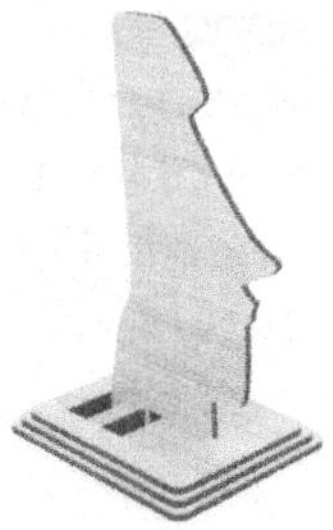

<u>Add vertical slots</u>
Building the head using a series of vertical slots, with a hole in the head which would serve as "hidden storage," seemed most appropriate for this design. It made the building process dead simple and also gave it a weight that would be difficult to achieve in any other way.

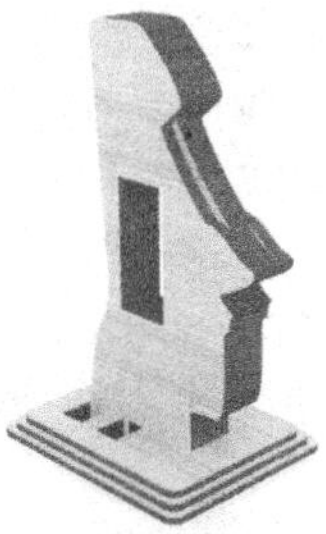

<u>Add eyes</u>
The eyes were the most difficult aspect of building this model -- I wanted them to be piercing, but at the same time require no additional design elements. Laser cutters produce burned edges, so in this case, using a flat piece provided the perfect automatic contrast I was looking for.

<u>Complete the head</u>
As I wanted the storage to be hidden, I made one side of the model solid and the other with an ear that you can grasp to pull out the tray to put your valuables onto. The internal tray accomplishes two goals. It gives the balance required to keep the ear side of the face from falling back and is an easy way to align things perfectly.

<u>Fine tune</u>
I later designed this model to be CNC cut as well, which caused the nice contrast for the eyes to be lost. There were a few internal guides added to the design to help with alignment issues, but the biggest challenge with this building system is that 6 mm plywood isn't always 6 mm, so you need to leave more/less room in the base.

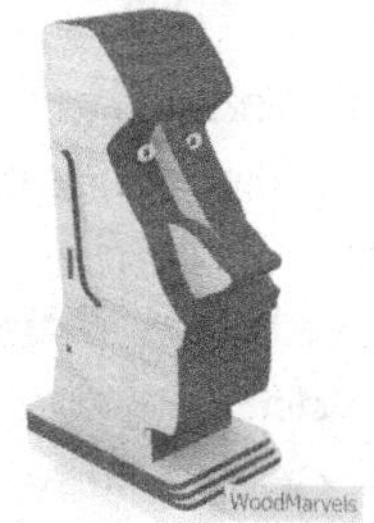

# Electronic Award

<u>Base dimensions</u>
I modeled this electronic award on a processor where you have the bottom plate and heat sinks above it to dissipate the heat. This model would feature lots of engravings to add details that would take too long to build piece by piece otherwise.

<u>Add a processor</u>
I decided to have the processor elevated purely for esthetic reasons. The nice thing about this was it also doubled as the perfect solid guide for all the pieces above and below. I also added lower side panels to the base to support the legs.

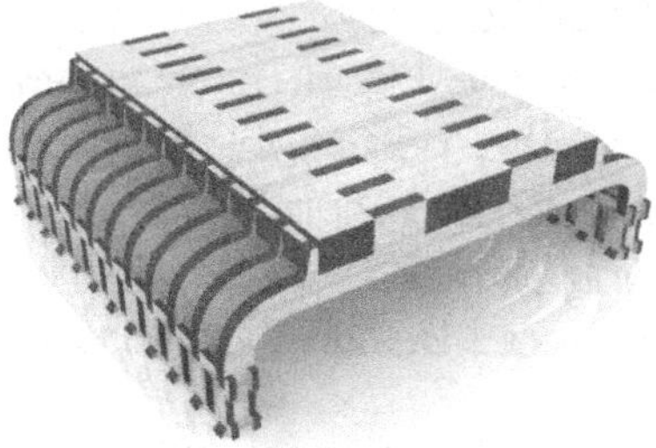

<u>Add fans</u>
The bottom worked so well that I decided that if this was working in real life, it would need more cooling, so I added fans (that work!) on both the front and back of the model along with additional heat sink fins on the top. This alone added a lot more detail to the model!

<u>Make refinements</u>
The rest of the model to bring it up to date was to flatten the front so you had an area to write somebody's name (this was an award after all) along with lots of additional internal components. I wanted this model to look "busy" even before the engravings were added.

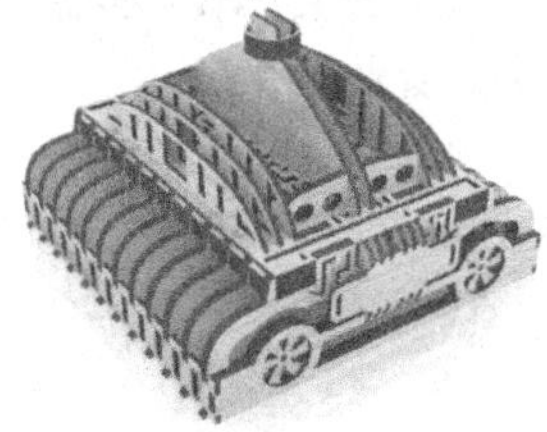

<u>Number 1</u>
Another aspect that I wanted transferred to this award was adding the #1 somewhere. The top made the most sense, and I figured out a way to actually have it spin on a column without falling down. Hollowing out the middle fins also added more detail to the model; in this case, perceived complexity increased perceived value.

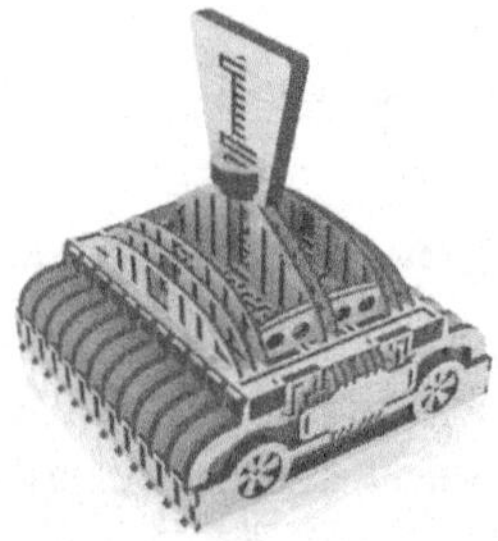

# Ferris Wheel B

<ins>Re-imagine a Ferris Wheel</ins>
The first model I designed on WoodMarvels.com that involved gears was a small Ferris Wheel. It worked all right, but looking at it after I had accummulated a few years of modeling experience, I decided it was time to bring the design up to date.

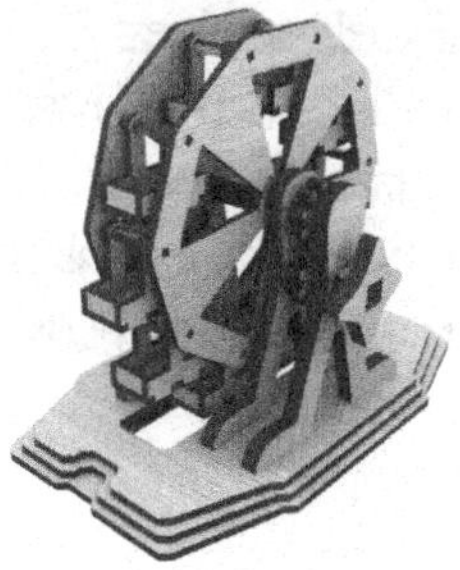

<ins>Start with the wheel</ins>
The largest part (and limiting factor cutting-wise) is the wheel, so I had to establish that at the very beginning of the building process. I also had to decide what aspects I was keeping from the original design and what needed to be redesigned.

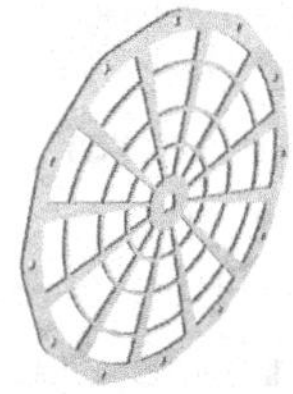

<ins>Establish width</ins>
To keep everything in proportion, after I had established a wheel design I liked, I had to start right away on the seats. The original model had lots of sticking issues due to a lack of clearance, so I made this Ferris Wheel with generous spacing and clearances all around.

<ins>Add gears</ins>
For the first Ferris Wheel, I designed the gears by eye, but this time around, I wanted something more precise. So I built the gears in simulation software and then exported the vectors as DXF files, which I then extruded and used for this design.

<ins>Make refinements</ins>
Once everything was finished, I added several rounds of refinements to the design and built the solid base that would hold everything together. The last thing you want is to be spinning the top while having to hold down the bottom ... creating a base with some weight to it helps alleviate this issue.

# 1911 Ford Model-T Torpedo Runabout

Build wheel spokes

This car is distinctive for two reasons: first, its spoked wheels and second, its long, blocky elongated body. As a result, I had to establish the proportions for each in order to successfully achieve the model accuracy I was looking for.

Build the frame

Once I had the wheel placement worked out, it was time to build the outer frame, again, as a way to make sure I kept within the proportions of the original. It's easier to build from the outside in, as your limitations are already set, than inside out where you are bound to go "over".

Add an engine block

The engine block of the Model-T is straightforward, but after fiddling around with a way to avoid build it without having to resort to a bunch of vertical slots, it was the only way to accomplish the look I was after. Each slot adds error as wood is rarely uniform in thickness, which was why I was trying to avoid layering.

Make refinements

The seating and layered back caused me lots of headaches. Again, I was overthinking this design, trying avoid building this model in the most efficient way. Sometimes the simplest path is the one to take, especially when you lack the resolution required to make it happen any other way.

Add final details

This model looks simple, and its construction in hindsight would have been equally so if I followed my gut feelings. Having the fenders "hang" with some glue isn't the strongest construction method, but then again, this Model-T isn't going to be going 100 miles an hour down the highway.

## Home Gym

Build the weight stack

Allow your passion for design to come from things you enjoy in everyday life. For instance, I really enjoy working out, so I decided to design a laser cut home gym!

I started with the weight stack to establish overall height of my model along, with width.

Build the frame

The weight stack is what holds everything else in place; get those proportions in place, and the rest is easy! You also have the additional weight created by layering wood, which helps when you need to cantilever other parts.

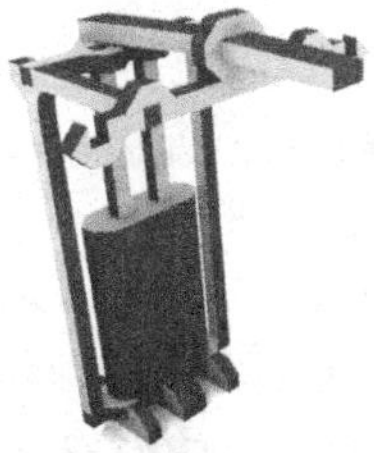

Add the Smith machine

A proper home gym needs a Smith machine, which allows for the safely control weight that your body really isn't used to yet. The barbell with laser-cut weights no less, helps to add to the visual appeal of this model.

Make refinements

Although I could have gone nuts adding details to the Smith machine alone, I decided to focus my attention on other more important aspects of the model, like the weight supports in the back, spotters, and of course, the bench!

Add final details

If you exercise, you know how weak you feel if your barbells or dumbells have no weight on them ... this model is LOADED with weight. Another unique feature is that the weight stack actually works -- you can do finger pulldowns with it! How's that for functionality?

# Mars Explorer

<u>Build the wheels</u>

I'm deeply passionate about space and science fiction -- so much so that I wrote my own science fiction book (see "other books"). As with other "vehicle" based models, I started with the wheels. I wanted to make these wheels extra tough and neat, so I doubled and tripled them up. If you want something to look tough, make it bigger!

<u>Build the frame</u>

I wanted this model to be elegant with all the weight in the back. If you are driving around Mars, you want the power to come from the back across a wide track for traction and the front just hovering over the loose sand. If you hit a bump, no big deal; you'll get a heads-up before something gets damaged.

<u>Add the rotating cockpit</u>

Another unique feature I wanted my Mars Explorer to have was a rotating cockpit ... you can't just open the door to your car on Mars, as you'd die or waste lots of valuable resources; hence, being able to look around if needed would come in very handy to check clearances before overcoming large obstacles.

<u>Make refinements</u>

If you are alone driving around Mars, you want to see everything around you so you don't get blindsided by a Martian with a bigger car! I also added a trailer in the back for groceries or interesting rocks picked up along the way, and plenty of storage on the cab itself ... with a skylight no less!

<u>Add final details</u>

As the gravity is lower on Mars than on Earth, you don't need an extensive ladder to get inside, but a small one to grab onto might come in handy. I also added lights mid-way up the front forks where they are well protected and along the cockpit. If I ever go to Mars, this is the vehicle I want to drive!

# Mars Rover: Opportunity

Start with the wheels

I'm a huge space enthusiast, and I wrote a science fiction model heavily focused on space as well, so making a laser-cut Mars Rover was a given in my mind. As this was a vehicle, as with other vehicles, I started with the wheels. They are among the most complex I've ever designed as they have internal slots.

Component by component

This model was going to be very complex, so I decided early on to build it one major component at a time, then bring them all together as the model was complete. In this case, the camera arm had lots of moving parts.

Add to the body

I envisioned, like the original, having all the electronics within the big box made up of the body. I wanted this model to be pretty accurate as well, so adding the unique leveraged axles and legs was something I had to figure out. If this was metal, it would have been much easier -- I had to double up many pieces for added strength.

Create the solar panels

There are massive "winged" solar panels on the top of the Mars Rover, so it was a great excuse to add engraving to the model. Once the major components were added to the body, developing a locking system to hold these panels was a requirement I had to build as well.

Make refinements

There were lots of refinements required for this model, as it was based on a real working model. The biggest challenge was building a wooden version of a metal model where you don't have the built-in material strength that would have been available otherwise. The easiest solution to this, again, was doubling up parts and adding guides.

# Mangonel Siege Weapon

Start with the wheels

I know this is getting repetitive but yet again, I formed the entire dimensions of this model based on the wheel base. For me, it's just so incredibly efficient to know this from the start and build up from there that I don't know how other designers build things any other way.

Build the outer structure

The outer structure of this Mangonel was the next obvious design step ... it builds up depth while giving you additional information about the upcoming catapult whose length (and power) is dictated by the height of the model and size of the wheel base.

Add supports

Having a powerful catapult is all fine but useless unless the structure supporting it is up to the task. Although I tend to wickedly overthink these things, if your superstructure is weak, then anything moving on it will both be innacurate and destroy the model due to unaligned forces over time. Yes, this uses elastic bands!

Add the catapult

Once everything was strengthened, it was time to build the catapult. The x, y and z forces in my mind were all aligned correctly and symmetrical from all angles. I wanted this model to be dead-easy to build and use, so I decided to just use a twisting action to power the floating catapult instead of a more complex assembly.

Make refinements

So, how do you twist a rubber band if it's within an enclosed area? The idea I came up with was having the supports on either side actually come out (they don't need to be glued as the elastic pulls them into the model) so you can just twist either side as you wish. This worked great when I tested it ... I was very surprised!

# Mangonel B Siege Weapon

Start with the wheels

Although this Mangonel shares the same name as the previous model, I built this one to be far larger and powerful. Instead of using an elastic twisting, I used a stronger elastic extending system to power the catapult. Because of this, everything had to be built much stronger to prevent twisting of the model.

Add additional details

This model had far more detailed wheels; as you can see, these are studded with a larger diameter. They also have caps that help to add additional esthetics and cross-members going across the body, especially in the middle. Overbuilding inside the body was necessary to produce maximum elastic power.

Create the front bumper

Where the original Mangonel was built using the minimal number of pieces, this one had a lot more details when it came to adding front and rear bumpers. This helps protect the axle and just gives it a more effective visual presence. I used the contrast produced by the laser cutting to create detail where in fact, there was none.

Create the catapult

In the previous model, the catapult floated in the middle, while for this one, it runs along its own axle. This means any power generated is more focused. It can shoot heavier things much farther away with a lot more precision as a result. The trigger system works in the same way, though this one is beefier.

Make refinements

This model was a very interesting design process as I already had a working model to follow; I just made it better in every way. I later improved on it further and made it even bigger and more powerful. Check out Mangonel B in the CNC table router section of this book -- can you spot the major differences?

# Stereo Bank

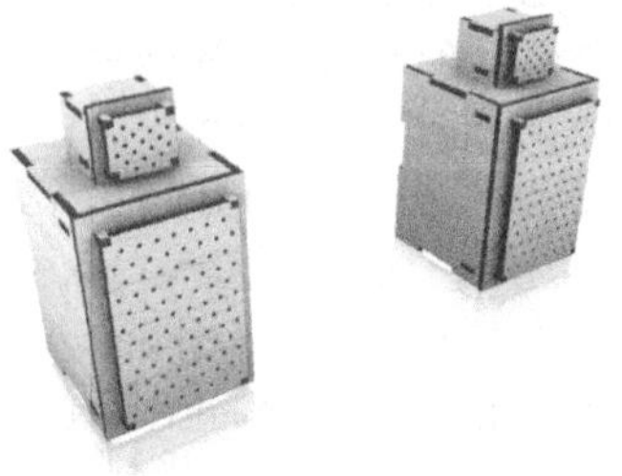

Figure out the length

If you look at a stereo system, the majority of the length comes from the controls in the center, but they are usually proportional to the speakers on either side. In this case, I designed the speakers first to establish the maximum length of the finished stereo bank I was trying to achieve.

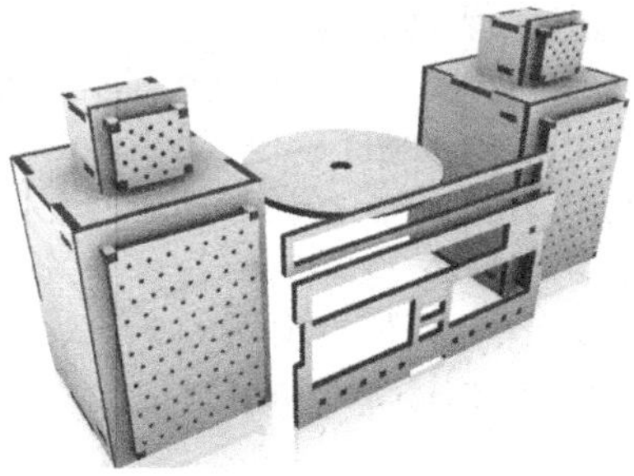

Build the control module

Once I had the length of the model established, I knew I had to start with the width -- most importantly, the maximum diameter of the vinyl disk on top. I wanted this to be a "traditional" stereo similar to the one I had while growing up.

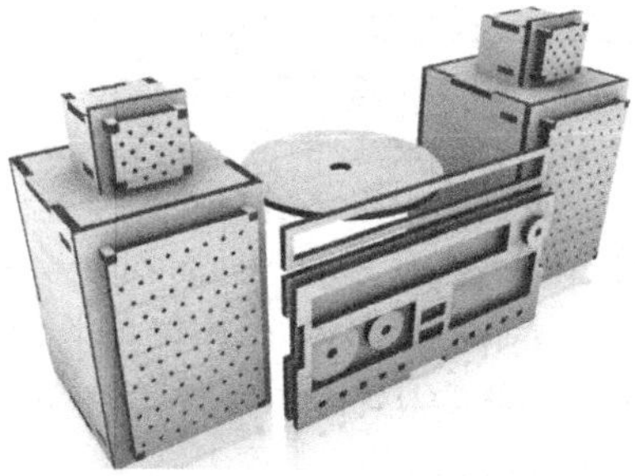

Add tapes

Not only was my goal to have a spinning record above, but I also wanted spinning tapes below! If the model is to have spinning parts, I find it easier to establish these parameters early in the model-building process, as it requires less modifications later in the building process.

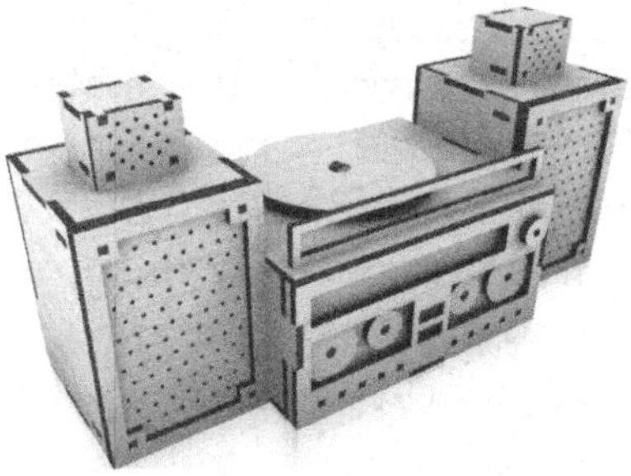

Add details

Once you have the initial model complete, it's time to accessorize it. This is probably the most interesting part of model making because you can let your imagination run wild to improve the look of the model..

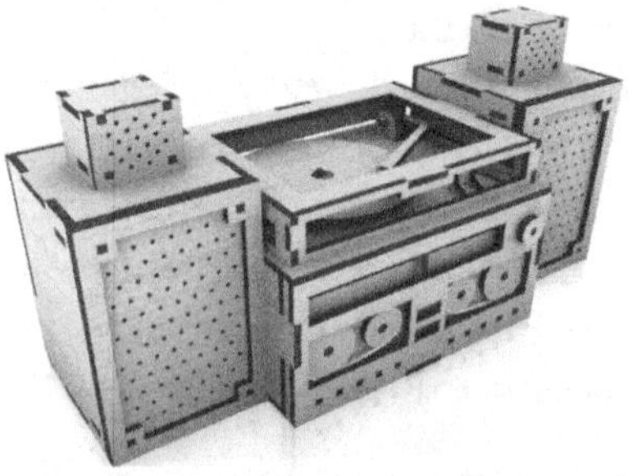

Make refinements

As I kept adding detail onto detail, I decided to build a lifting case on top of the record player. This made the inner vinyl hard to see, so I added additional "windows" all around and above it. The mechanism to open and close it was a simple hinge cut at the same time as the rest of the model.

# Rocket Coaster

<u>Sketch the design</u>
I've gotten into the habit of making a quick sketch of what I plan to build, as I find it simplifies the design process considerably. Although I build everything in my mind first, it's still highly malleable until it's put down on paper.

<u>Complexity first!</u>
Usually I begin a new design with the base first so I can establish the total size of an item from the very start. But in this case, I built the arm first. I knew this would be the most complex part of the Rocket Coaster, so best to put that away first. Bcause it was meant to spin, the arm is what defines the overall model size.

<u>Add rockets</u>
As you see in the picture, I built the rocket holders at the ends of the arms in the wrong direction, so I had to rotate them 90 degrees and add the rockets. I wanted the rockets to be able to spin freely as the arm spun, which added to the complexity and part count considerably.

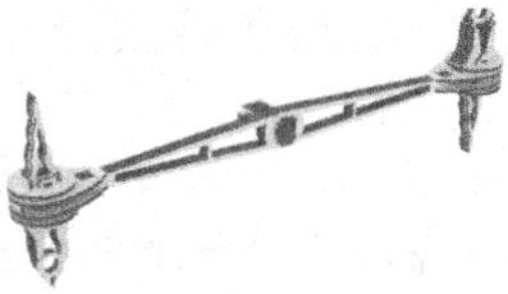

<u>Add some gears</u>
My goal with this design was to use gears to rotate the arm, as I built the model, I realized I could build it so that only one open wheel in the back would turn lots of gears in the front at the same time. Again, this adds to the complexity and piece count, but I gained a new level of novelty with the model, making it wonderfully unique.

<u>Add the base</u>
I had to make sure there was enough clearance at the bottom for the arm not to hit the base. Afterwards, I wanted the base to be wide and strong enough not to move around while a person spun the wheel in the back. I made it two levels to elevate the model higher and provide more weight for the bottom.

# Rusk Plane

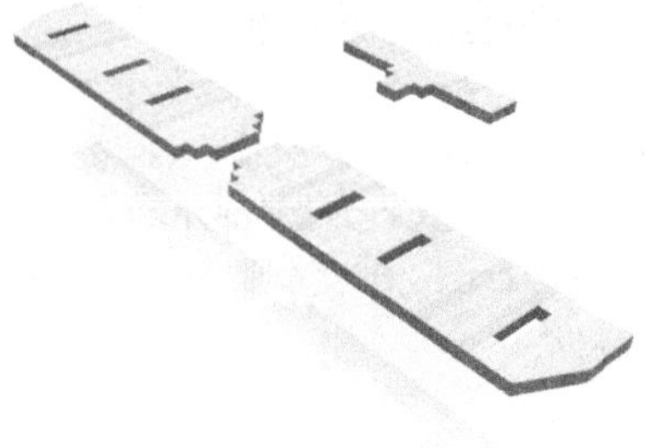

Sketch the design
Again, this plane idea was from my science fiction book (page 432). The fun part of writing a book is that your only limitation is imagination, which leads to a great brainstorming session. Here, I knew my limiting factor was going to be the wings, so I started with that first.

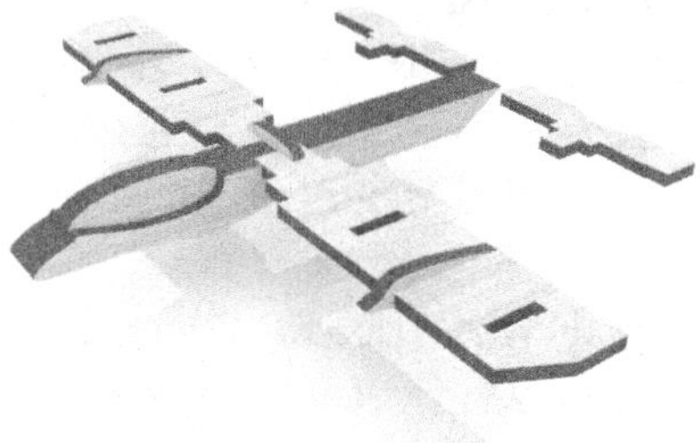

Create the cockpit
Once I established the wingspan of my plane, it was time to build the cockpit! Again, my limiting factor here was the length of the cockpit, which would dictate the maximum wingspan I could have while still staying in proportion. As you can see, I went from a single to twin tail at the end.

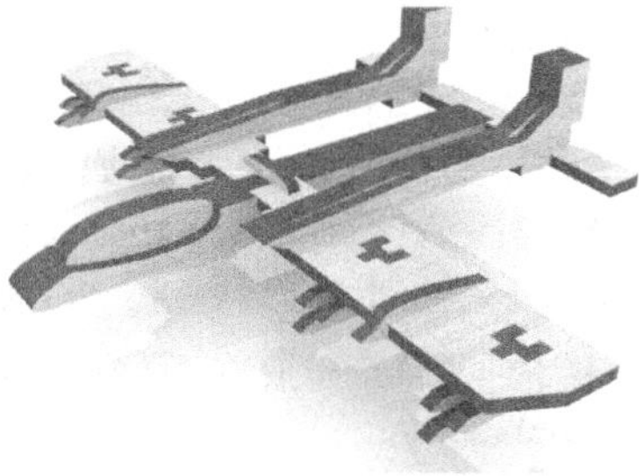

Work on the twinning and ammo
I liked the look of the twin tail, so I decided to go a different direction with this aircraft, by having basically the entire back interlocked. I created not only a neat-looking design but something that's INCREDIBLY STRONG! I didn't want it just to be a massive stack of wood though, so I knew I'd have to add holes to air it out a bit.

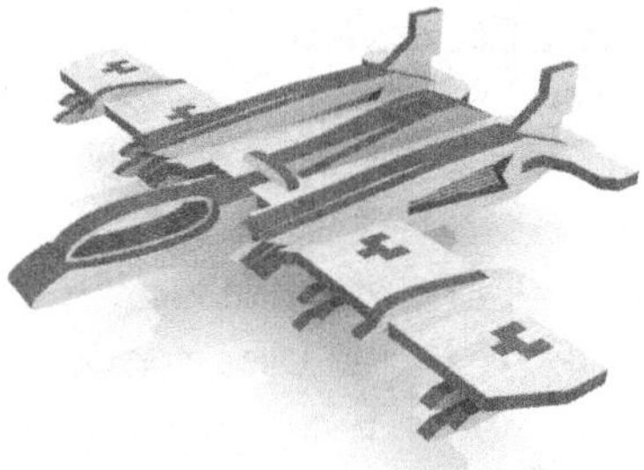

Add some holes
Adding holes throughout the back and the cockpit really made the model more visually appealing, but then the wings looked out of place. Having a straight wing doesn't convey stength, mass and influence. I wanted this model to be very tough and stocky -- not long and sleek (weak).

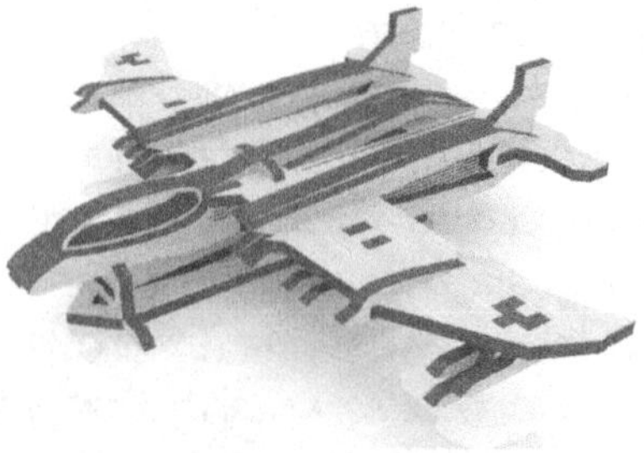

Finish the model
After an incredible number of refinements, the model was finished. I swept the wings back and created a "holder" to keep it in place on a desk. This model is very strong and could be thrown around quite easily, so I wanted all the components to be well-supported and not stick out too much.

# Science Award

<u>Start with the equation</u>

The hardest part of this award was going to be writing letters into the award. I played with several ideas until I found one that worked best and was fully supportable within the framework I had established.

<u>Build from the ground up</u>

The next step was building the superstructure to support the equation and also be strong enough so the entire award didn't fall over.

<u>Add holes</u>

One of my favorite movies is "Stargate," and I wanted this award to have a similar persona but more simplified. Once you know science, you can do amazing things, and I wanted this award to represent what was possible not only in woodworking but knowledge as well.

<u>Add details</u>

In science, you either achieve your goal or you don't -- there isn't much gray area in between for experimentation. I wanted this award to be given to the winner, not second or third place, though I added the flexibility to do so by giving myself lots of room in the round design.

<u>Make refinements</u>

I wanted this award to really be a showcase, so adding additional flaring along the top, an area for a plate on the base, and multiple levels of layered detail helped me achieve the goal I had set.

# Shopping Cart

<u>Start with the wheels</u>
I got a request from a laser shop owner in the United States who wanted a gift he could cut for his daughter's birthday present. A shopping cart! Well, that came way out of left field (I hate shopping,) but then I challenged myself to built it for him to his specs.

<u>Build from the ground up</u>
A shopping cart is essentially a car that is way too high, so after establishing the wheelbase, I began working on the cart end of things. It needed to be strong enough to carry some stuff but not so out of balance as to fall over. I ran simulations in my head and came up with a design where the weight was pretty even all over.

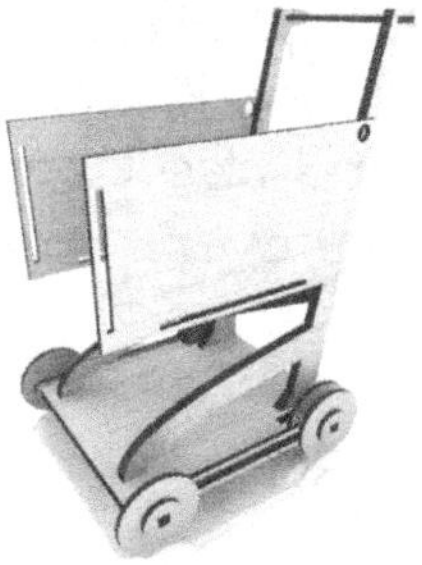

<u>Add holes</u>
The purpose of a shopping cart isn't to just carry things, it's to show others the loot you found so they can find it too! As a result, I had to make the model as "transparent" as possible, with plenty of holes so everybody could easily see what the little girl had in her shopping cart.

<u>Add details</u>
Although rarely used, most shopping carts have a back that can swivel so they easily fit inside of another cart for storage. I built this into the design as well -- it doesn't have any real purpose as you can't stack one into the other, but just something that moves is always fun.

<u>Make refinements</u>
I added quite a few small refinements to this model to make the assembly extra-easy and to remove as many pieces as possible (especially small ones). Overall, it was a good project that challenged me to think of modeling for a whole other demographic.

# Sydney ANZAC Monument

Start with the center

I've had a fascination with Australia since I was a kid -- after designing the Sydney Harbour Bridge, I wanted to design another Sydney monument. The ANZAC Monument was only my second real "architectural" model, and my board size was the only limiting factor I was concerned with. I wanted this to look awesome! Screw cost!

Create columns

As this monument is world-famous, I decided to work extra hard to be as accurate as possible. After building the superstructure, I had to figure out how to build columns with little guys sitting on each one, all of which had to interlock with both the floor and the wall. I spent HOURS on this aspect of the model alone.

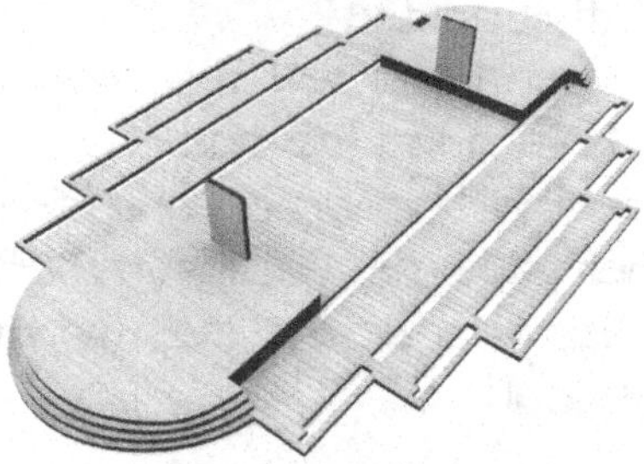

Build the foundation

I had a very tough time on the top, and then realized I had to have it all supported on an extra-strong foundation due to the weight of everything! This foundation is massive, almost a meter long; I needed the resolution because I wanted the little guys on the pillars to have some detail instead of just being small sticks.

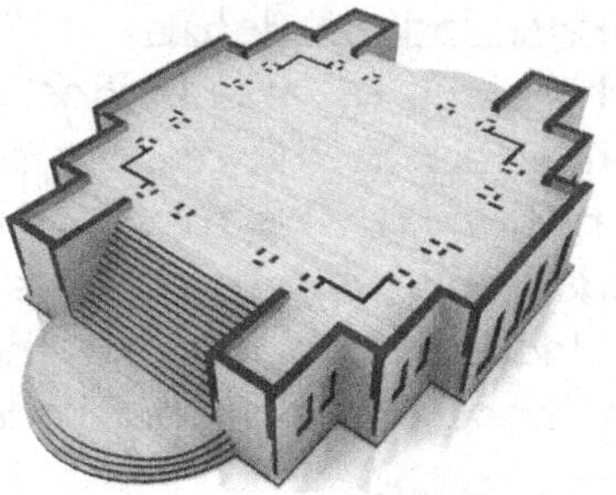

Finish the foundation

I spent days getting the foundation to work properly with the highly detailed top ... weight was an issue, but I wanted to make sure I was as accurate as possible from every angle possible. Building a monument is easy; building one so well known is quite another.

Refinements abound

Looking at this model, it looks pretty damn good! There are endless layers supporting every aspect of the model hidden throughout. As of this writing, I've designed over 200 models, and this one was among the most challenging and dense I ever built. It has 296 pieces fitting in a space that's about 60 cm square!

# Sydney Harbour Bridge

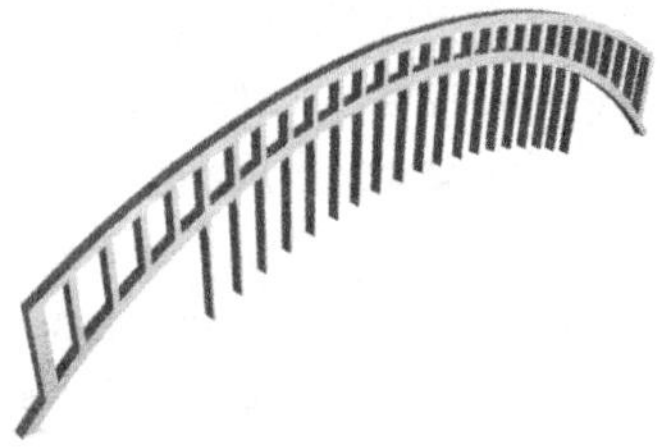

Start with the arch

This bridge is basically one huge arch with an anchor on each side. The arch would make or break this model, so I had to spend a lot of time getting this part down and as accurate as possible given the material limitations I had to contend with.

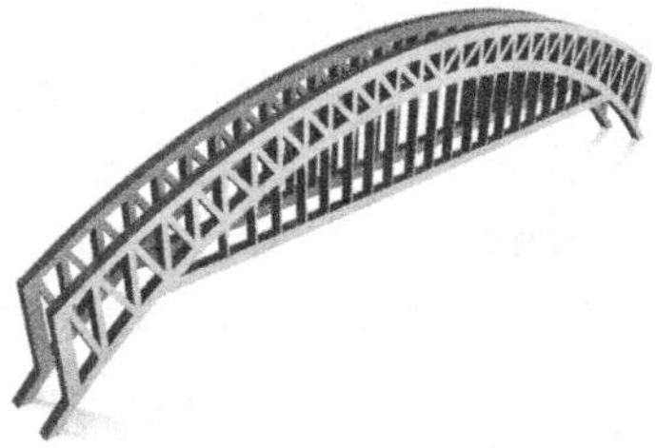

Double up the arch

Once I had built an arch that was as accurate as possible, the easiest way to begin adding depth was by doubling up the piece. This established dimensions and also the maximum width of the road, so a little more thinking was involved than at first sight. Proportionality must be along the x, y and z axes to work.

Anchor the bridge

I knew when I started this model that I was basically building half a bridge and then cloning the other half. Because the anchor on one side has to fit perfectly on the other side, perfect symmetry makes this model rather easy to design even if the real one isn't perfectly symmetrical.

Make under-bridge details

Since the sides of this bridge are well known, the bottom where it fits into the anchor points is also a detail that I had to nail down. The most difficult aspect was making the model as accurate as possible on such a small scale where I lacked resolution, as the arch was so large (making these parts proportionally smaller).

Take shape

Now that I had two anchor points, I decided it was time to revisit the arch to make additional changes to it. There were going to be cross members going all the way across the top; as a result, additional clearances were required to make this model more accurate.

# Tower Bank

<u>Start with the sides</u>
A tower is essentially a vertical cylinder or elongated box, so making something proportional is pretty easy. The challenge comes in trying to make something asymmetrical when it's screaming to be symmetrical.

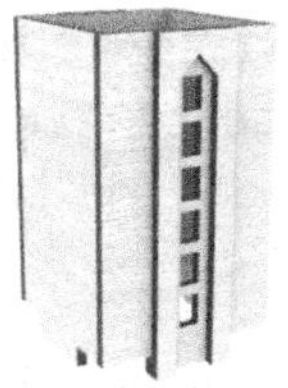

<u>Add branches</u>
What I really like about medieval buildings is how over-built they really are -- they are incredibly massive and strong, and I wanted this look to follow through with this model. A great way to accomplish this is by building vertical "fins" that run along the side of a building. If you have something heavy, it needs these for stability and support.

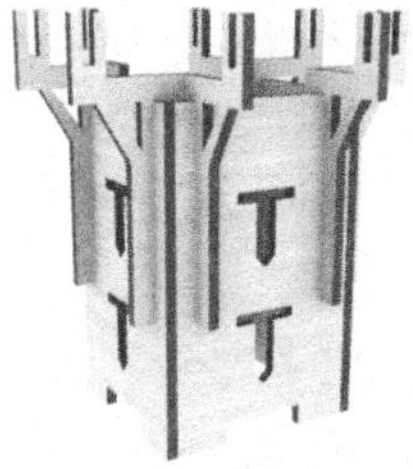

<u>Add medieval characteristics</u>
Medieval buildings are easy to spot architecturally speaking, as they have the blocky roofs and small slit windows (which come in handy when you are trying to defend a building while also providing cover for offensive tactics). In this case, the bank needed slits to show how much money you had and prevent it from falling out.

<u>Create support structures</u>
In the picture, you can see the support structures. These were re-built several times as I tried to bring the piece count down and also make them more distinctive. Nothing is worse than trying to build a model where all the pieces look very similar!

<u>Flatten the roof</u>
After going along with the traditional roof style, I decided to flatten it. This required further rebuilding, but the look gives the roof more thickness and brings a more dynamic look to the model.

# TV Bank

<u>Start with the sides</u>
My goal with this project was to create a bank that could be customized easily for any child -- the screen would be a piece of paper they can draw on and "insert" into the front of the bank ... making it a TV bank!

<u>Add sides</u>
I knew that to keep everything proportional from the start, I'd have to establish the dimensions of the TV first. I also wanted it to be easy to carry, so the sides needed handles.

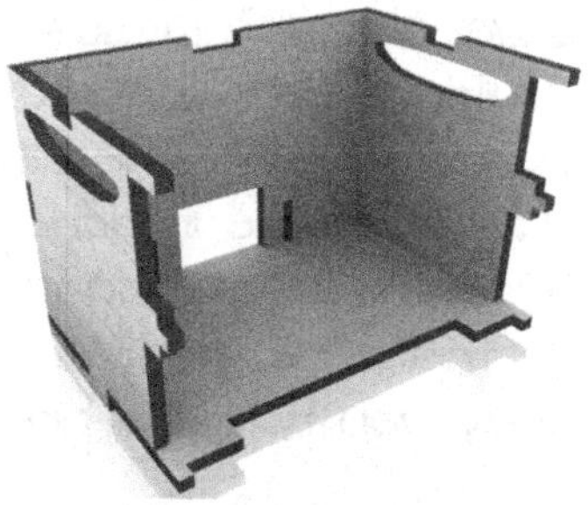

<u>Money retrieval</u>
One thing I hated about my banks growing up was how hard it was to get the money out of them. It seems so simple, but turning the bank upside down (reversing the process you used to put coin in the first place) always seemed unimaginative and silly. A design flaw I didn't want in this bank!

<u>Support structures</u>
As this was to be a TV bank, I needed to make the front screen "hover" over the front facade so there was plenty of room to slide a piece of paper in.

Creating simple extended handles was the easiest thing I could come up with.

<u>Add details</u>
After I added knobs and made holes under them (speaker,) I achieved my goal in creating a bank that's easy to build, fun to use and customizable whenever a child wants to use it. Taking the money out via a back slot is EASY!

## Jonathan Engel's Design Philosophy

Jonathan is a medical device research scientist by trade and inventor/tinkerer by avocation. His interests include renewable energy, embedded systems, zymurgy, and the outdoors, to which a recent foray into fatherhood has added toymaking. You can contact him by e-mail at: jonathan.engel@gmail.com. As you can see, Jonathan has a far more methodical approach to his designs; they look fantastic, and he was generous enough to share his design knowledge with the world!

The first step in designing a wooden laser-cut toy is finding an inspiration. My designs have focused on construction equipment due to my son's proclivity for such vehicles, as well as their natural suitability for the medium owing to their chunky and utilitarian designs. When I need an inspiration for converting a real-world piece of construction equipment into a robust laser-cut wooden toy, I turn to the world of scale models.

For some reason, there is a market for highly accurate, highly detailed, and very expensive die-cast scale models of construction equipment. Not one to look a gift horse in the mouth, I make use of online photos and reviews of these models in my efforts. It is much easier to capture photos of a 100,000 pound bulldozer from all angles when you can pick it up and flip it over. This makes understanding how the machine works and which details should be captured in the wooden toy much easier. One of my favorite places to go, which has outstanding photos of a wide range of well-made models, is Cranes Etc out of the UK (www.cranesetc.co.uk). Manufacturer websites are also a good place to look, with Caterpillar (www.cat.com) having videos on most of their equipment in action so you can see how they move.

If you compare my model to the ones seen on these sites, you'll see some compromises were required, but overall the designs remain true to the original machines.

**Material Limitations**

Once you've made some headway into deciding what you'd like to build, you should start to think about the limitations of your medium. I design and build wooden toys pretty much exclusively for one-eighth-inch (3.2 mm) thick baltic birch plywood. It is available, uniform, attractive, and strong.

A well-designed wooden toy can be scaled to be cut from other material thicknesses, but the scale of the toy will change as well. So if you design a toy to be 12" (~300 mm) long in eighth-inch plywood, it will be 24" (~600 mm) in quarter-inch (6.4 mm) ply. Additionally, the thickness of your material cannot be easily altered by laser cutting, and the strength of your material will dictate how small of a feature you can tolerate on your design. With all those considerations in mind I have found that for making a wooden toy for a child in the range of 2-6 years of age that will be strong yet light, as well as handheld and easy to store when not in use, eighth-inch baltic birch plywood is the best compromise.

Many projects are designed for quarter-inch plywood, but I find that not much detail can be achieved without making the toys far too large. Other commonly laser-cut materials such as acrylic may be used as well, but they represent trade-offs in terms of strength, ease of cutting, ease of gluing, etc. Depending on the material, the quality of the cut edge may also be of concern. You'll be doing a lot of gluing or other assembly method, and having consistent and predictable joints will go a long way towards protecting your sanity and the success of your design.

**Process Limitations**

In addition to material selection concerns, you should evaluate how the parts for your design will be cut.

If you are going to be using a hobbyist type low power laser cutter versus an industrial scale laser cutter with higher power, better optics and a larger cutting bed, then you'll need to factor that into your design.

Carefully evaluate the ability of your chosen cutting system to cut your chosen material. On lower power laser cutters for example, it may take multiple passes to cut through thicker material, which can be a real pain and time suck. Thicker material will impact the kerf and parallelism of the sidewalls of the cut on pretty much any laser system.

You will also need to accommodate some of these limitations into your design. For example, cutting eighth-inch thick baltic birch plywood on my laser system requires that my design account for about 0.1 mm of kerf on a feature such that when I want a 5 mm slot I design a 4.9 mm slot, and when I want a 5 mm tab I design a 5.1 mm tab.

These design rules give me parts that fit together with a friction fit that makes glue assembly much easier as the parts stay together. However, this can be problematic when transferring designs to other laser systems or scaling for other materials, as the kerf relationship may not transfer.

If you think your design will be scaled extensively, you may need to re-design for the various materials.

For example, in quarter-inch material, simply doubling the slot and tab dimensions mentioned above for eighth-inch plywood will yield a slot cut at 9.8 mm and tab at 10.2 mm, which may no longer fit together as the laser kerf will not necessarily scale linearly with material thickness. Avoiding such a tight fit by allowing more slop between parts may be a solution as well.

Another obvious limitation of laser cutting is that you can only cut out 2D parts, and any 3D body must be built up of slices or boxed in by interlocking faces. In designing a laser cut version of a construction vehicle like a mining dump truck, this limitation comes to the fore when designing the large dump bed as well as the large wheels.

In the case of the dump bed, I chose to use interlocking faces to build a large functional 3D part, while choosing stacking slices for the wheels. The images below show examples of these approaches for building 3D parts out of 2D pieces.

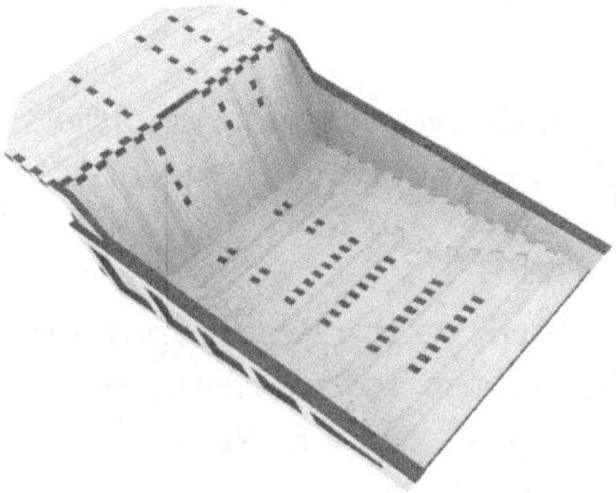

Particular to the design of laser cut vehicles and mechanisms is the limitation that laser cutters cannot easily cut out a cylinder or axle with a length much greater than the thickness of your material.

You can get around this limitation by a number of methods. You can choose to make use of commonly available wooden dowels in your design, you can use square axles in round holes and accept the sloppy fit and action, or you can approximate dowels by using a combination approach where round washers are glued onto square laser-cut axles at the critical points.

The following image shows all three approaches. Methods for making axles or pivots for moving models include round dowels, square axles, and hybrid washer-on-square-axles.

My personal preference is to use wooden dowels, as the square axles in round holes method feels like a kludge, and the hybrid approach is complex to design and sometimes difficult to assemble.

However, dowels are not laser cut, and if you plan to sell your design as a laser-cut kit it is difficult to include dowels in the kit. People may balk at having to source their own dowels or even be unable to source the proper size dowel due to the inconsistencies of local units of measure.

As a result, most laser-cut kits out there use the square axles in round holes method. Shudder.

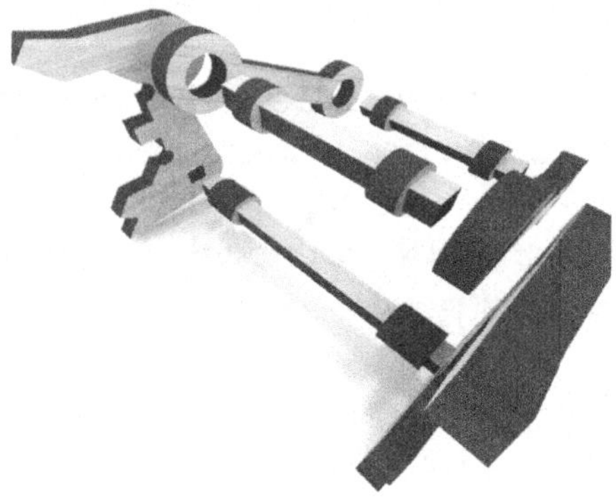

The image to the right shows another detail exploded view that details how the hybrid method uses a square axle and laser cut washers to make a smoothly rolling axle with a combination of round holes for moving parts and square holes for stationary parts.

Focus is another process limitation that you may need to consider when designing your model. Depending on the flatness of the material stock you have in mind for your design, the ability of the laser to cut uniformly through the material as well as to cut features of uniform size may be impacted. For example, thin plywoods often bow or warp and are difficult to hold flat on the laser cutter bed.

The result is that if you focus the laser on a flat area, you may find the laser out of focus on areas that stick up depending on the focal length of your laser system. This can result in increased kerf as the laser is less focused on the material, and if the cutting power is spread over too large an area you may not cut all the way through the material.

This is a real pain which can require hand trimming to free the parts in mild cases and complete scrap in severe cases. Applying additional cutting power to compensate is an option but may throw off your kerf estimates in the flat areas, as extra power is transferred to the workpiece and increasing the material burned out.

You can intentionally defocus the laser, choosing to focus in between the flat and bowed areas so that the laser spot is more uniformly out of focus to get a more uniform cut characteristic. You can also increase the slop in your design to allow for increased variation in kerf. Trial and error may be the best method with your material and laser system of choice.

**Design Challenge**

Every model design presents its own set of challenges. Usually, each machine has its own key design challenge that must be solved to yield a satisfactory result. In most cases, it doesn't make sense to work on the design very much at all until the key challenge is solved.

For example, in my front end loader design, there were two design challenges. One was the jointed body, which needed to be very strong and allow the front and back halves of the machine to pivot freely. This was rather easily solved by adding additional stacked wood layers.

The most difficult challenge was to come up with a method of approximating the hydraulic cylinders that allow the front end loader bucket and arm to lift and tilt.

I mulled over this problem for a couple of weeks before having a flash of inspiration that led to the design pictured in the first picture below. The moving mechanisms of the front end loader are basically multi-bar linkages with links of variable length. The lengths are variable but limited. Thus by building a box like "cylinder" with a captive "shaft" the hydraulics can be approximated.

An added bonus is that the friction of the rectangular shaft in the box cylinder allows the mechanism to stay in position rather than flop limply about when in use. Within mere weeks of publishing this front end loader design on the web, other designers had already appropriated this solution into their own vehicle designs.

The other pictures below show the hydraulic cylinder installed in the front bucket and arm assembly for the front end loader design. The variable length of the hydraulic cylinders allows the bucket to tilt realistically.

I mocked up these linkages with cardboard pieces in order to get the relative lengths and pivot points just right so that the bucket tilted through a realistic range of motion. This can be quite helpful and is a lot easier to change at this stage in the design process.

The design challenge for the dump truck was a method to build a large yet strong lift bed with a hinge mechanism that could stand up to the abuse a child can subject a toy to. The solution was to design a series of polygonal faces that linked together with interdigitated box joints and a parallel array of stiffeners with an integrated hinge pivot.

As seen in the image below, the resulting dump bed looks similar to the real thing. Although not up to true roughhouse play, the design is robust enough for most situations. The large number of closely spaced hinge ribs gives a very strong joint with only a small dowel. You can also see the supports for the top of the dump bed and the additional "cosmetic" ribbing on the sides of the bed.

Watching kids play with these kinds of toys, they often put their entire weight on the sides or top of the dump bed while "driving" the trucks. As a results I also designed the top of the dump bed to rest on the cab when in normal position and the bottom of the bed rests on the strong frame box under the bed to help support it during play. So far so good.

For tracked vehicles like excavators and bulldozers, I puzzled over how to make a wooden belt for tracks for months. In the end, I came up with a three piece solution that creates a pretty realistic, flexible, and scalable wooden belt for vehicle tracks. The design uses square axles in round holes, which is not ideal but was necessitated by the small scale of the parts and the sheer number of rotational joints needed to make a complete set of tracks. No one wants to cut and glue a few hundred small dowels into small laser cut parts unless they have to.

The image below shows a close up of the wooden belt design I arrived at. In the middle of the belt plate you can see a pair of slots that allow gear teeth to engage and rotate the belt. The side pieces with axles holes need to be glued on as the next belt plate is added to capture the square axle on the next plate.

This process is to build up the belts one link at a time. Care should be taken to keep glue out of the axle holes to allow the belt to flex freely when assembled. One limitation of this design is that the belts can't get much smaller than shown, which limits the size of machine you can make while maintaining realistic proportion between the tracks and the rest of the machine.

Prior to arriving at this wooden solution, I was ready to use stepper motor rubber belts on wooden pulleys as tracks. These belts are available in a number of widths and a large array of circumferences to accommodate almost any vehicle design.

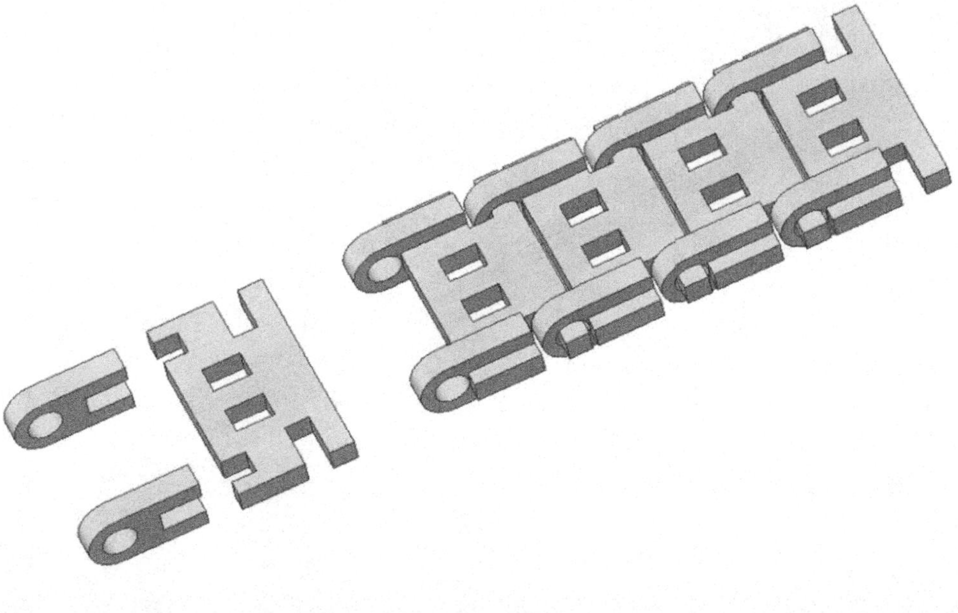

### Remainder of the Design

With the primary design challenge solved on your model, the rest of the design centers on creating pieces that will glue together to give the structure desired, with details that make the whole thing look good.

Box joints like those pictured below are perfect for building laser-cut 3D objects. Glued together, the increased surface area and interlocking nature of these box joints gives them great strength. As discussed in the limitations section, you'll need to characterize your material and laser system to get a set of design rules for achieving consistent snug fits of your box joints.

My design rule for 3.2 mm baltic birch plywood on a 30W laser system is to design 4.9 mm by 3.2 mm slots and 5.1 mm by 3.2 mm tabs. I repeat these slots and tabs the length of the part to be joined.

The basic design process is to draw out the shape of the final part desired, and then put in the details of the slots and tabs.

As you add new parts you may have to go back and tweak the slots to match up, and have odd lengths on parts that are not multiples of 5 mm long. Building assemblies and subassemblies in CAD is invaluable at this stage, allowing you to check the fit of each part in the design and correct as needed.

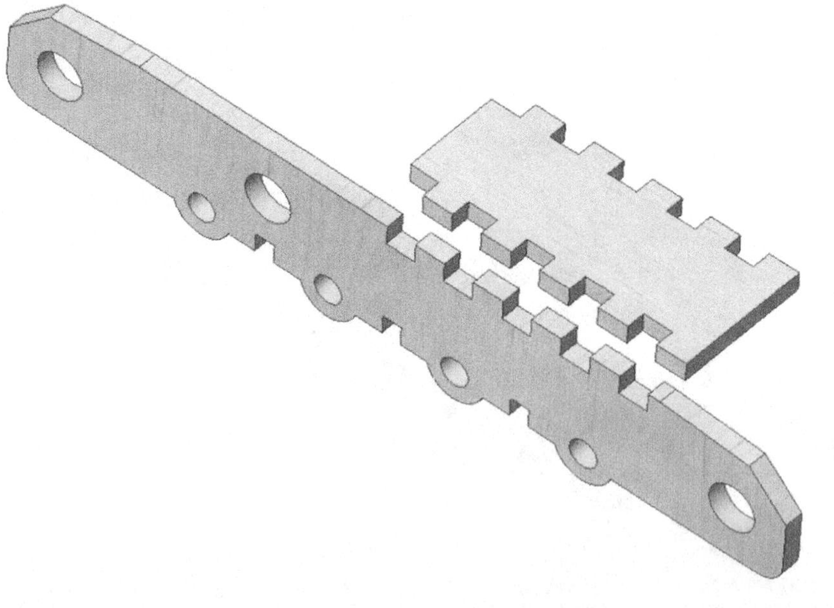

As you continue to flesh out the design of your toy, don't get too hung up on capturing every detail. For example, I thought long and hard about how to make the front wheels on the wooden dump truck steer like the real machine.

I just couldn't come up with a way that I thought would be robust enough in wood to allow kids to play with. So I settled for no steering.

Additionally, a lot of the bigger machines have guard rails and lights and other fine details that would be cool to capture but are just not feasible to cut out of wood and have last through a single play session. You don't want these toys to be shelf princesses; these are made to play with.

Using laser engraving rather than cutting you can design a lot of details like textures, weathering, mud, nuts and bolts, grills, exhaust ports, etc. For example, if you look closely at the wheel picture at the beginning of this article, you'll notice the lug nuts, tread blocks, and rim details I cut into the wheels for the vehicles.

**Laying out for cutting**

Depending on your CAD system, you can maintain and propagate design changes through a number of methods. Ultimately, the CAD is used to create a 2D outline of the parts for cutting. You should learn how to use your CAD software to maintain links between the 2D drawing for cutting and the 3D model so that you aren't constantly having to iterate your cut files as changes are made to parts. Changes will likely be needed as you do your first few builds from cut parts.

This is fairly simply done in Solidworks, where the 2D part view drawing can be updated as the 3D part linked to the view is changed. You can make a large master drawing that has views of all the parts needed for your project, and any changes will appear in the master cut file as if by magic.

In practice though, I've found that it becomes necessary to use other vector art software like Illustrator or Inkscape to manipulate the master cut file into sub files for cutting and to do color mapping for vector engraving or other details that are not cut through the material.

These vector files are not updated by changes to CAD and can be a pain to keep up-to-date.

One of the reasons for the need of a vector art program is the need to fit various sized material blanks or recut single parts that break or fail to cut properly. Using Ponoko, for example, they have uniform material sizes, but when you are sourcing and cutting your own material you might find that you have a handful of 12" x 14" pieces and some 10" x 16" pieces. In any case, you'll want to rotate the parts and fit them together as tightly as feasible to reduce the material wasted and the time spent prepping and loading material into the laser cutter.

The image below shows a sample cut sheet for the front end loader design. As you can see, fitting the parts in can be a challenge. Round parts are especially difficult to fit together tightly. A lot of the material used on the dump truck is for the six wheels.

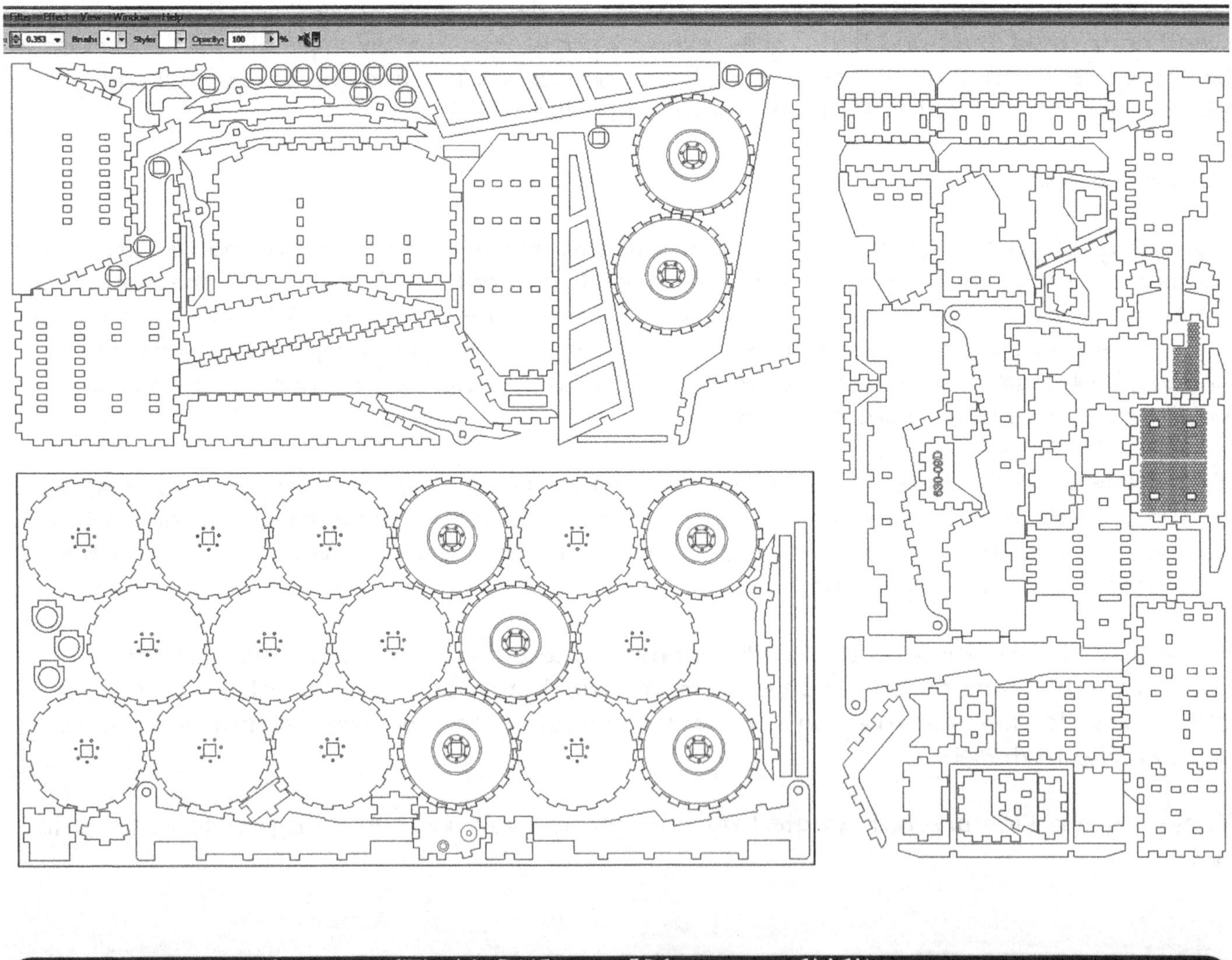

## Assembly

When the time comes for assembly, your hard work, design prowess, and attention to detail will all pay off in one big expletive-laced tirade as you berate the designers of such an ill-conceived product ... Oh wait. In reality though, the assembly of a laser-cut wooden toy is often more difficult than "Onsert Tab A into Slot B." What can appear foolproof in CAD can be tough to realize in reality.

All those interlocking box joints may only be able to be assembled in CAD, and the angles on certain parts prevent fitting the real parts together despite your best efforts. For example, the assembly of the deck of the wooden dump truck toy is a pain since there are tabs on slanted parts that need to come together with tabs on horizontal parts at the same time. When everything is slathered in glue, this can get frustrating.

Keep calm and think each step through, as order can be quite important when assembling these toys. Dry assembly is a good idea to help plan and visualize the assembly, as wood glue can very quickly make these parts impossible to separate.

When you have thought through a plan of assembly, get out a small paintbrush and some wood glue and start assembling pieces into logical units that can be allowed to dry before making bigger assemblies.

I brush the glue into the slots on both sides of the box joint being assembled. Use masking tape or small hobby clamps to hold pieces together or in alignment if needed. When using dowels, you can leave the dowels long and trim to length using a flush cut saw once the glue has dried.

Take care to keep glue off of moving parts and mechanisms so they can move after assembly. Keep a supply of paper towels handy and water to moisten the towels with to help wipe up excess or spilled glue.

Do your best but don't be too alarmed when you leave glue fingerprints on the assembly, as it is really tough to avoid.

Once complete, resist the urge to cycle any mechanisms or move parts too vigorously until the glue has dried overnight. Once dry, you can sand, stain, paint, seal, or otherwise finish your toy as you see fit. I typically leave them in as-assembled finish in order to highlight the cuts and other details of the design. That and I am lazy. Enjoy.

## Abacus B

Completed 78 piece model.
Finished dimensions using 10 mm (25/64 inch) thick wood.
L: 390 W: 30 H: 170 (mm) or L: 15-23/64 W: 1-3/16 H: 6-11/16 (in)

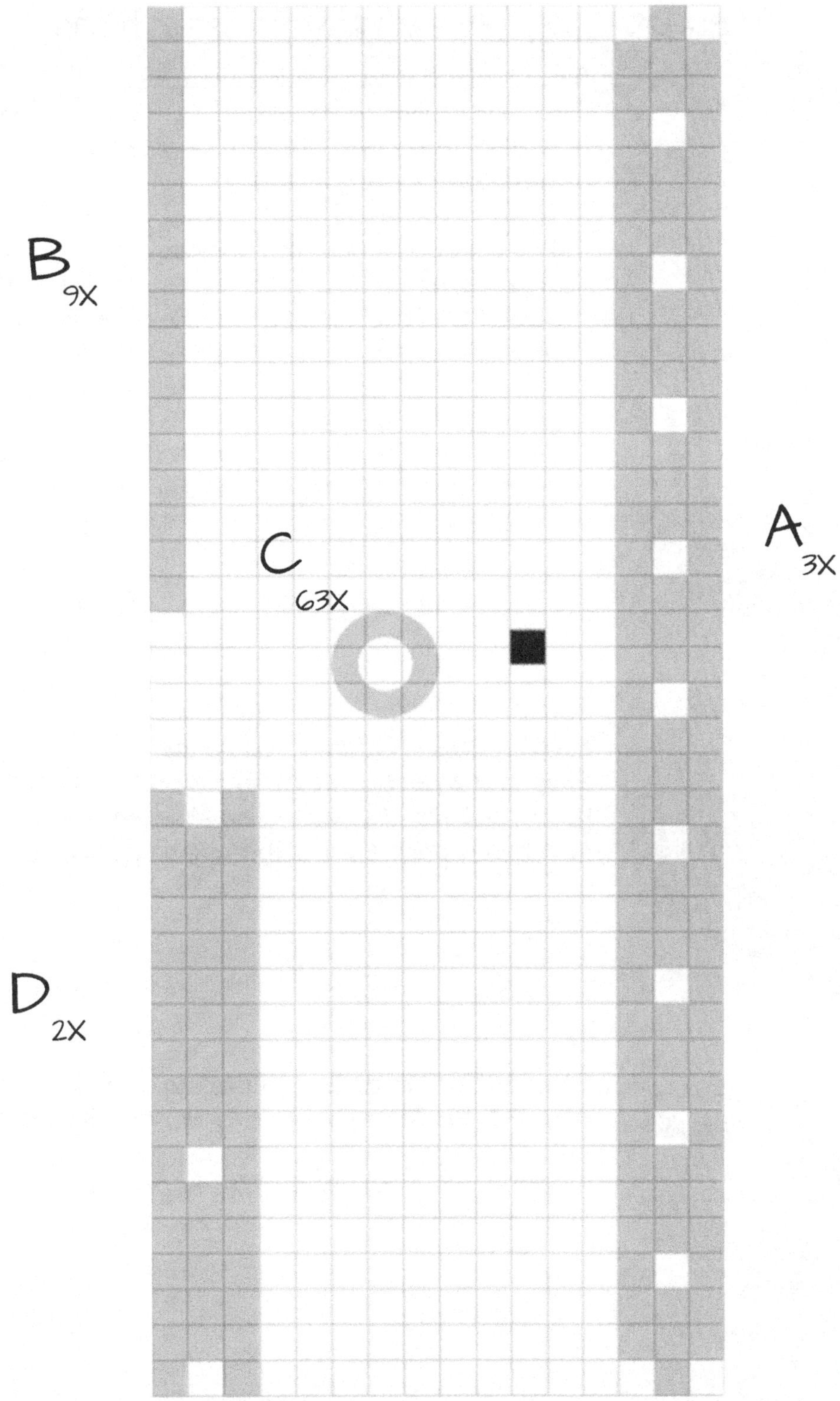
B
9X
C
63X
A
3X
D
2X

# Step-by-Step Assembly Instructions

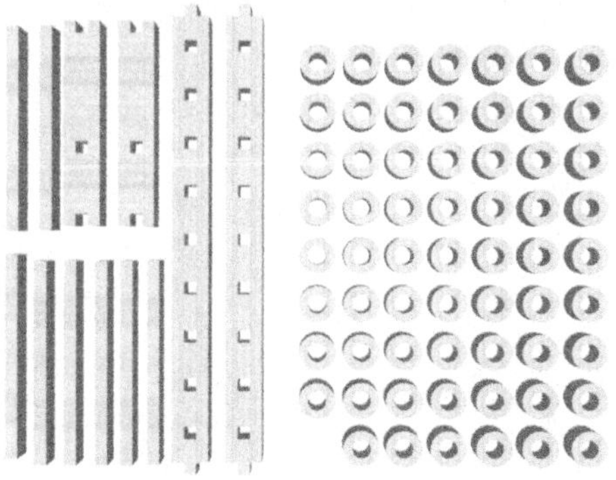

Step 1: Label and check inventory

Once you have all the pieces in front of you, the process of putting them together is no different from assembling pieces of Lego.

Label, using a pencil, the various parts and double-check your inventory. Take your time and enjoy the building process! Before jumping to the glue, be sure to do a dry run.

Step 2: Start the base

Slide and glue together part A (base) followed by all nine part Bs (rods). Once this is dry, slide onto each rod five part Cs (disks).

Step 3: Add the upper numerals

Slide and glue into place the second part A, followed by both part Ds. You can then slide into place, without glue, two more part Cs on the upper portion of the abacus.

Step 4: Perform some finishing touches

Add the final part A with glue and allow it to dry.

There is no need to apply any preservatives or paint; feel free to sand any sharp edges down. ENJOY!

# Bird Feeder

Completed 11 piece model.
Finished dimensions using 10 mm (25/64 inch) thick wood.
L: 300 W: 170 H: 200 (mm) or L: 11-13/16 W: 6-11/16 H: 7-7/8 (in)

# Product Renders

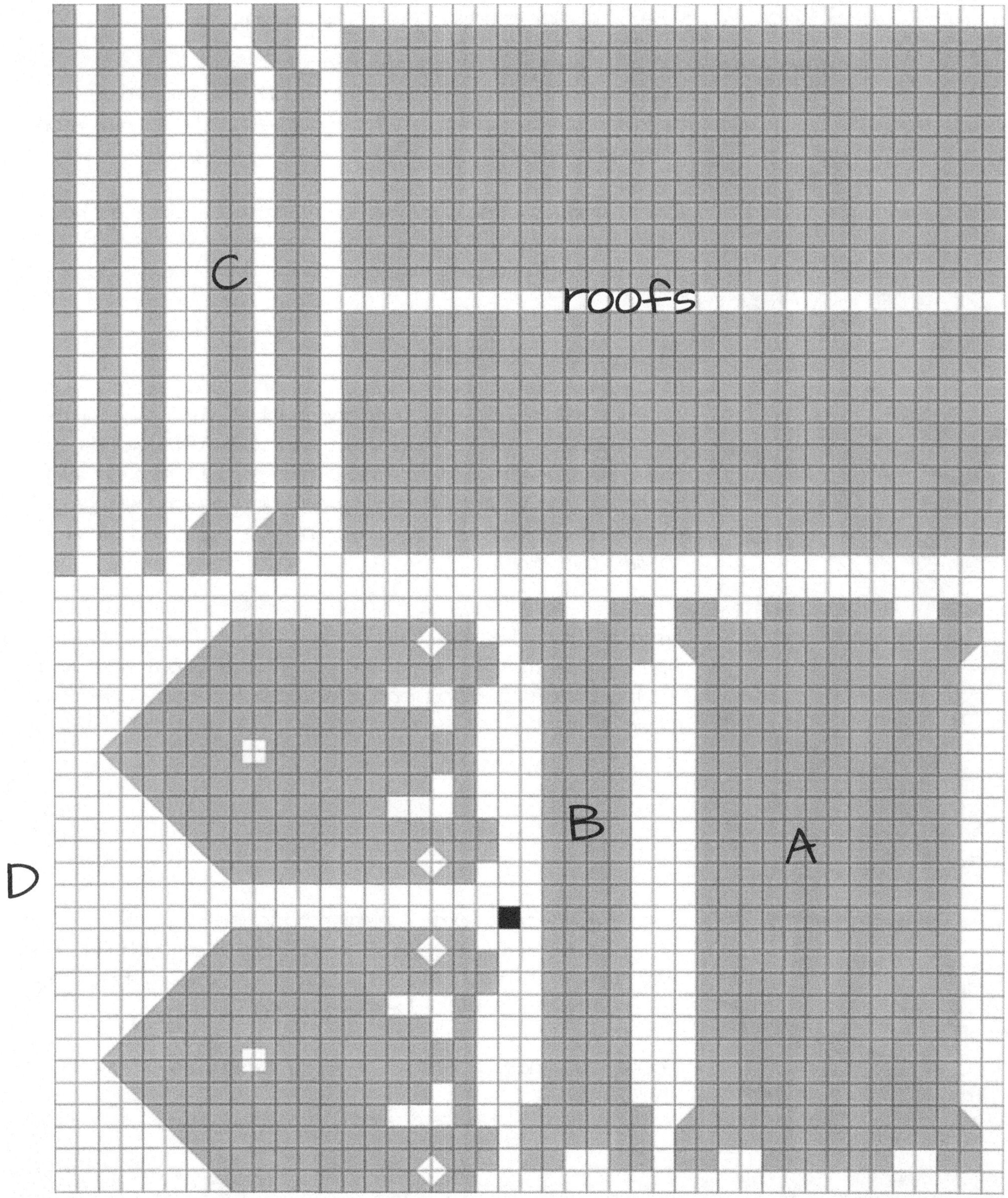
C
roofs
B
A
D

# Step-by-Step Assembly Instructions

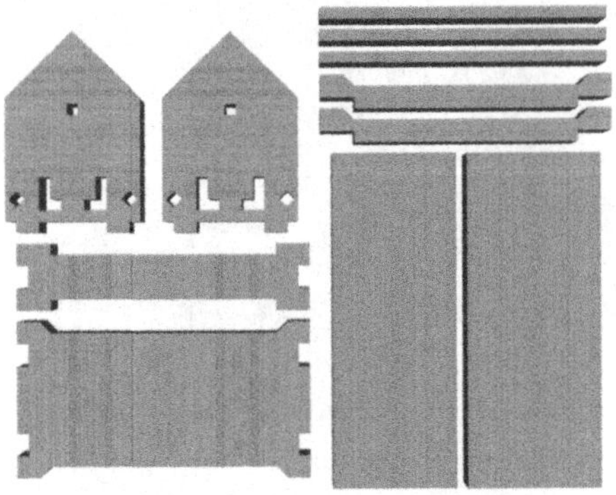

Step 1: Label and check inventory
Once you have all the pieces in front of you, the process of putting them together is no different from assembling pieces of Lego.

Label, using a pencil, the various parts and double-check your inventory. Take your time and enjoy the building process! Before jumping to the glue, be sure to do a dry run.

Step 2: Align the walls
Place wax paper under the model and place part A flat on its surface. Slide in only one part D, then slide in the three dowels, part B and both part Cs. Afterwards, cap the entire assembly with the other part D, making sure to apply glue where all these parts intersect. Using a pair of clamps, make sure the assembly is tight and allow it to dry overnight.

Step 3: Drop the roof
Apply some glue along the edges of the front and back of the bird house. You can now drop the roof onto the model.

Step 4: Perform some finishing touches
After allowing the pieces to bond together with the glue overnight, you may now sand any protruding parts flush, along with the outer edges if you so wish.

There is no need to apply any preservatives or paint. ENJOY!

# Bird House B

Completed 8 piece model.
Finished dimensions using 10 mm (25/64 inch) thick wood.
L: 200 W: 170 H: 200 (mm) or L: 7-7/8 W: 6-11/16 H: 7-7/8 (in)

Product Renders

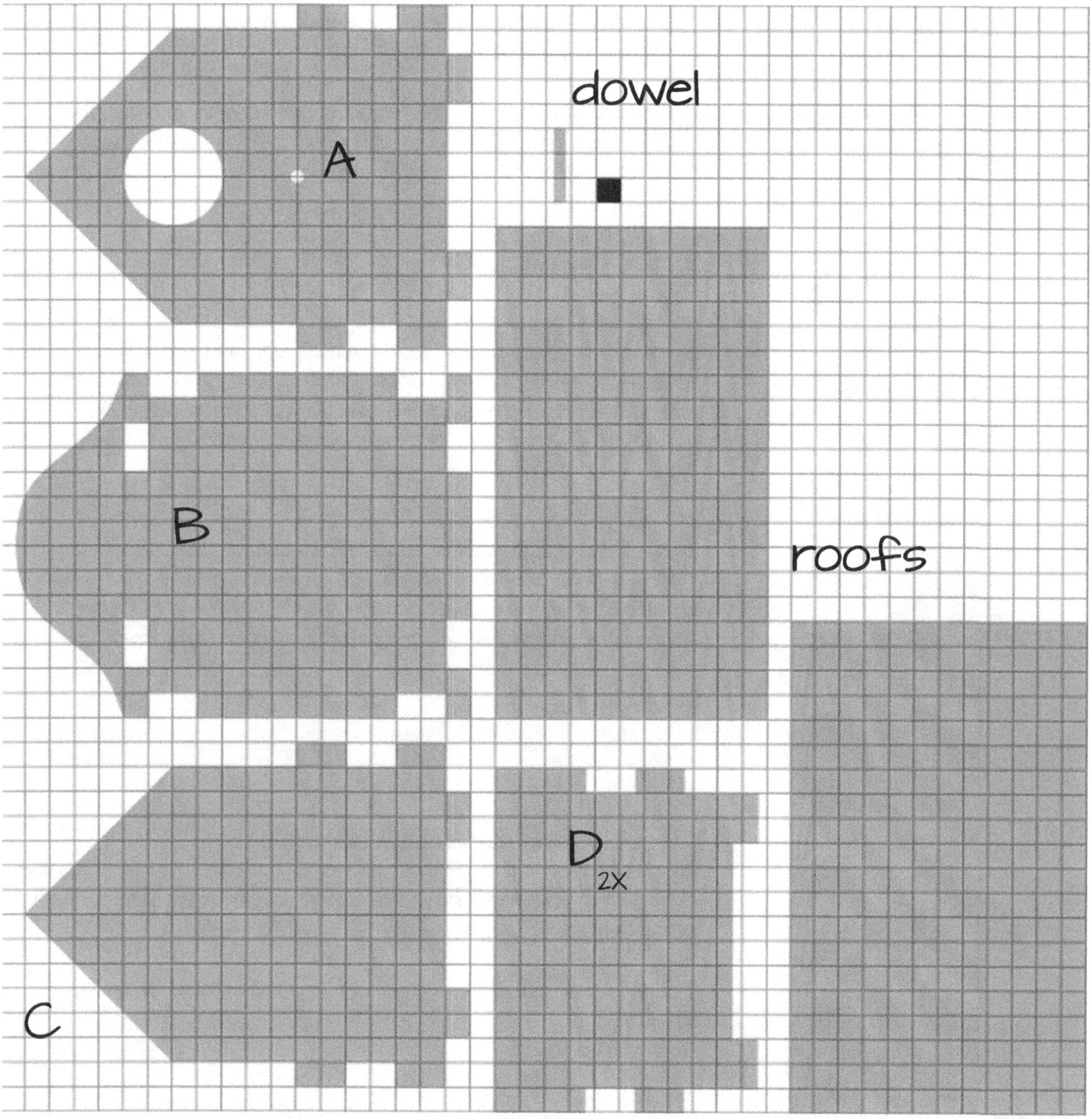
dowel
A
B
roofs
D
2X
C

# Step-by-Step Assembly Instructions

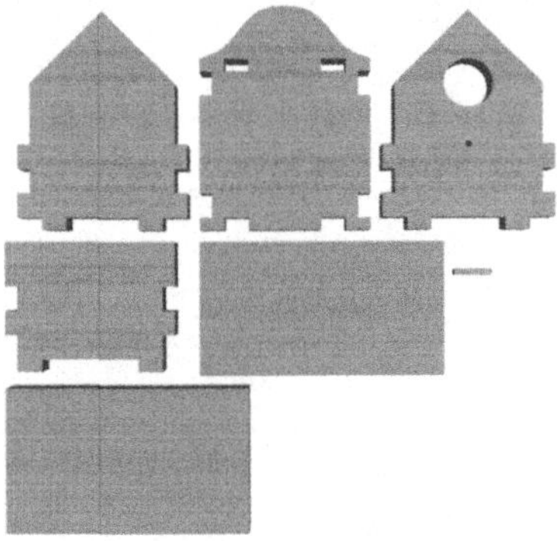

Step 1: Label and check inventory

Once you have all the pieces in front of you, the process of putting them together is no different from assembling pieces of Lego.

Label, using a pencil, the various parts and double-check your inventory. Take your time and enjoy the building process! Before jumping to the glue, be sure to do a dry run.

Step 2: Align the walls

Place wax paper under the model and place part A flat on its surface. Apply glue on the inside edges where each of the parts A, B, C and D meet up, and slide them together. Afterwards, apply a set of clamps to keep everything together in the middle of the model.

Allow the model to dry overnight.

Step 3: Insert the front dowel and drop the roof

Apply a little glue on the tip of the dowel and insert it into the front of the bird house, followed by applying some glue along the edges of the front and back of the bird house. You can now drop the roof onto the model, applying glue where the edges intersect.

Step 4: Perform some finishing touches

After allowing the pieces to bond together with the glue overnight, you may now sand any protruding parts flush, along with the outer edges if you so wish.

There is no need to apply any preservatives or paint. ENJOY!

# Bird House C

Completed 21 piece model.
Finished dimensions using 4 mm (5/32 inch) thick wood.
L: 210 W: 150 H: 320 (mm) or L: 8-17/64 W: 5-29/32 H: 12-19/32 (in)

## Product Renders

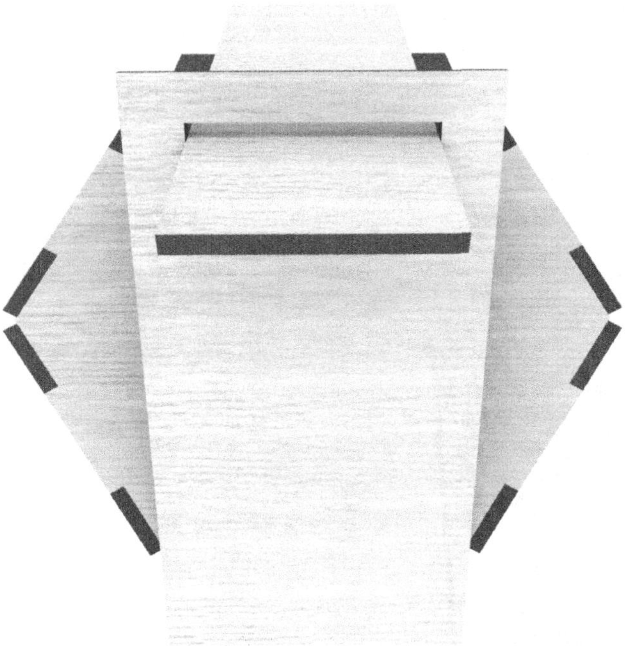

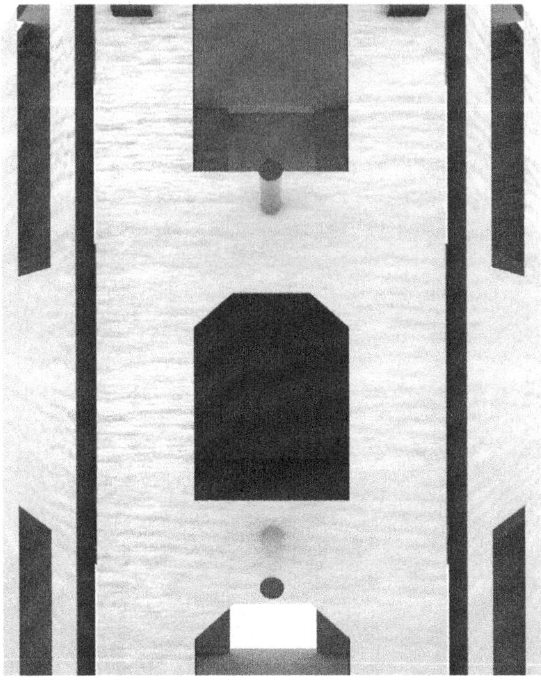

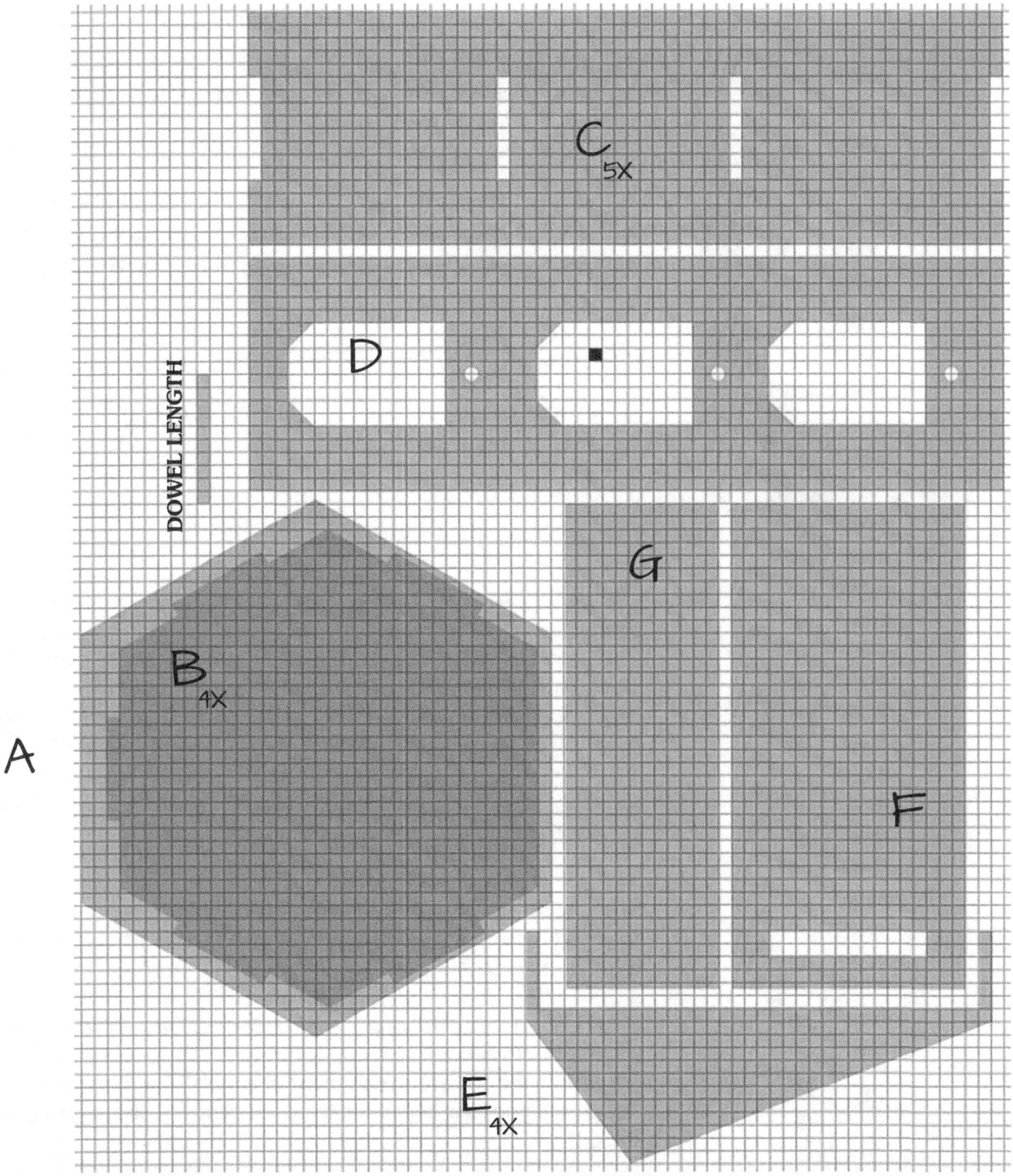
C
5X
D
DOWEL LENGTH
G
B
4X
A
F
E
4X

# Step-by-Step Assembly Instructions

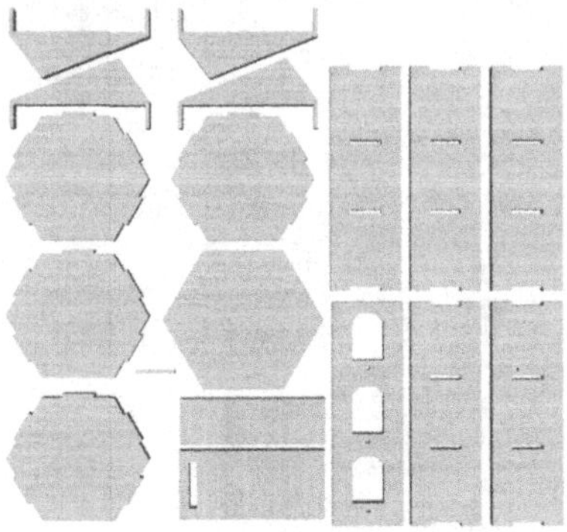

**Step 1: Label and check inventory**

Once you have all the pieces in front of you, the process of putting them together is no different from assembling pieces of Lego.

Label, using a pencil, the various parts and double-check your inventory. Take your time and enjoy the building process! Before jumping to the glue, be sure to do a dry run.

**Step 2: Assemble the foundation**

Glue part A under a part B. You can now glue into place all five part Cs with the part Bs being used as a guide.

**Step 3: Add supports**

Glue into place, at the front of the bird house, part D along with the small dowels that form the perches. Glue into place the four top supports (part E) on the top of the roof.

You can then glue the roof in place, first part F followed by part G.

**Step 4: Perform some finishing touches**

You may now sand any protruding parts flush, along with the outer edges if you so wish.

There is no need to apply any preservatives or paint if this is an ornamental piece; otherwise, apply caulking on ALL exterior seams (especially the roof) and apply weather-proofing paint to the exterior of the bird house. ENJOY!

## Eiffel Tower

Completed 58 piece model.
Finished dimensions using 4 mm (5/32 inch) thick material.
L: 500 W: 500 H: 1500 (mm) or 19-11/16 W: 19-11/16 H: 59-1/16 (in)

## Product Renders

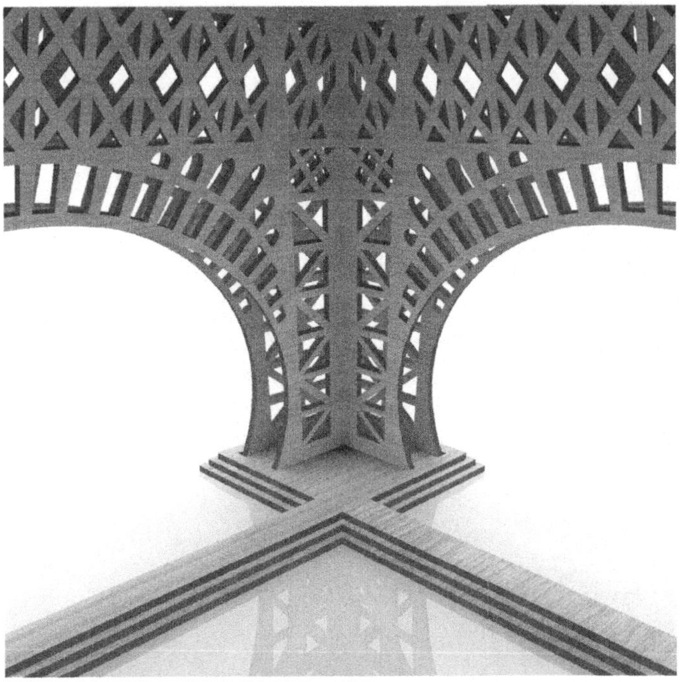

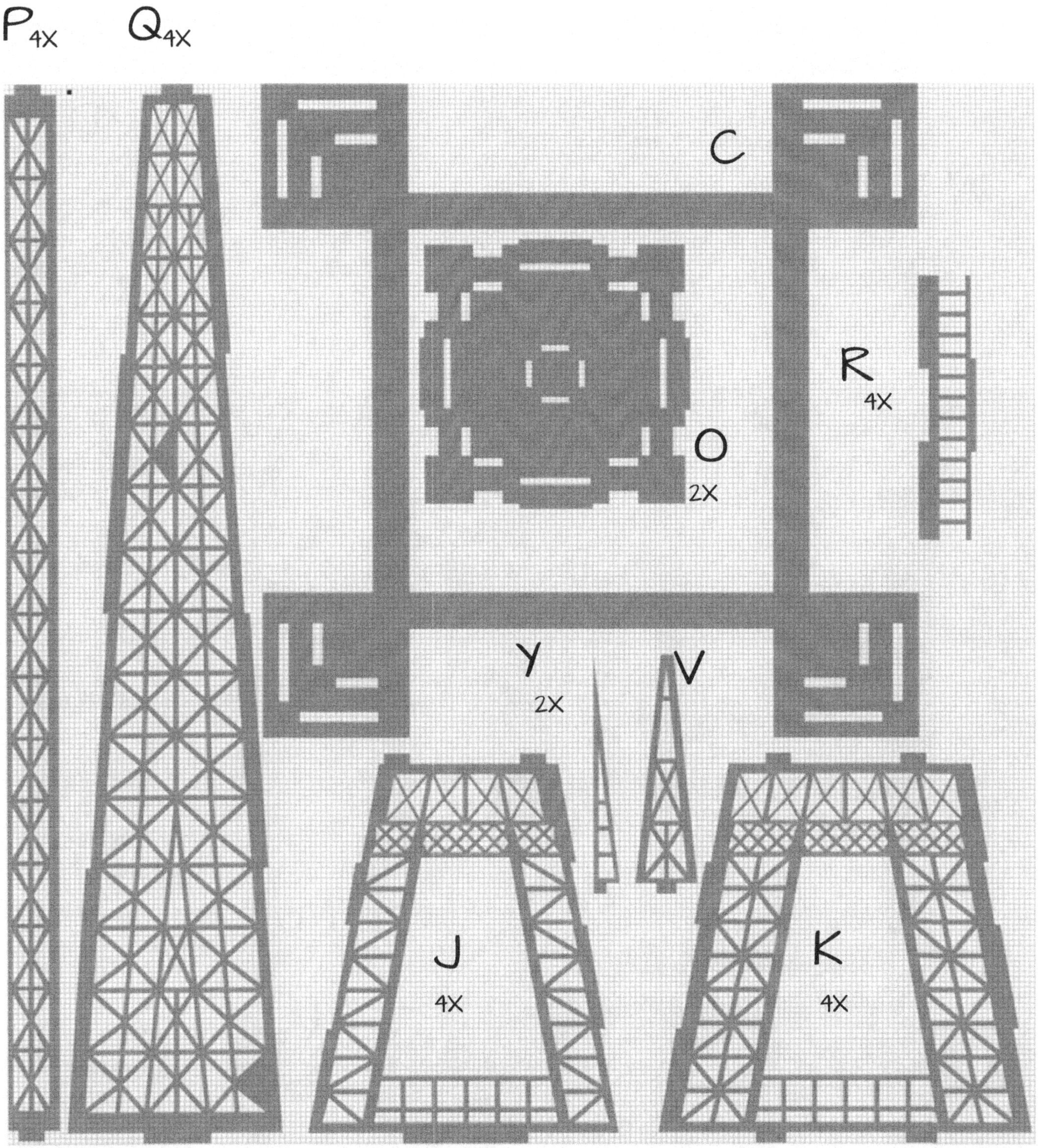
P 4X
Q 4X
C
R
4X
O
2X
Y
2X
V
J
4X
K
4X

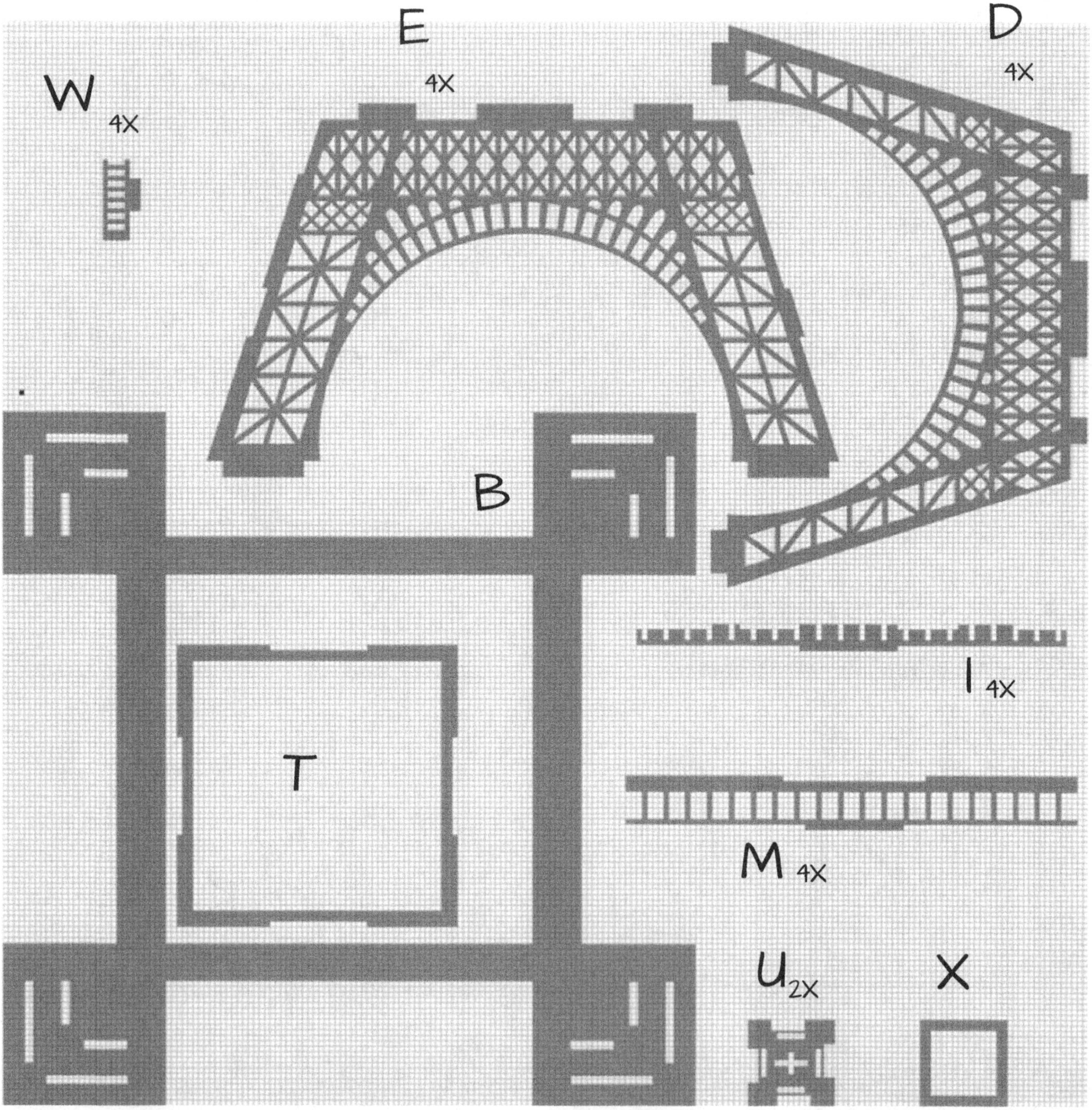
E
4X
D
4X
W
4X
B
I
4X
T
M
4X
U
2X
X

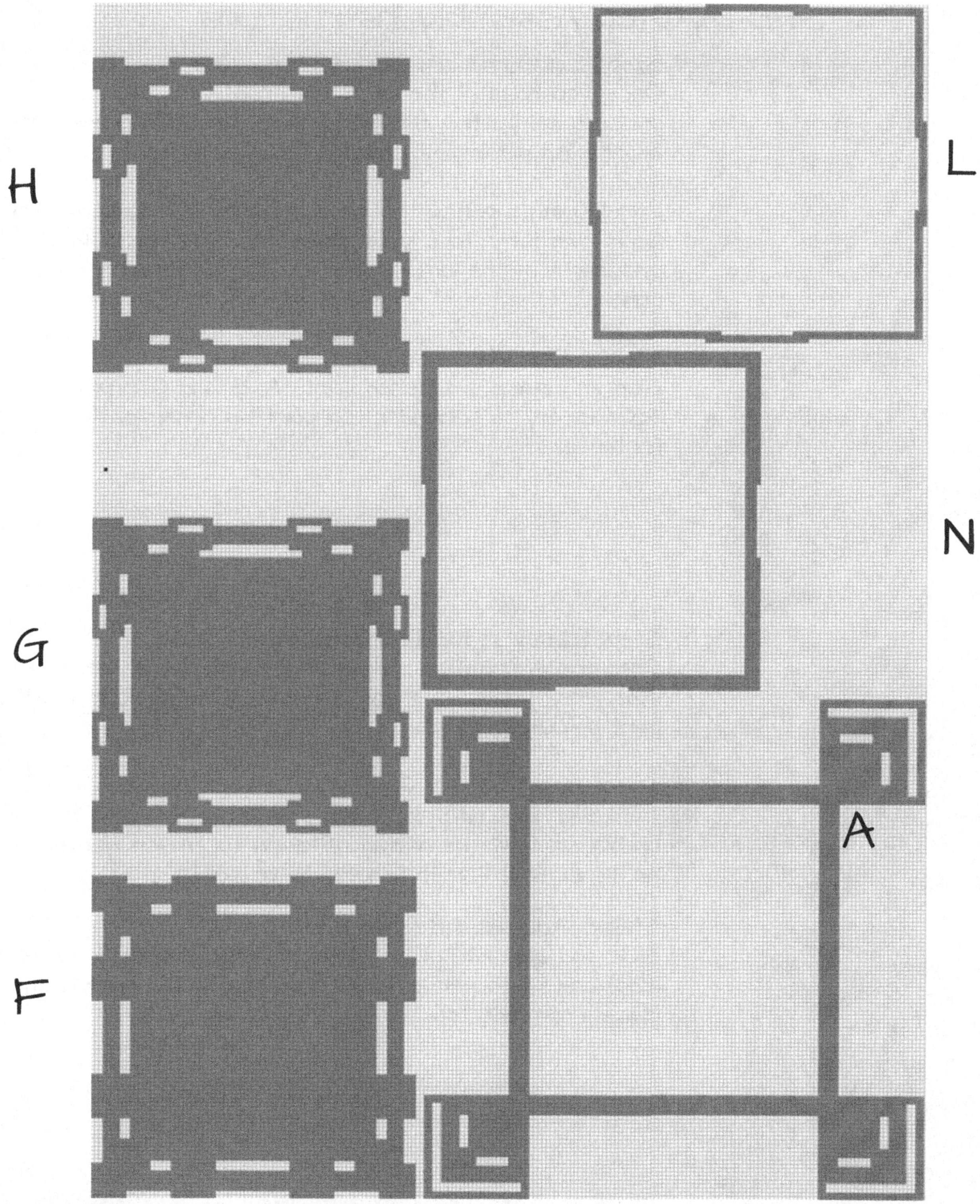
H
L
N
G
A
F

# Step-by-Step Assembly Instructions

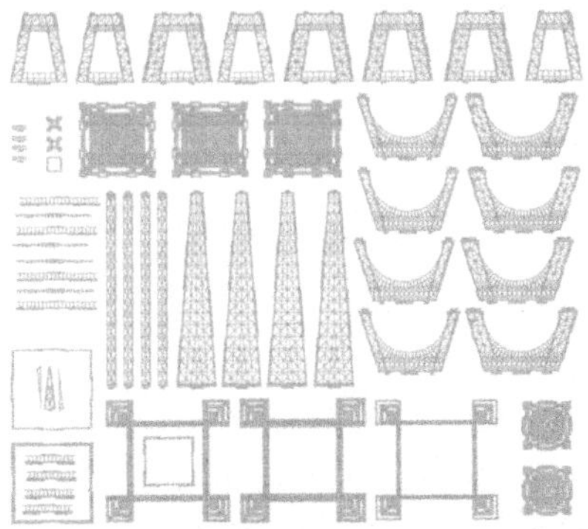

Step 1: Label and check inventory
Once you have all the pieces in front of you, the process of putting them together is no different from assembling pieces of Lego.

Label, using a pencil, the various parts and double-check your inventory. Take your time and enjoy the building process! Before jumping to the glue, be sure to do a dry run.

Step 2: Build the foundation
Glue into place staggered parts A, B and C in place followed by all four identical part Ds (inner legs) and Es (outer legs).

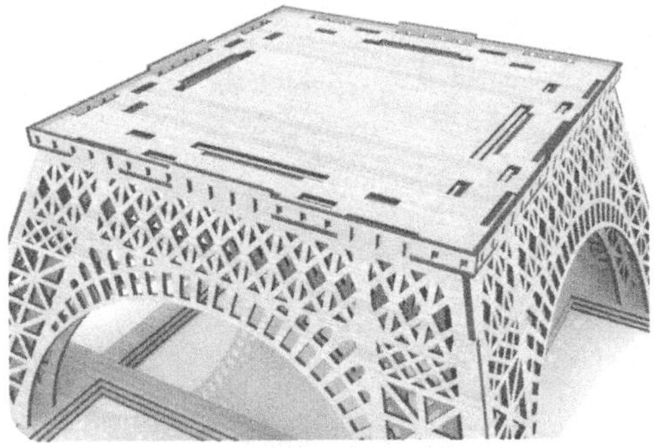

Step 3: Lock in the foundation
Onto the foundation, slide and glue into place the staggered parts F, G and H followed by all four identical part Is (outer flank).

The purpose of the first floor of this model is to lock the lower legs into place and provide a platform to build up higher. Don't be afraid of using too much glue here, as it's critical for the model to be solid.

Step 4: Add second floor legs
Slide and glue into place all four identical part Js (inner legs) and Ks (outer legs,) followed by sliding part L into place from the top.

Step 5: Finish the second floor

Slide and glue into place all part Ms (window panes,) followed by part N (roof).

You can now slide and glue into place part O (third floor,) which will lock the legs into place.

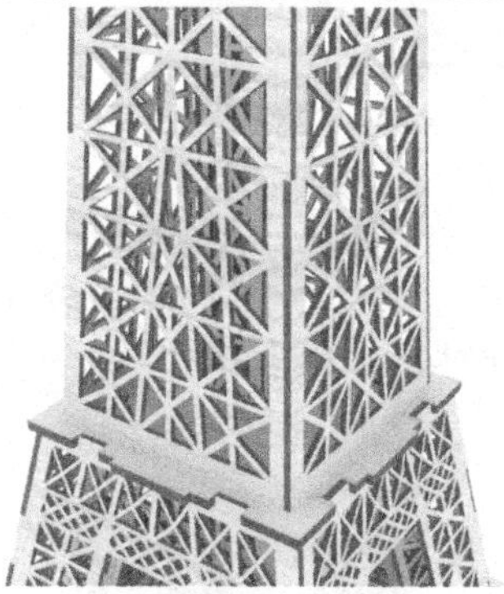

Step 6: Add the elevator shaft

Slide and glue into place all four part Ps (inner shaft) followed by all four part Qs (outer shaft).

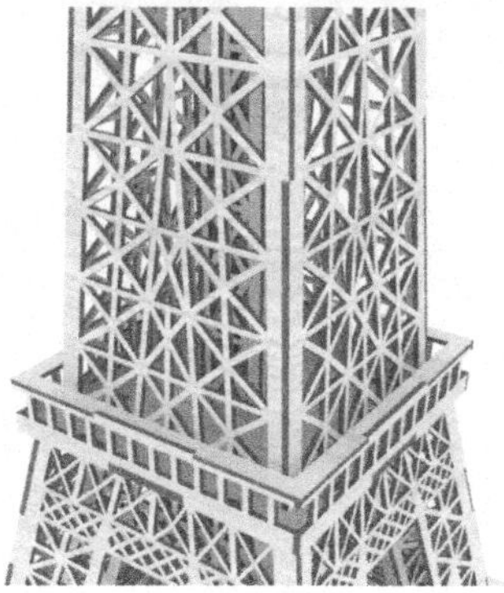

Step 7: Finish the third floor

Slide and glue into place all four part Rs (window panes) followed by part T (roof) slid from the top.

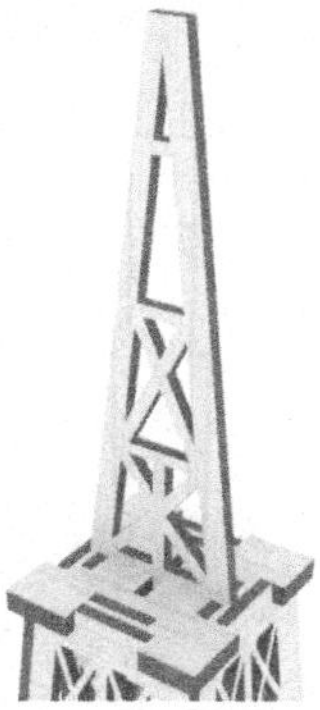

Step 8: Start the fourth floor

Slide and glue into place both part Us (fourth floor platforms) followed by part V (antenna).

Step 9: Finish the fourth floor

Slide and glue into place all four part Ws (window panes) followed by part X (roof) and both part Ys (antenna supports).

Step 12: Perform some finishing touches

Allow everything to dry overnight with clamps to hold things together.

There is no need to apply any preservatives or paint; feel free to sand any sharp edges down. ENJOY!

# Electronics Award

Completed 82 piece model.
Finished dimensions using 6 mm (1/4 inch) thick material.
L: 445 W: 445 H: 445 (mm) or 17-33/64 W: 17-33/64 H: 17-33/64 (in)

## Product Renders

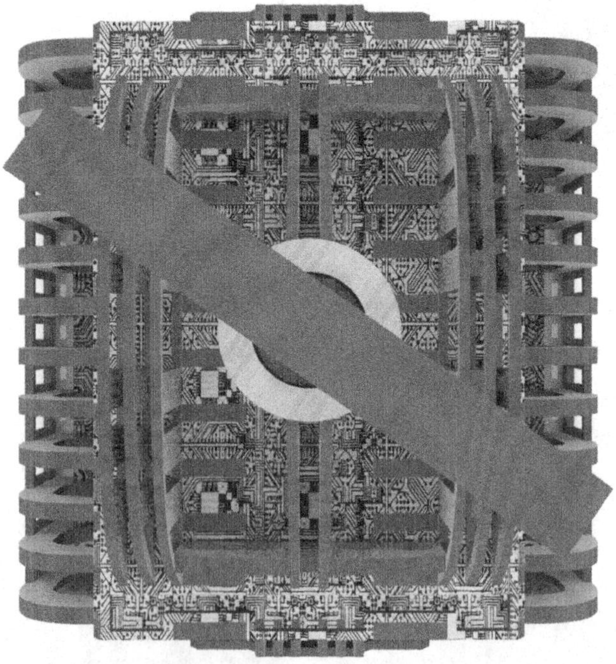

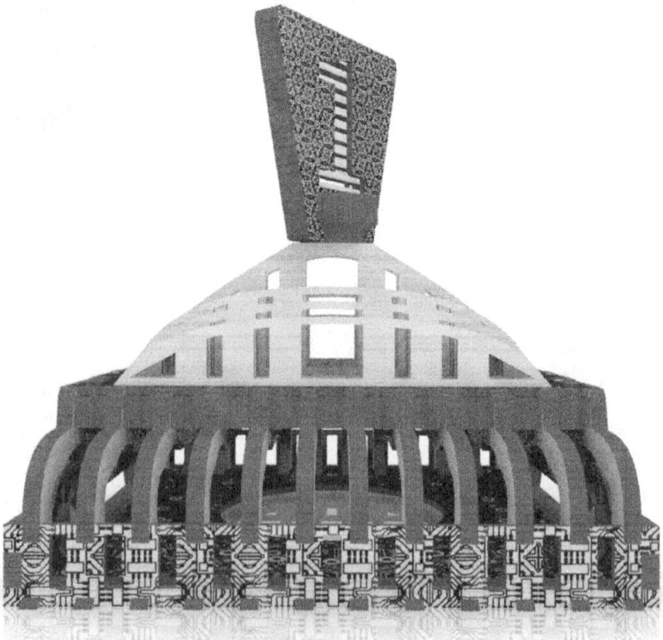

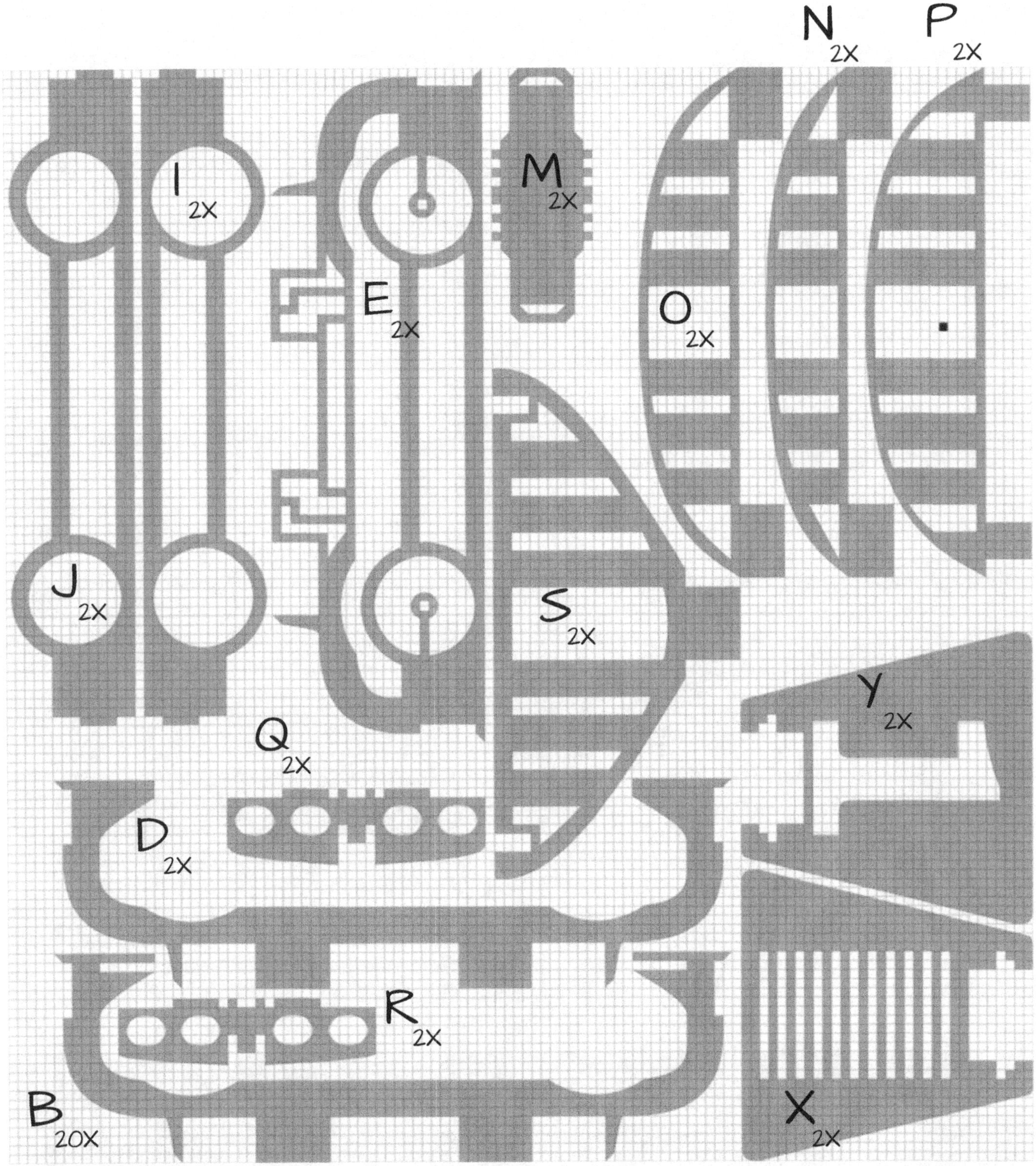
N 2X
P 2X
I 2X
M 2X
E 2X
O 2X
J 2X
S 2X
Y 2X
Q 2X
D 2X
R 2X
B 20X
X 2X

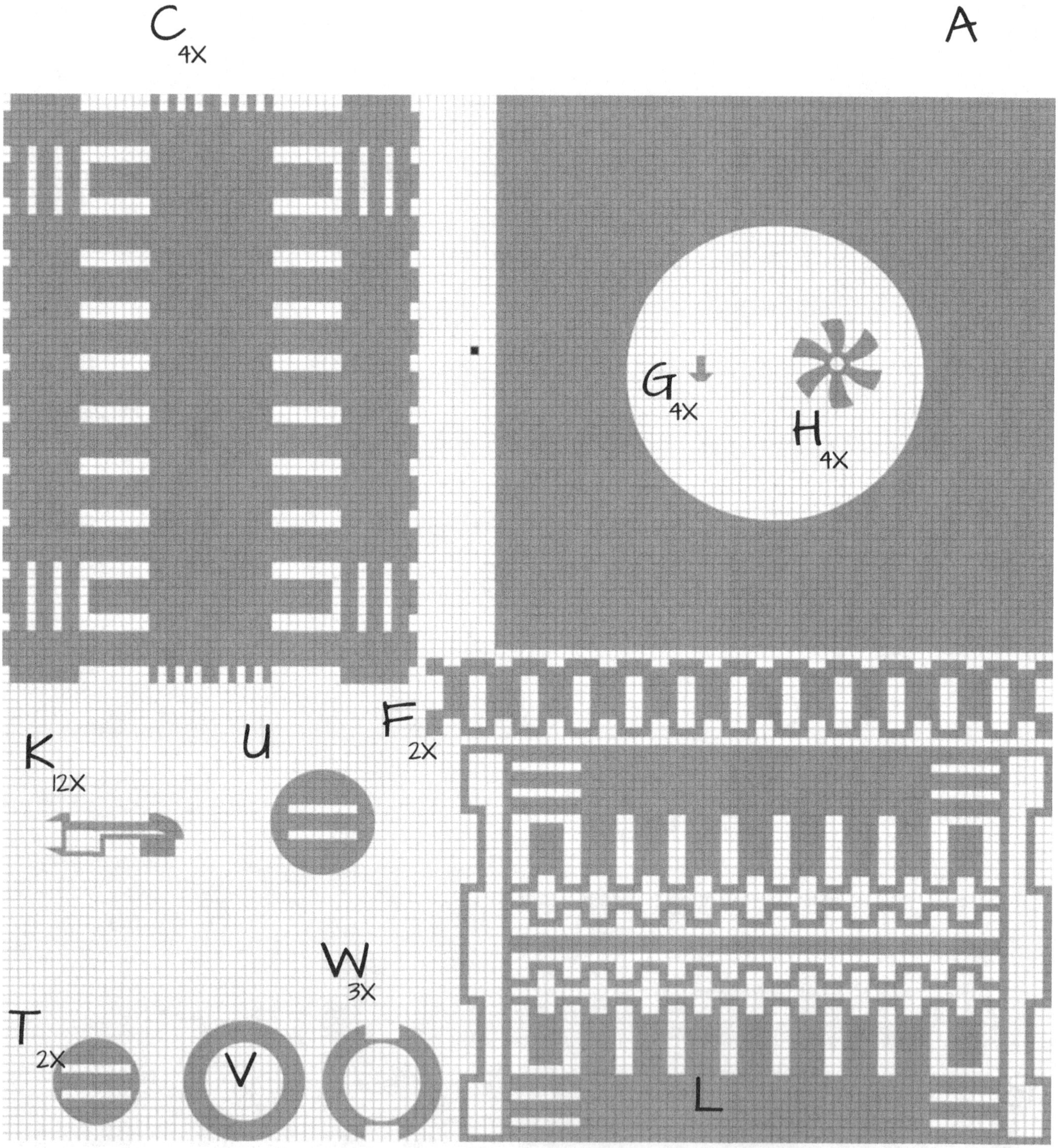
C 4X
A
G 4X
H 4X
K 12X
U
F 2X
W 3X
T 2X
V
L

# Step-by-Step Assembly Instructions

Step 1: Label and check inventory

Once you have all the pieces in front of you, the process of putting them together is no different from assembling pieces of Lego.

Label, using a pencil, the various parts and double-check your inventory. Take your time and enjoy the building process! Before jumping to the glue, be sure to do a dry run.

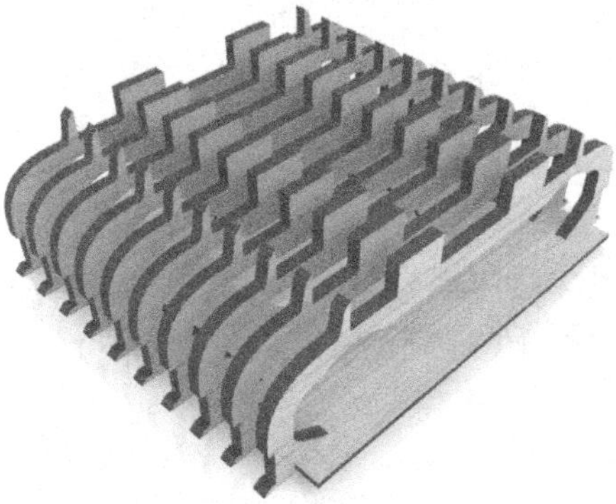

Step 2: Create the foundation

Slide, without glue, all 20 part Bs onto part A.

Note: Part A is square, so it doesn't matter which side you glue them onto.

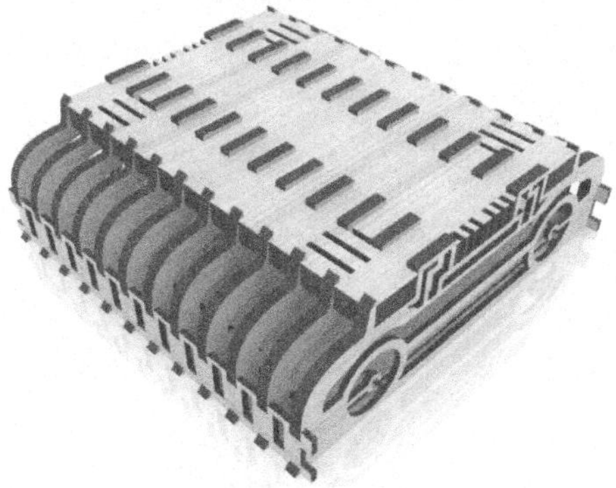

Step 3: Buid up and out

Slide and glue together all four part Cs on the top of the trophy. On the front/back, slide into place part D, then E on both ends. On both sides, glue into place a part F.

Inside the small circular holes of part F, slide into place, with some glue, part Gs.

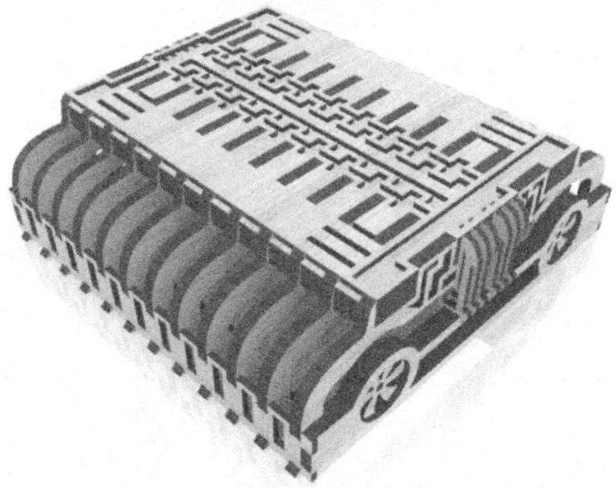

Step 4: Assemble the fans

Onto the four part Gs, slide into place, without glue, part Hs (fan) on each. You can then slide and glue into place the part Is (large hole) then Js (smaller hole). While you are doing this, glue and slide into place the part Ks (vertical heat sinks).

On the top, glue and slide into place part L.

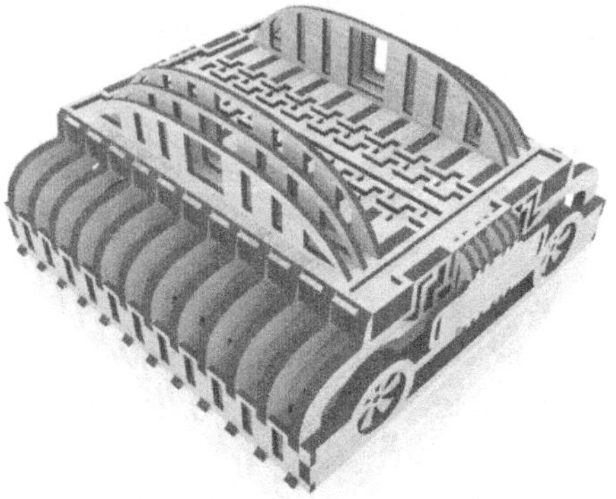

Step 5: Add vertical fins

On both the front and back, slide and glue into place part M (face plates). Slide and glue into place vertical fins from the shortest to tallest, outside to inside, parts N, O and P.

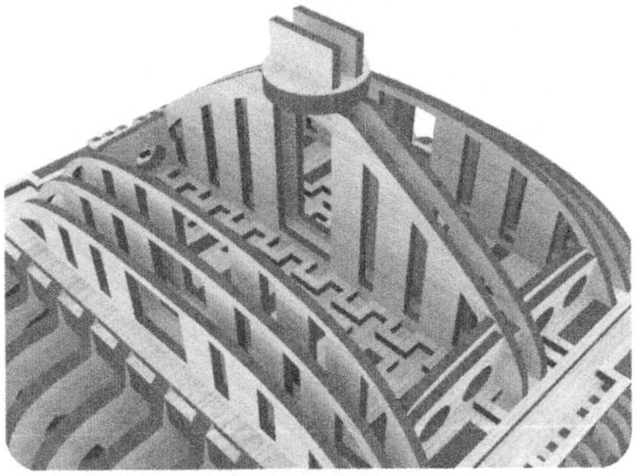

Step 6: Add banner supports

In the top slots, slide and glue into place parts Q and R in both the front and the back of the trophy. Afterwards, glue and slide into place both part Ss.

You can now slide and glue into place both part Ts, which will help keep everything spaced properly.

Step 7: Add the swivel

Drop into place part U, then part V.

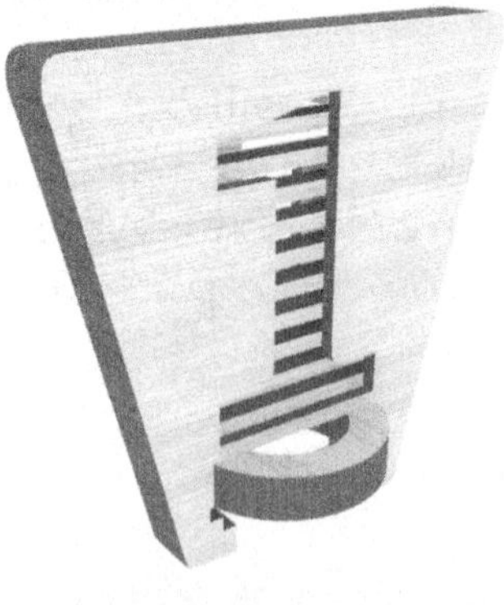

Step 9: Build the banner

Glue together parts W, X and Y.

Step 9: Put it all together

You can now slide the assembly you built in step 9 onto the base you built earlier. It should freely rotate.

Step 10: Perform some finishing touches

There is no need to apply any preservatives or paint; feel free to sand any sharp edges down. ENJOY!

## Fish Pad

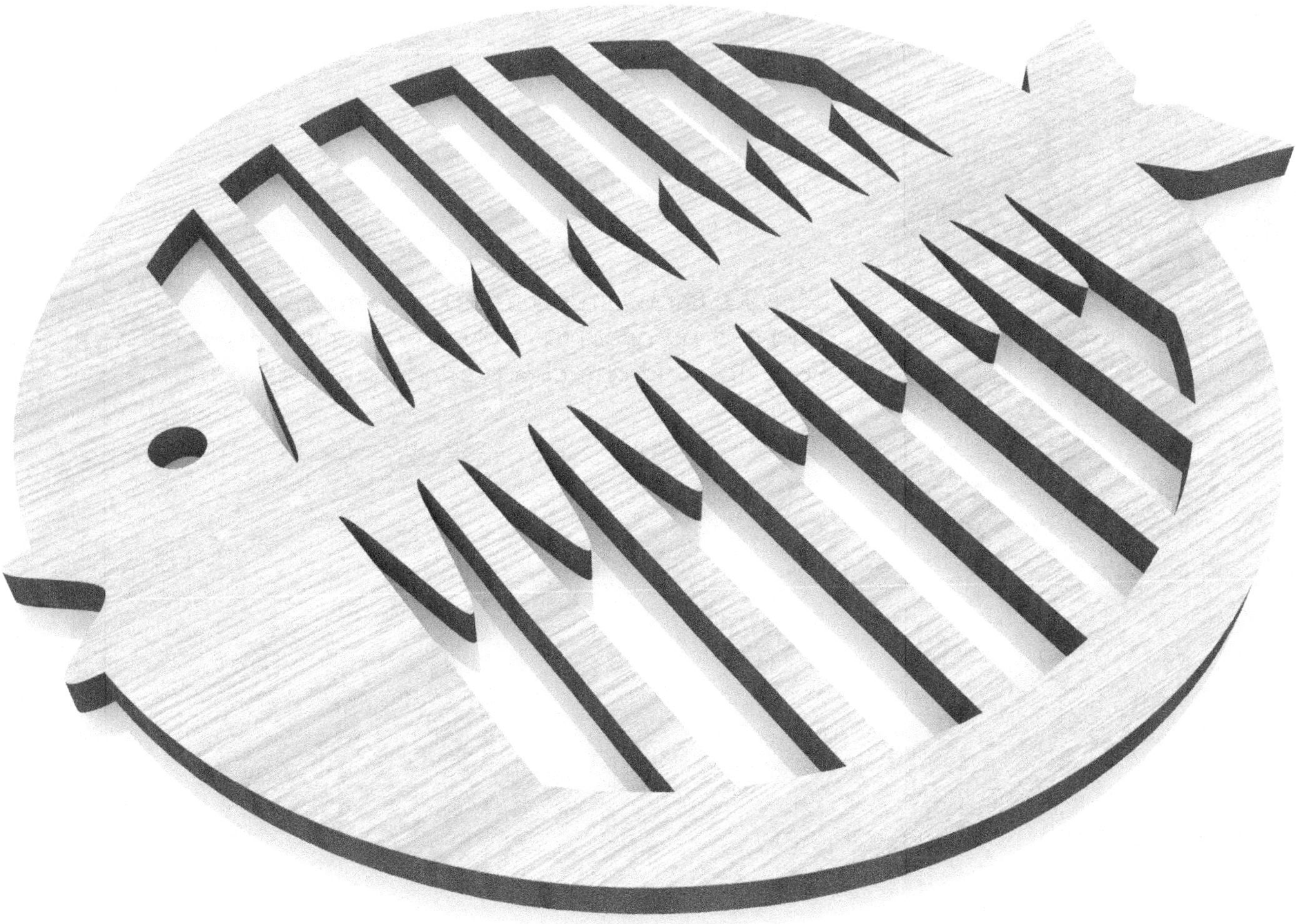

Completed 1 piece model.
Finished dimensions using 6 mm (1/4 inch) thick material.
L: 200 W: 180 H: 60 (mm) or L: 7-7/8, W: 7-3/32, H: 2-23/64 (in)

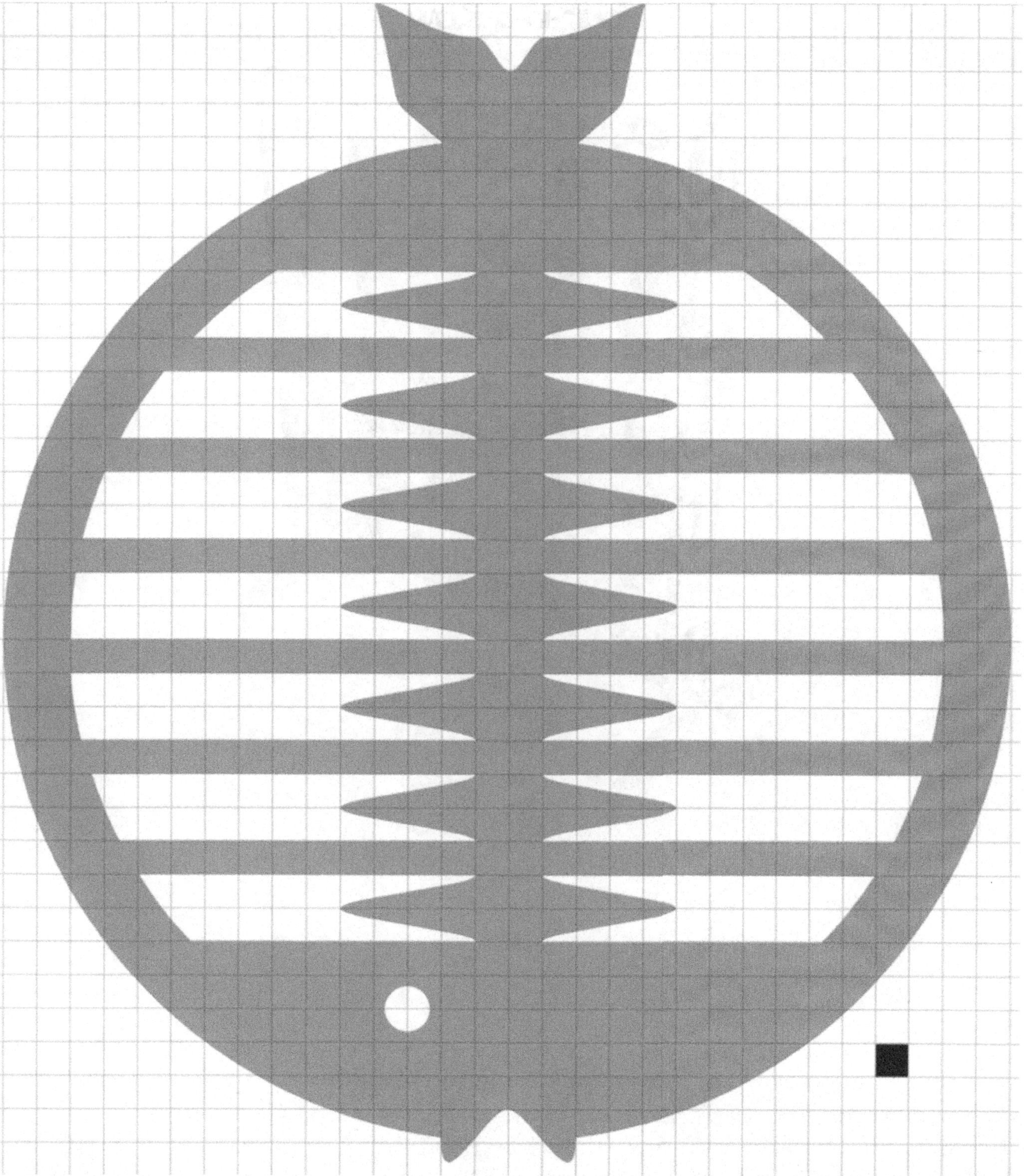

# Four-in-a-Row

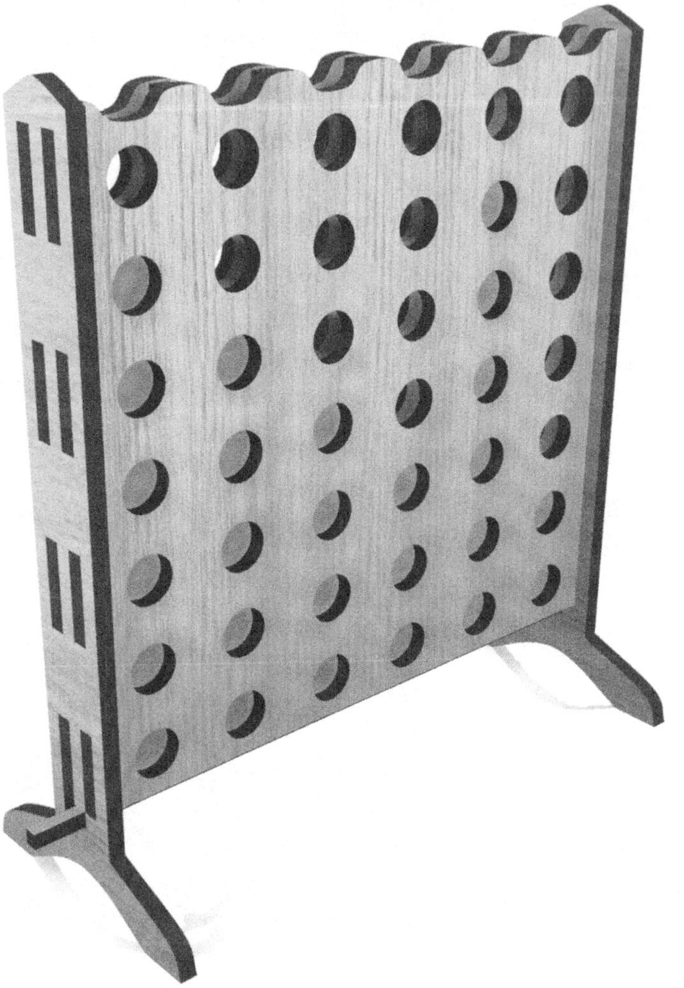

Completed 52 piece model.
Finished dimensions using 10 mm (25/64 inch) thick wood.
L: 430 W: 170 H: 430 (mm) or L: 16-59/64, W: 6-11/16, H: 16-59/64 (in)

## Product Renders

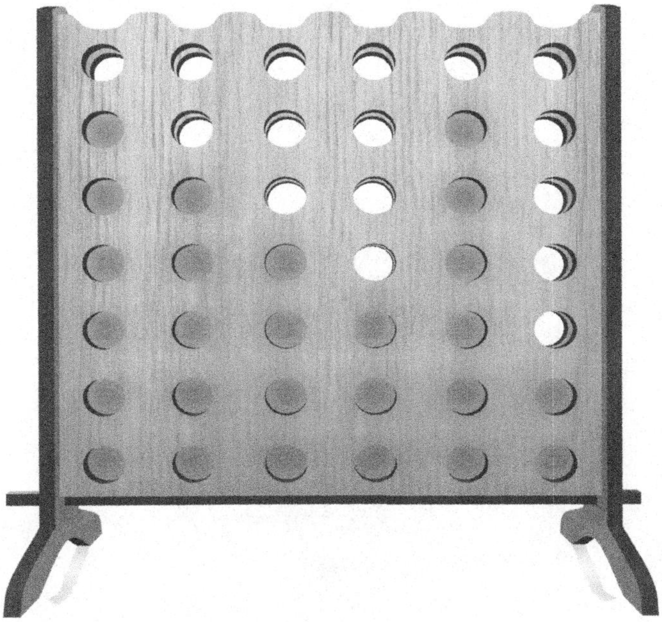

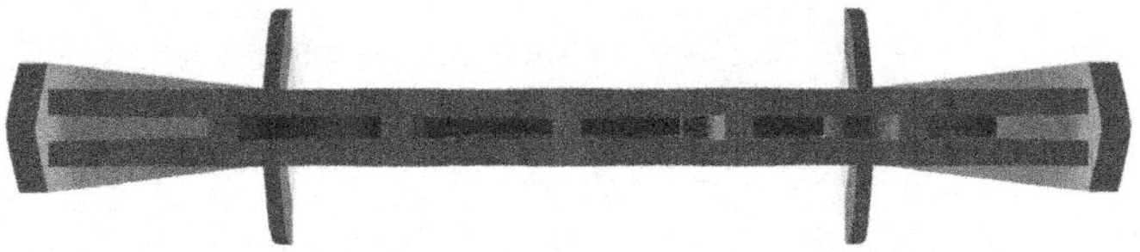

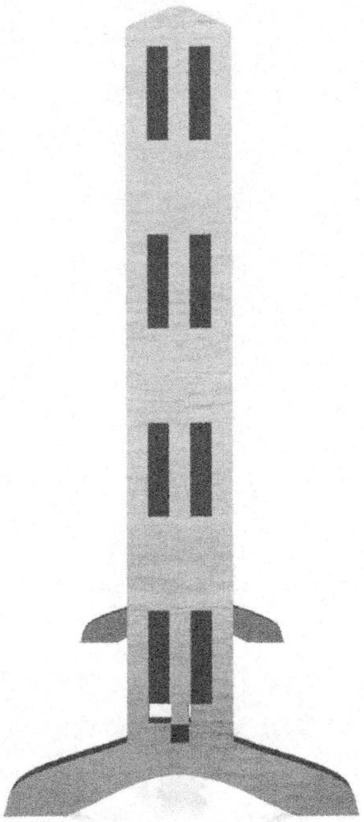

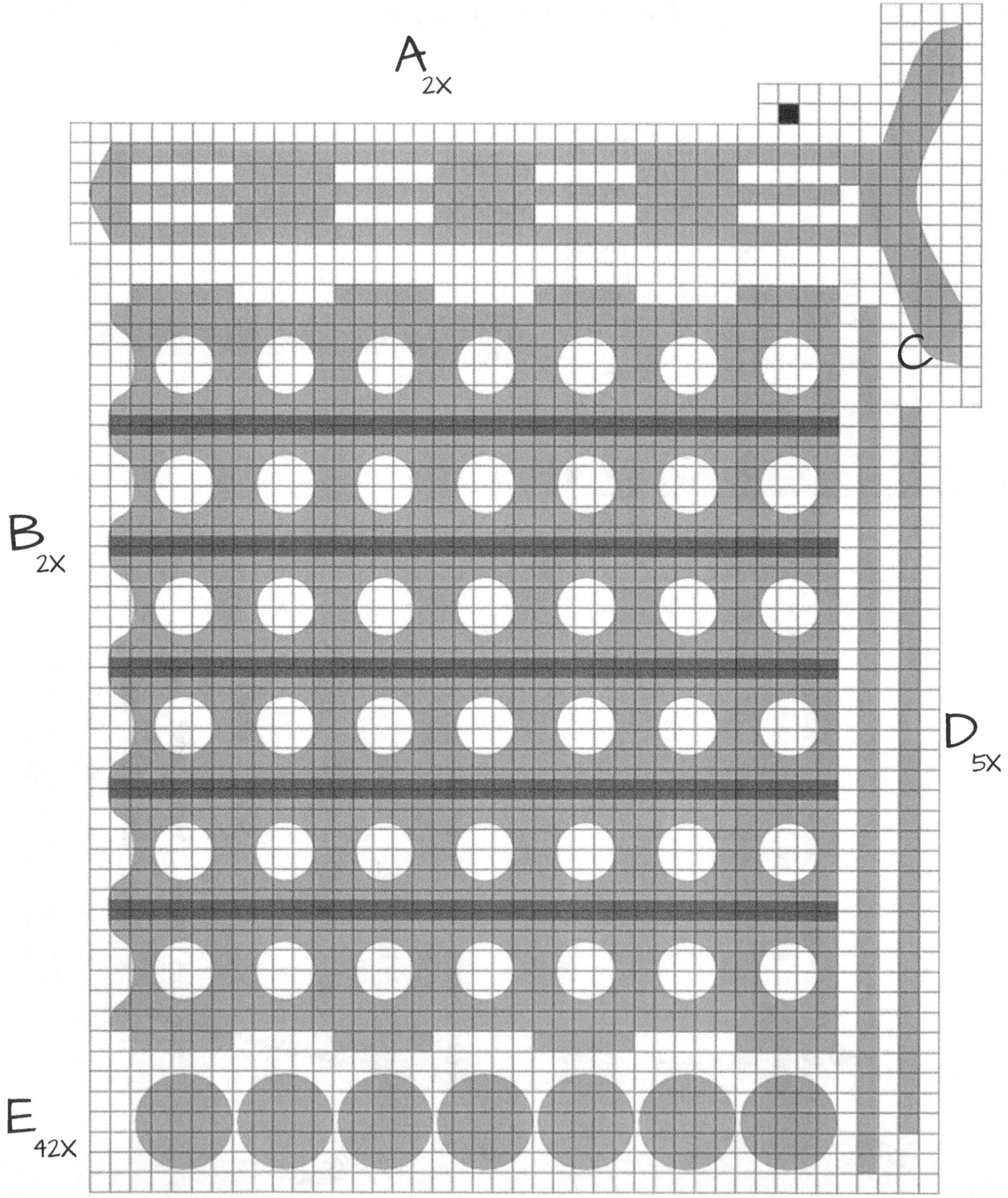
A
2X
B
2X
C
D
5X
E
42X

## Step-by-Step Assembly Instructions

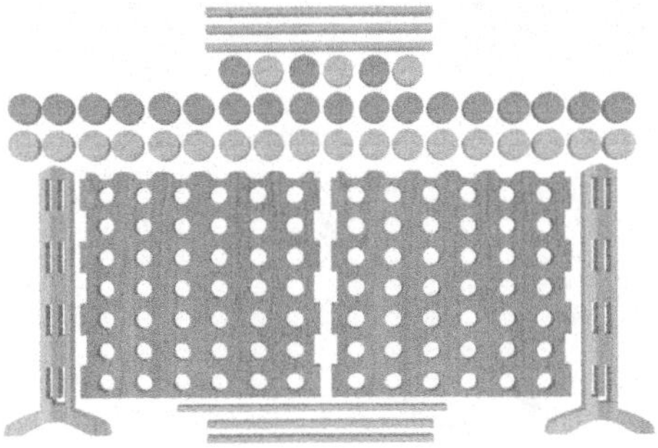

Step 1: Label and check inventory
Once you have all the pieces in front of you, the process of putting them together is no different from assembling pieces of Lego.

Label, using a pencil, the various parts and double-check your inventory. Take your time and enjoy the building process! Before jumping to the glue, be sure to do a dry run.

Step 2: Add backing
Apply the rods with glue in the center of each column of holes.

Step 3: Glue the rest of the pieces together
Once the part B sandwich is completely dry, stick both part As on either side, applying glue where the parts intersect to hold them in place.

Step 4: Perform some finishing touches
After allowing the pieces to bond together with the glue overnight, you may now sand any protruding parts, flush along with the outer edges if you so wish.

There is no need to apply any preservatives or paint. ENJOY!

# Japanese Miyajima Gate

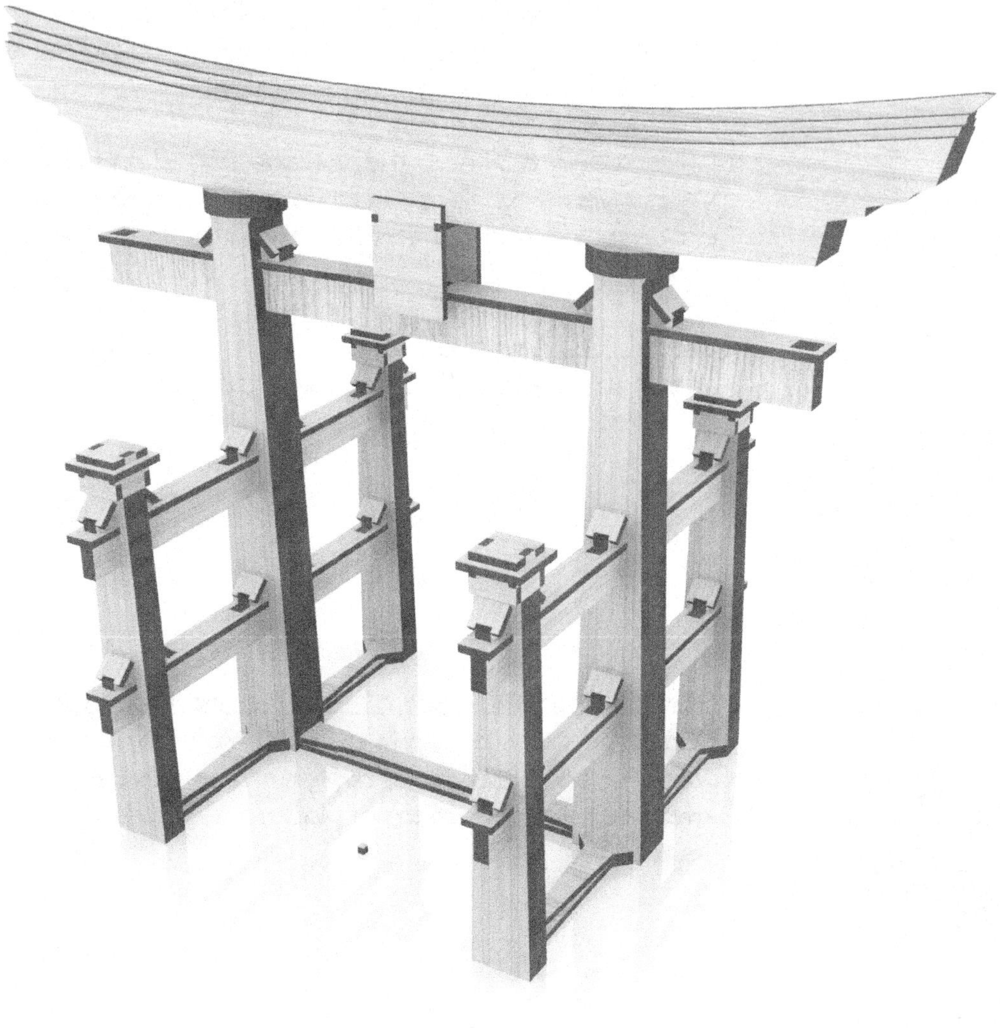

Completed 125 piece model.
Finished dimensions using 4 mm (5/32 inch) thick material.
L: 540 W: 280 H: 410 (mm) or L: 21-17/64 W: 11-1/32 H: 16-9/64 (in)

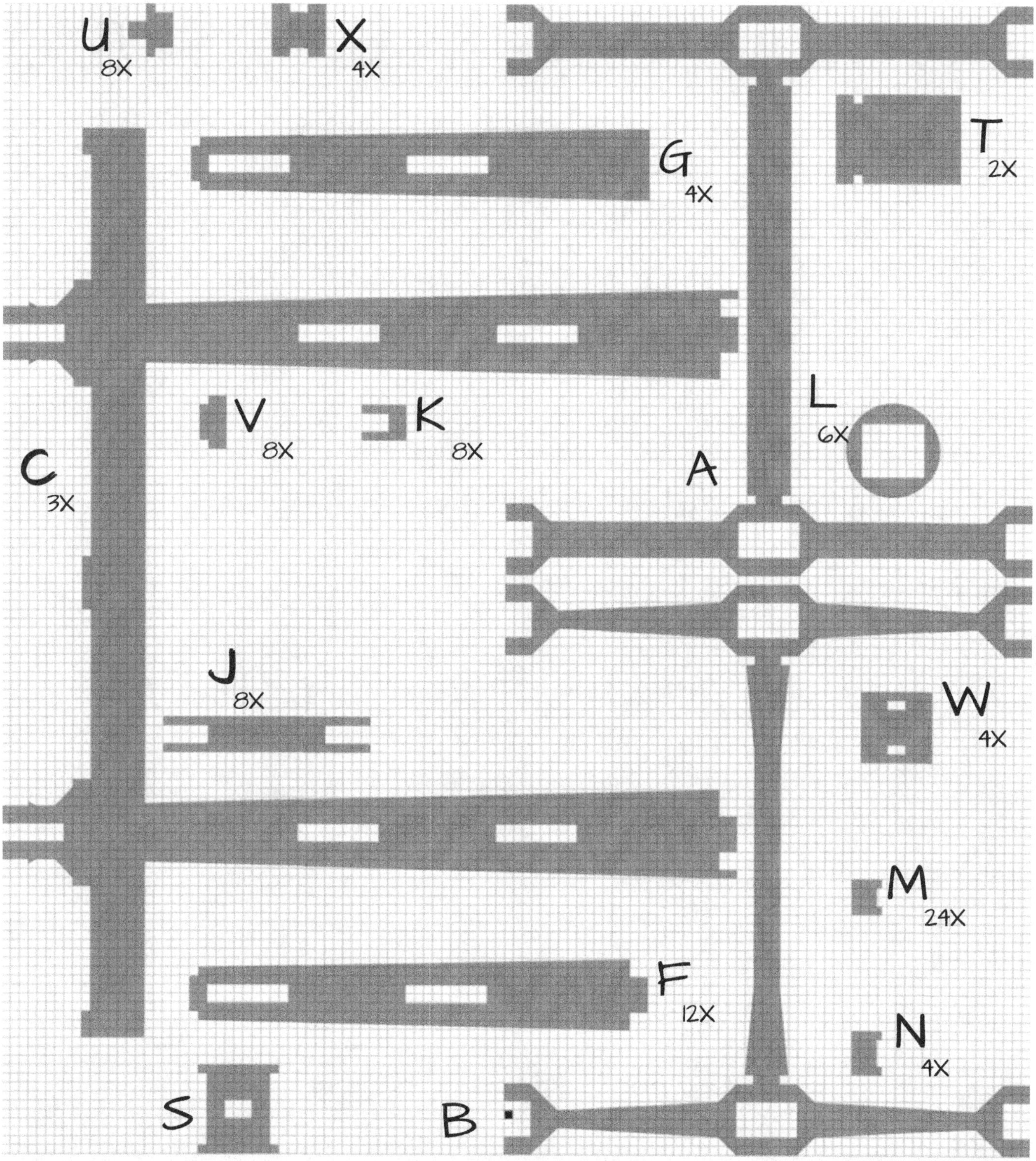

U
8X
X
4X
T
2X
G
4X
L
6X
V
8X
K
8X
C
3X
A
J
8X
W
4X
M
24X
F
12X
N
4X
S
B

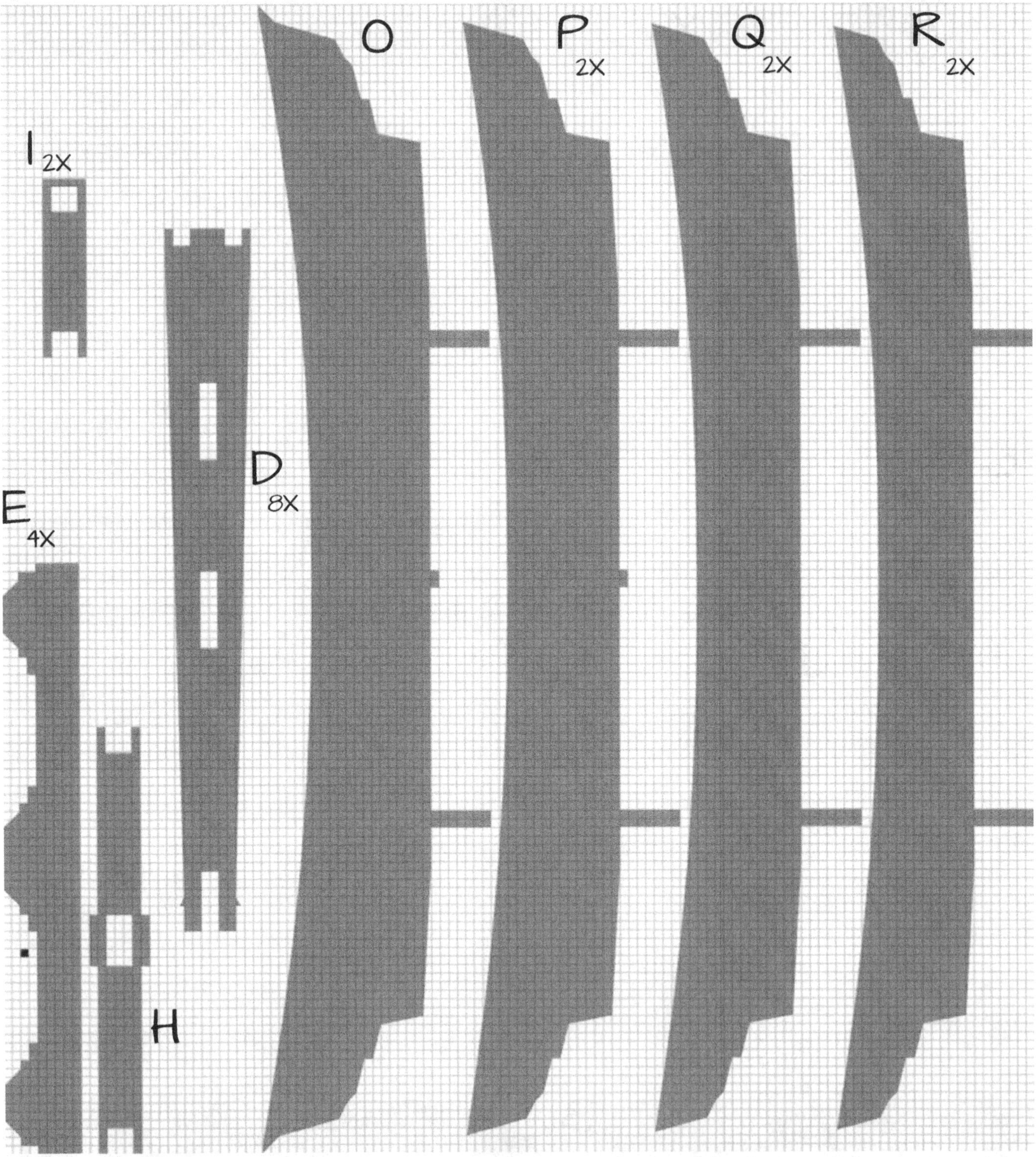
O
P 2X
Q 2X
R 2X
I 2X
D 8X
E 4X
H

# Step-by-Step Assembly Instructions

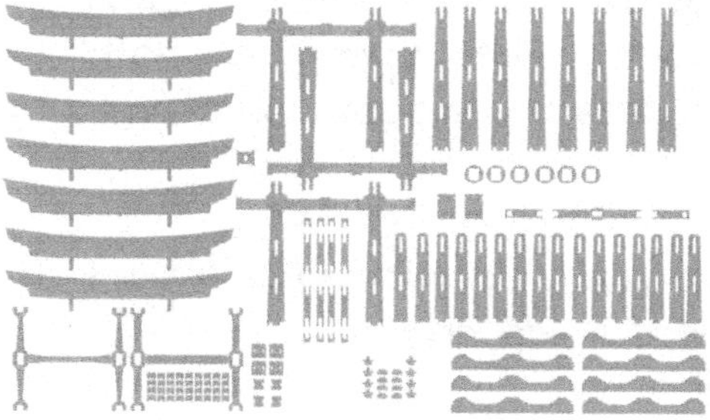

**Step 1: Label and check inventory**

Once you have all the pieces in front of you, the process of putting them together is no different from assembling pieces of Lego.

Label, using a pencil, the various parts and double-check your inventory. Take your time and enjoy the building process! Before jumping to the glue, be sure to do a dry run.

**Step 2: Start with the center**

Start by laying down part A, followed by part B, on a flat surface. Then glue into place three part Cs (supporting columns).

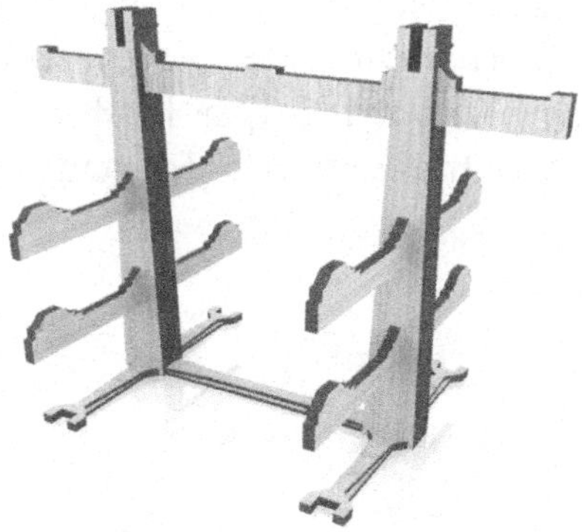

**Step 3: Build outwards**

Onto the foundation, slide and glue onto the part Cs, two part Ds (major columns) on both sides of each column.

Horizontally, slide and glue into place doubled up part Es into all four holes within the columns.

**Step 4: Add minor columns**

At the end of each part E, slide and glue into place three part Fs (minor column with peg bottom) followed by a part G (minor column with straight bottom) on the outside.

Step 5: Add horizontal flashing
Starting on the top surface of the supporting column, slide and glue into place a part H with a part I on both ends. On the top of each supporting column, slide and glue into place three part Ls.

For the minor columns, slide and glue into place a part J followed by a part K at the outside end,.

Step 6: Add flashing caps
Add all 24 identical part Ms on both edges of each minor column, both the upper and lower ones.

Onto the supporting columns, do the same with four part Ns on either side of them.

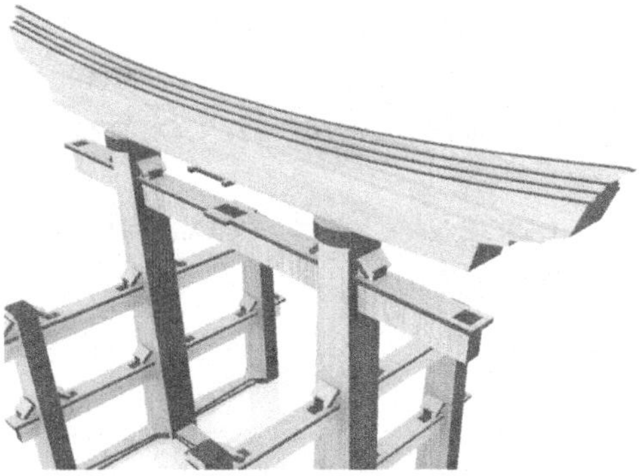

Step 7: Add the ornamental roof
Slide and glue into place part O in the middle, followed by parts P, Q, R on both sides with a part S (portrait holder) in the middle, underneath the ornamental roof.

Step 8: Add the tablet
Slide and glue into place a part T on both sides of the ornamental roof -- be a little creative here; you have lots of room to put whatever message you want.

Step 9: Build end-caps

Glue together two part Us with part Vs, using a W to cap the assembly. Follow this with a part X.

Repeat this step three additional times until you have four completed end-caps.

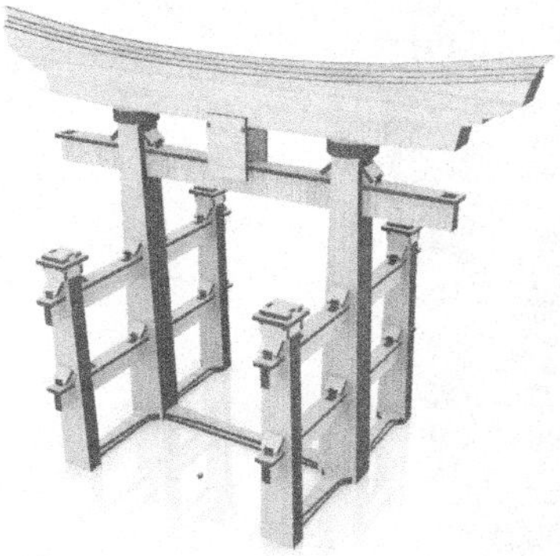

Step 12: Perform some finishing touches

Slide the end-caps onto the ends of all the minor column assemblies and allow everything to dry overnight with clamps to hold things together.

There is no need to apply any preservatives or paint; feel free to sand any sharp edges down. ENJOY!

# Pen and Pencil Holder

Completed 17 piece model.
Finished dimensions using 6 mm (1/4 inch) thick material.
L: 130 W: 72 H: 90 (mm) or L: 5-1/8 W: 2-53/64 H: 3-35/64 (in)

## Product Renders

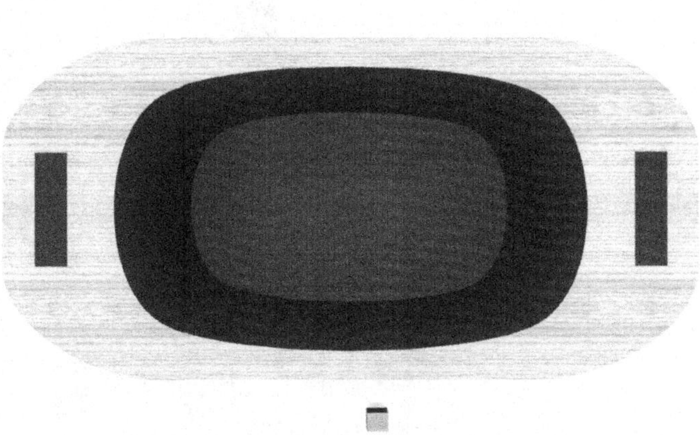

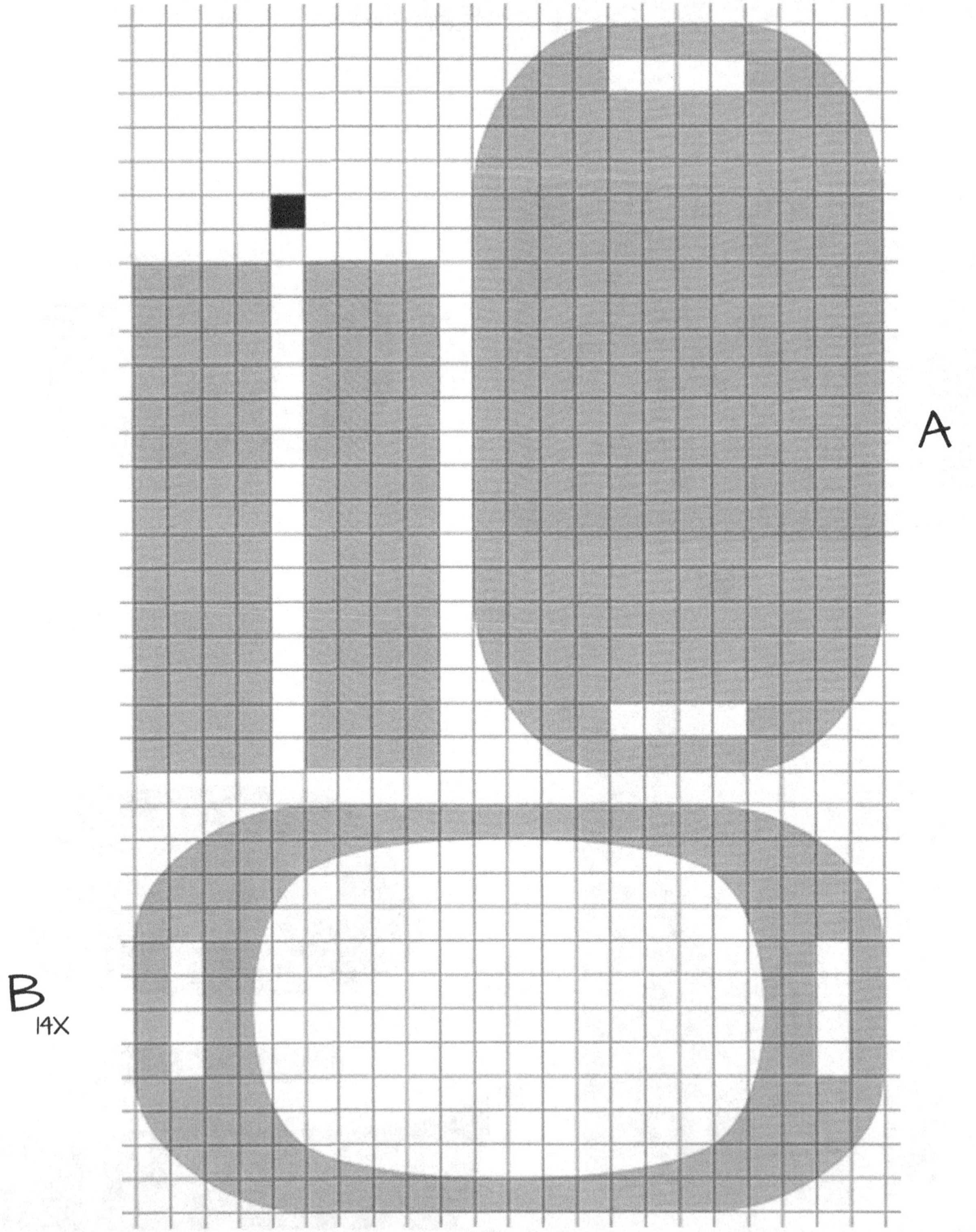
A
B
14X

# Step-by-Step Assembly Instructions

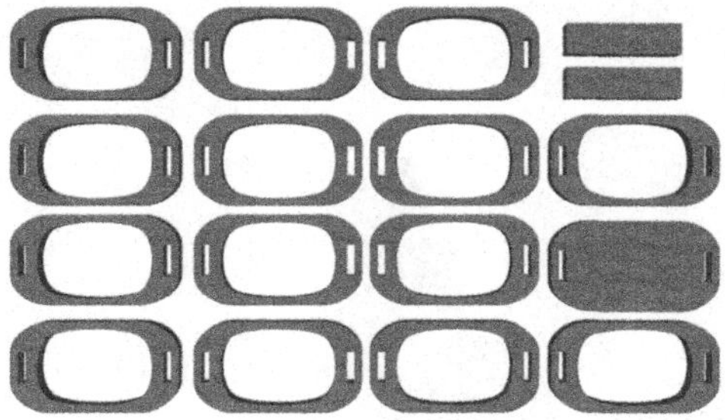

Step 1: Label and check inventory

Once you have all the pieces in front of you, the process of putting them together is no different from assembling pieces of Lego.

Label, using a pencil, the various parts and double-check your inventory. Take your time and enjoy the building process! Before jumping to the glue, be sure to do a dry run.

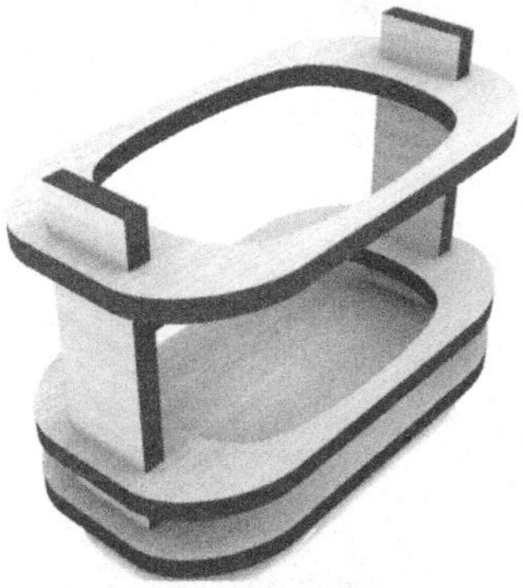

Step 2: Glue and slide

Place part A on some wax paper and glue the dowels into its holes. Afterwards, slide on alternating cut wood strips, adding a little glue where the parts intersect.

Once everything is stacked, place something heavy on top while everything dries overnight.

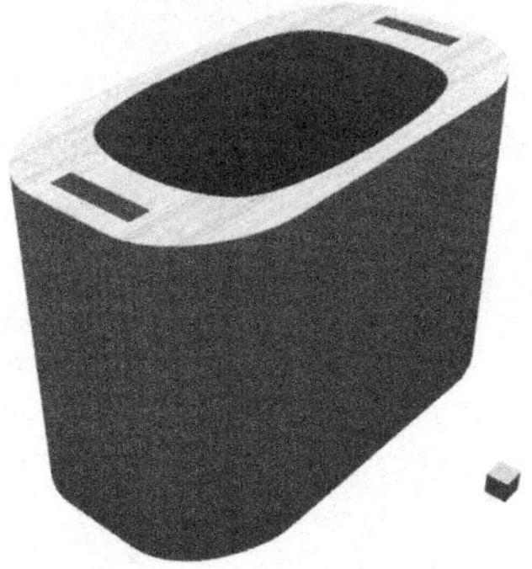

Step 3: Perform some finishing touches

Sand any protruding parts flush, along with the outer edges if you so wish.

There is no need to apply any preservatives or paint. ENJOY!

# Rocket Coaster

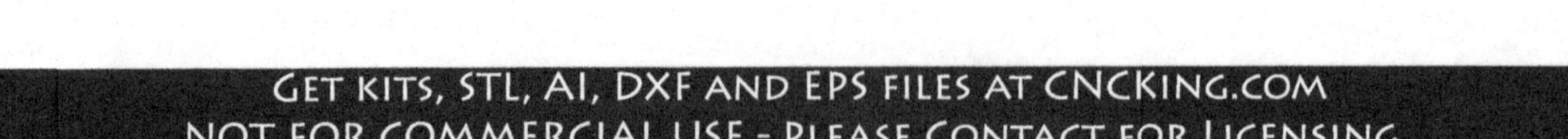

Completed 126 piece model.
Finished dimensions using 6 mm (1/4 inch) thick material.
L: 130 W: 72 H: 90 (mm) or L: 35-7/16 W: 13-25/32 H: 32-9/32 (in)

Product Renders

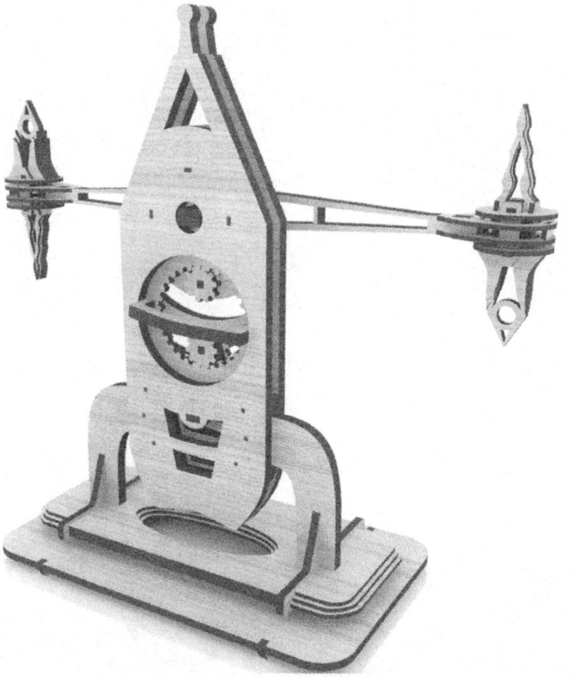

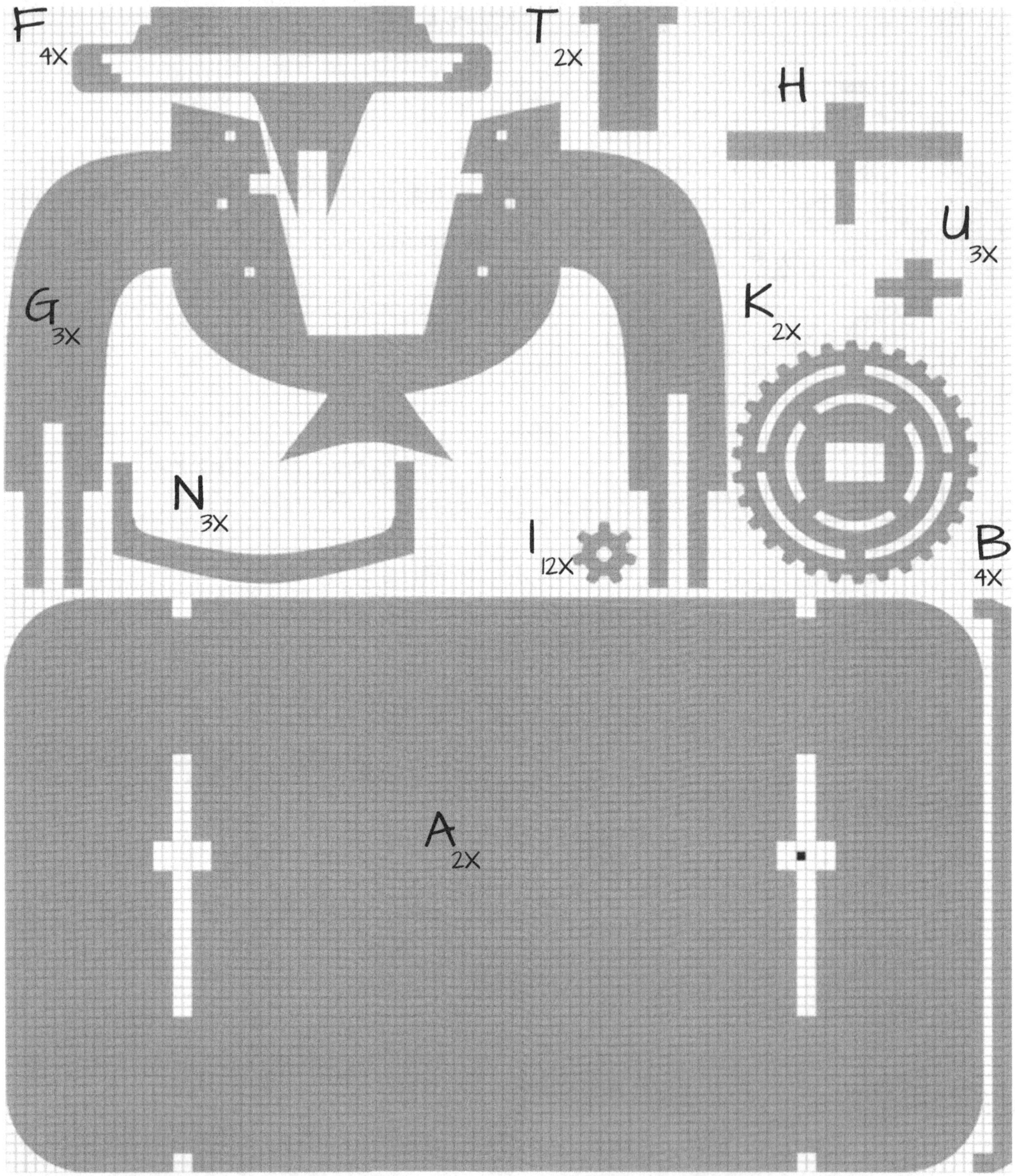
F 4X
T 2X
H
U 3X
G 3X
K 2X
N 3X
I 12X
B 4X
A 2X

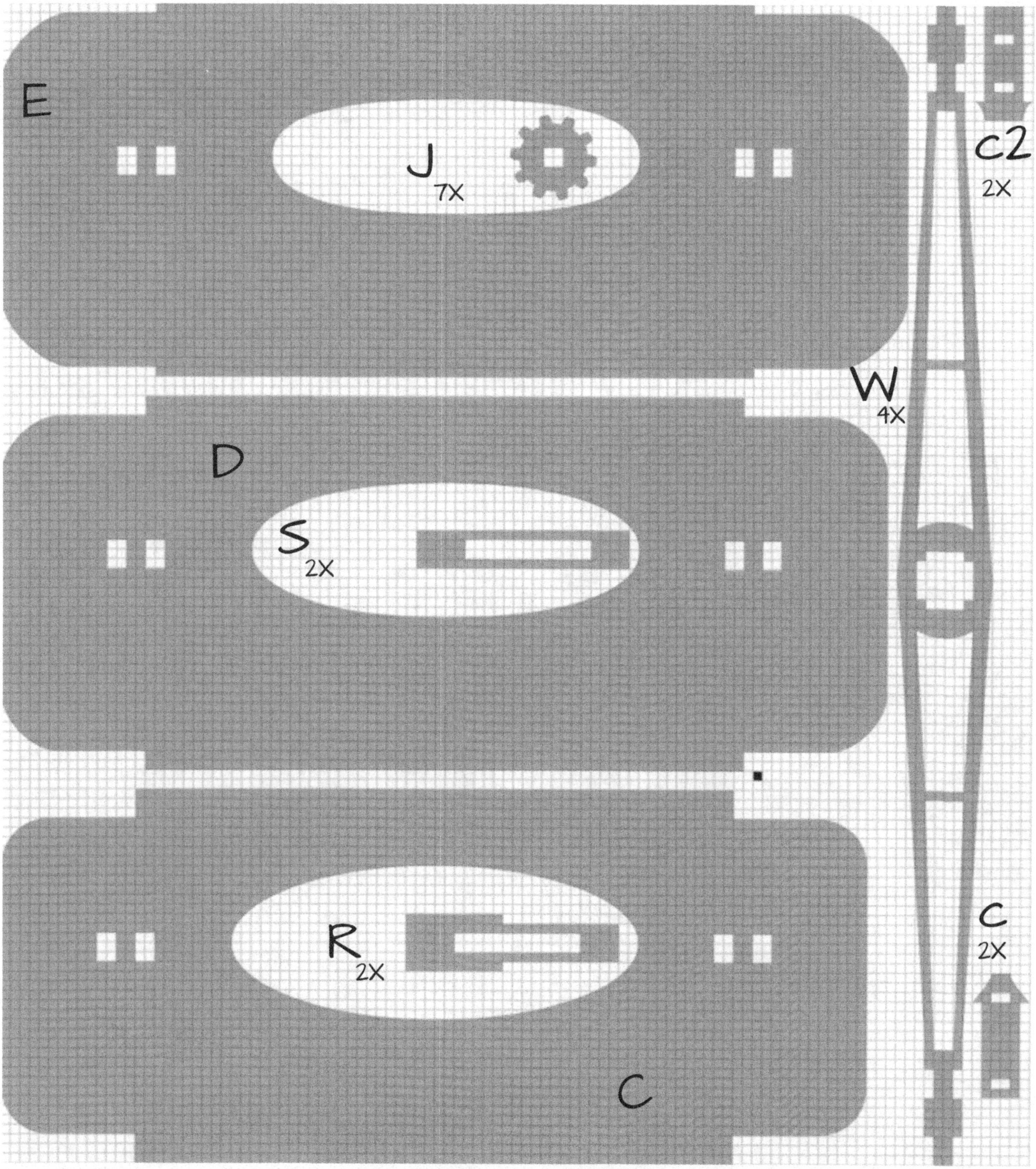
E
J
7X
C2
2X
W
4X
D
S
2X
C
2X
R
2X
C

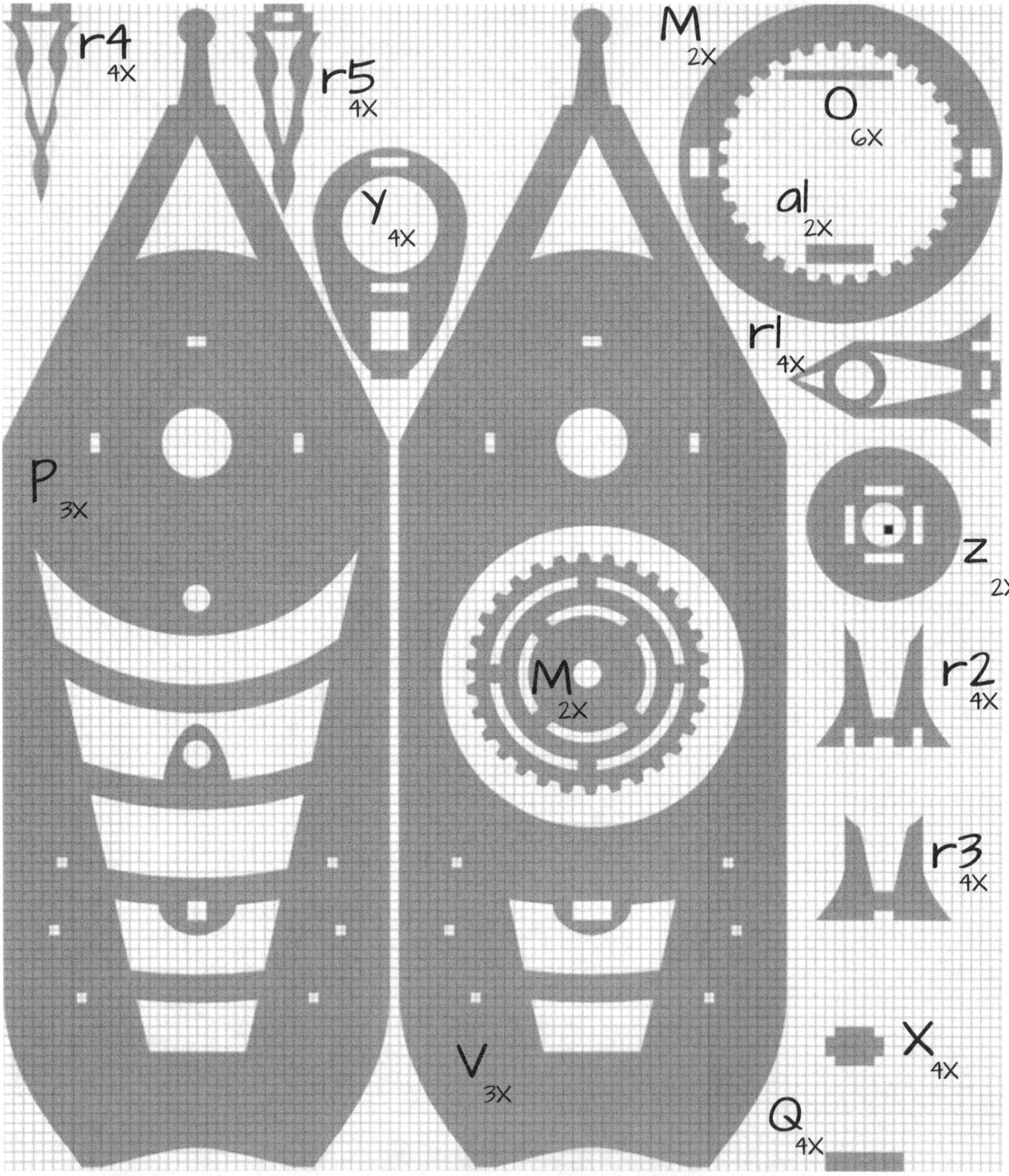
r4
4X
r5
4X
M
2X
O
6X
Y
4X
al
2X
rl
4X
P
3X
Z
2X
M
2X
r2
4X
r3
4X
X
4X
V
3X
Q
4X

# Step-by-Step Assembly Instructions

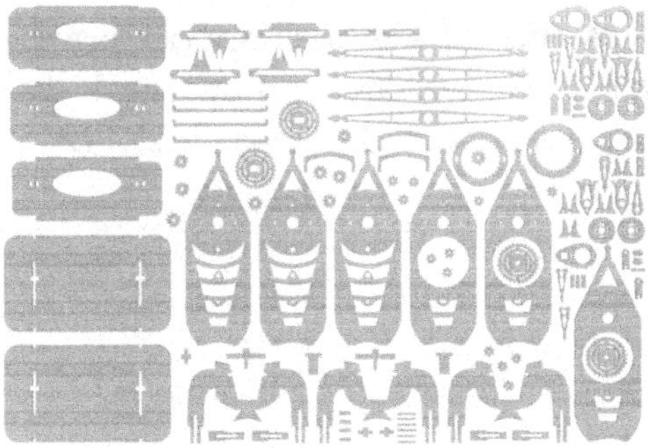

Step 1: Label and check inventory

Once you have all the pieces in front of you, the process of putting them together is no different from assembling pieces of Lego.

Label, using a pencil, the various parts and double-check your inventory. Take your time and enjoy the building process! Before jumping to the glue, be sure to do a dry run.

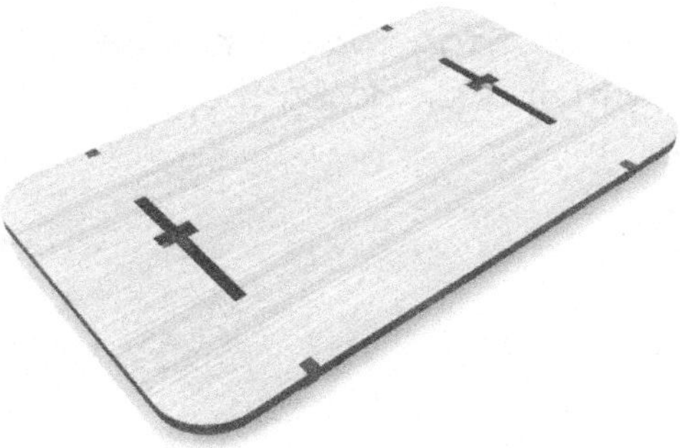

Step 2: Create the foundation

Slide and glue into place both part As (platform) and Bs (rails).

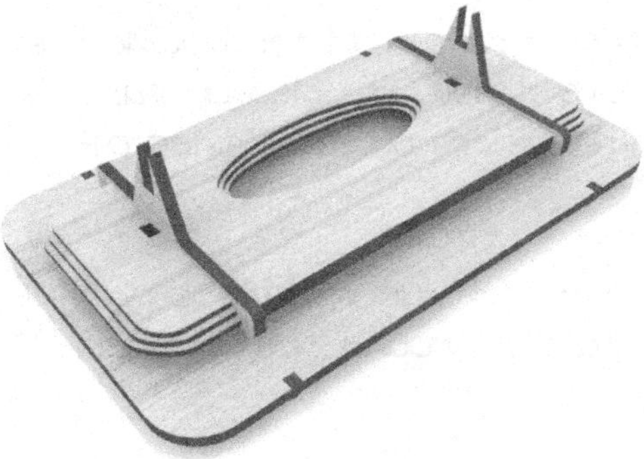

Step 3: Complete the foundation

Slide and glue together parts C, D and E; then on both edges, glue and slide into place part Fs (support rails).

Step 4: Add the rocket's legs

Slide and glue into place all three part Gs (rocket legs) and then glue into place part H (supporting cross member).

Step 5: Double up gears

Glue together to double up all the gears. You should have a total of 6 doubled up part Is, three doubled up part Js and a doubled up K, L and M (with part Ns inserted for the handles).

Step 6: Start adding gears

Using part O (pegs) to keep everything together, slide into place all three part Ps with the six doubled up Is.

Step 7: Add the central gear

In the middle of the six smaller gears, add the doubled up part L (large gear) then above it, two doubled up part Js with doubled up part Qs uniting them on both sides of part Ps.

Ahead of the large wheel, slide and glue into place part J. The gears should all freely rotate around one another.

Step 8: Add the main gear

In the back of the model is the main gear, which is hollowed-out with a handle so that you can easily rotate all the gears in the front from one place.

As you did in the bottom, on the top, add two doubled up part Js with doubled up part Qs uniting them on both sides of part Ps. Slide the main gear between the top and bottom. Don't apply any glue.

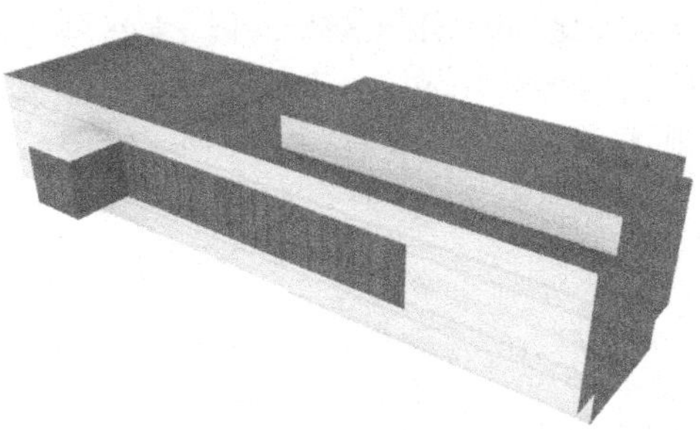

Step 9: Create the arm axle

What turns the main arms is a large axle through the top. It's built by combining all part Rs, Ss and Ts together to form one solid axle.

Allow it to dry before inserting it into the model so that it doesn't glue itself to other components.

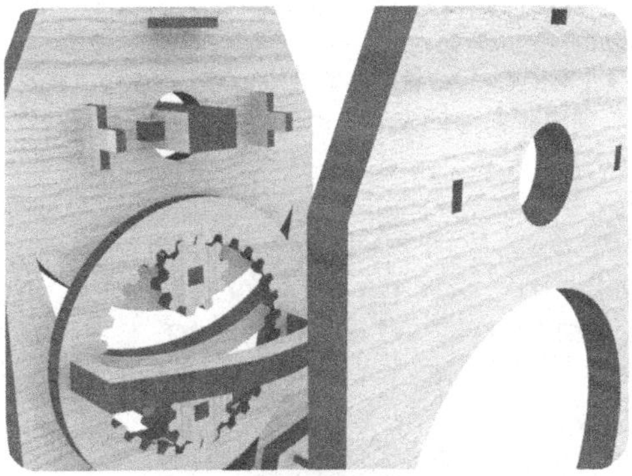

Step 10: Complete the back

From the back, slide but don't glue into place the axle assembly you made in step 9, so that the pegs that stick out prevent it from falling forward towards the front into the large hole in the top. Around it, put three part Us (small spacers) followed by part Vs (back pannels). Glue the back panels in place, but make sure the gears still freely rotate (you may have to sand their edges down).

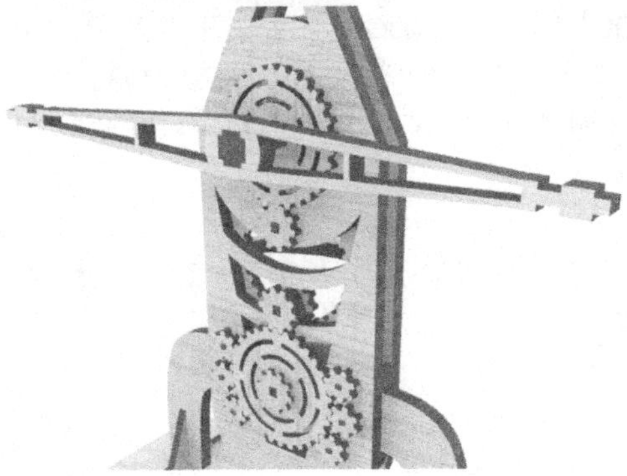

Step 11: Add the front arms

Slide and glue in the front of the rocket coaster the doubled up part Ks along the back. Now that the back gear is freely spinning, allow everything to dry overnight.

At the front edge of the axle, glue and slide into place all four part Ws.

Step 12: Add the rocket mounts

Although only one end of the arm is shown in the picture, do the same in reverse at the end of the other arm to complete both rockets.

Slide and glue into place two part Xs (spacers) in between two part Ys to form the central support for the rockets.

Step 13: Start with the top
Place a part Z first on top place. Next, slide and glue into place two part r1s, followed by r2s on both sides and then r3s. Once that's all together, drop into place a part c1 on both sides, then put a part Q through all the parts.

Step 14: End with the bottom
From the bottom, slide into place both part c2s, inserting a part a1 to keep it from falling just under the first part Z you placed.

Then glue and slide into place another part Z (from the bottom again,) followed by both part r4s in the middle and the r5s on either side, using a part a1 to keep everything in place. You want the rocket to freely rotate but not fall out.

Step 10: Perform some finishing touches
Test the gears. Everything should spin freely; you may have to sand down a few parts where too much friction is occurring.

There is no need to apply any preservatives or paint; feel free to sand any sharp edges down. ENJOY!

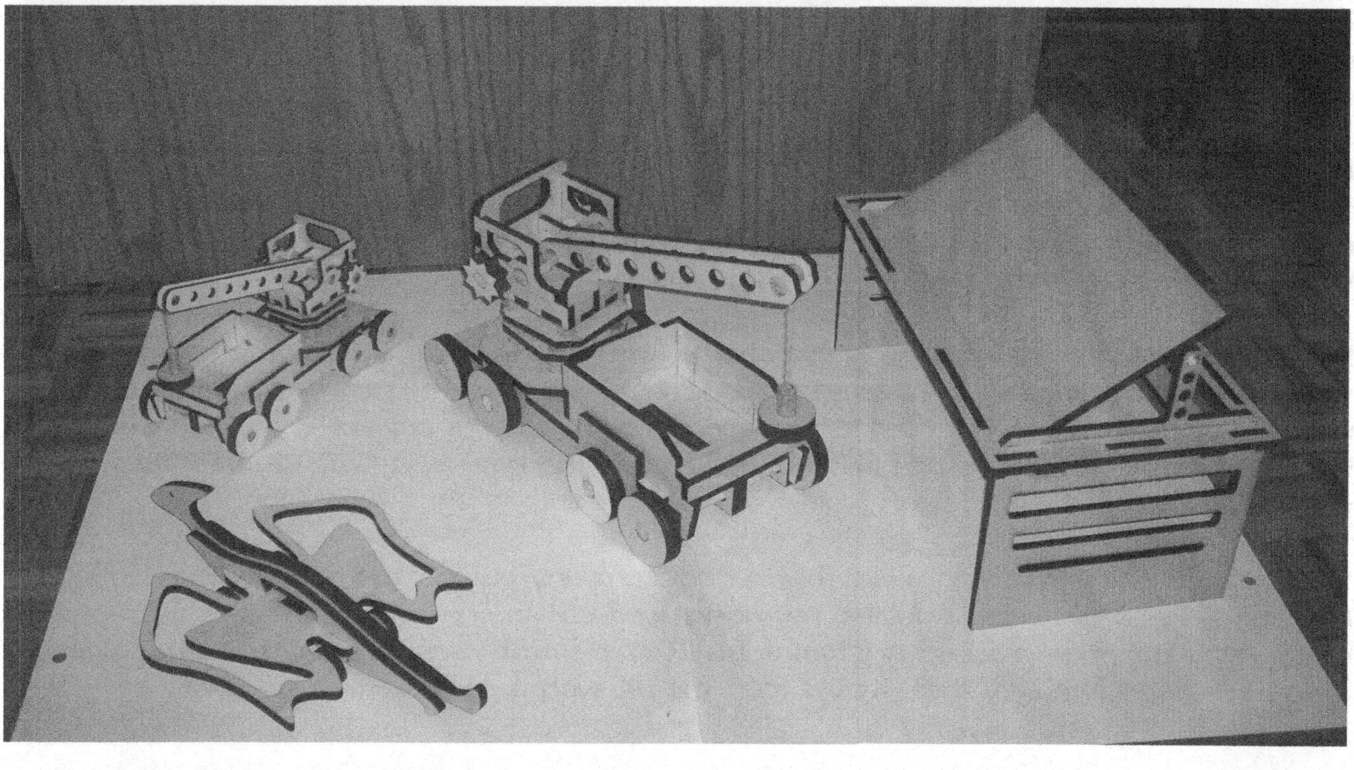

# Eiffel Tower and Japanese Miyajima Gate Behind the Designs

Many of the laser designs you've seen thus far in the previous section were all done YEARS ago (hence their simplicity, they stand-out!). I've been so focused on learning and designing for my ShopBot Desktop (my CNC table router) that I neglected my laser designing skills for an extended period of time. The result? Even better designs and production processes! How is that? I focused entirely on redesigning a lot of my previous laser designs to be adapted for table routers - 30 of them so far - and that made me realize first-hand some aspects of my designs that needed to be improved upon.

As with anything in life, if you focus on one thing, it tends to bleed into other areas - design is definitely one of them!

My first design after an extended laser sabbatical was the Eiffel Tower. I had been wanting to make a laser version of it for a very long time - they are really a dime a dozen but they weren't built the way I wanted it to be built. With all the fine details and intricacies along with major design compromises made during the design process, I'm extremely happy with the results. My focus was two-fold.

First, create something that was instantly recognisable YET required as few parts as possible. I didn't need massive part counts on this model because the complexity of the superstructure itself accomplished this.

My second goal was to make it BIG. I've seen these Eiffel Towers at trade shows around the world cut by lasers and they were always TOO SMALL. Fresh from having to rescale my laser models to be 2-3x larger - I wanted to make this have presence and the easiest way was to make it big.

Due to the size and material thickness choice (4 mm instead of 6 mm), my biggest fear was strength. It's all good to have a big model but I didn't want it to be easily pushed over nor did I want the base to crumble. This meant, oddly like the original, the bottom would have two sets of supports and I had to make sure the base was FAR HEAVIER than the top.

What I never imagined though was that a cool company out of the UK called Radecal Machine Sales & Service would see my design and cut it - not using a laser cutter or out of plywood but their massive Aristomat GL 2032-8 which uses an oscillating blade with laminate instead! Not only that, they'd tour the UK with it at their trade shows!

RADECA
calmachinesales.com | Print, Cut
ARISTO
AGFA
Roland's NEW
Imagine.
Roland

I knew ahead of time that the Eiffel Tower would be a no-go for most shops, it has absolutely no ROI potential when you consider that it takes about 12 hours to cut at 3/4 size using a 60 watt laser. 200$ x 12 = 2400$US in cutting excluding labor and materials but my focus for this design wasn't to generate sales but to generate interest - and a personal challenge to myself.

The other Eiffel Tower designs I saw out there were made to be cut quick with minimal assembly, though I shared the minimal assembly frame of mind, I didn't want to compromise size. What I found most interesting is that I sold a few of these digital files in the first few weeks of launch!

The following was my design process behind the Eiffel Tower - the total design time for this project I believe was 20-30 hours spread out across a few weeks with the vast majority of it spent on figuring-out why my 3D mesh was falling apart. 3DS Max is great but I've noticed once I start having mesh with lots of holes in it (similar situation happened with my Sydney Harbor Bridge), the mesh degrates creating lots of holes that shouldn't be there. This is purely cosmetic as the output file is fine, it only shows-up during rendering time which makes it all the more frustrating as clients see this before making a purchasing decision.

Angle the legs
The most important aspect of the Eiffel tower is the base, the legs have to be strong (considering each hole introduces a new weakness), wide and as accurate as possible to the original. The model was made or broken based on the design decisions I had here. To make matters worse, I had to double this so there was both an inner and outer shell to the base. Lots of angles and calculations needed to be dead-on!

Let support
After I build the base, I still had a fear of the legs, because they were angled out, would be too easily broken as there wasn't anything holding them back other than glue. As a result, I decided to create a base support structure (not part of original Eiffel Tower) in order to maintain stability and add much needed support strength to the entire assembly. It also made the base more solid as a result.

First Observation Deck

Once I had all my angles and support structures figured-out (with all the related built-in guides helping assembly), it was time to make the first observation deck. The biggest challeng here was giving myself enough room for the tower to keep growing upwards while having a deck that was somewhat accurate to the original. The material at scale is far thicker than the original so some compromises needed to be made.

Fine Tuning

As with all my models, this one especially, there is quite a bit of fine tuning after the initial calculations are made. This included looking at it from several different angles to make sure it wasn't asymmetrical, that parts weren't going through others (which is impossible outside the virtual world) and that visually, everything looked strong enough. I'm not an engineer but looking at a model, I can tell if it's strong enough.

I was very happy with the results and after making this landmark, I decided to make another - I REALLY LOVE JAPAN - the people are amazing, the entire country is spotless and incredibly efficient (public transport is incredible!)

Unlike France or Australia, I didn't know of any major landmarks within Japan - there are towers, temples and a host of other major buildings - but something that was known outside of Japan to most foreigners is something that is harder. My criteria for a laser model are pretty flexible and short.

1) I must be able to simplify the model to be under a hundred pieces yet still make it accurate and interesting to the eye. With lasers you have "two colors", that of the burned edge which is usually light or dark brown for plywood and the surface which is beige to yellow. A wall is boring to look at as you only see the surface without any contrasting edge (I try to limit my engraving which would be one workaround).

2) The model must be a challenge yet be easy enough to assemble. There is no point making a model that's a thousand pieces just because it can be a thousand pieces... under a hundred pieces, repeatable yet unique pieces and most important, challenging but not so much that the client (or me) would give-up building it.

3) The model must be of interest for me to build to begin with. If I have no passion for the subject matter, regardless of money or other benefits, I just can't bring myself to make it. I get quite a bit of requests from people asking me to design XYZ and they'll wire me some money right away but I usually turn them down. I want to make things that are fun, interesting and challenging and I find that if I have no passion for what I'm designing, all three of those go out the window and the results WILL SUCK!

After searching for a few nights, I found a model that was itching to be made laser cutter - the Japanese Miyajima Gate! It's stunning visually, it's the only gate IN THE WATER and most importantly, it's massive and world renown... a UNESCO World Heritage Site.

My first challenge with this model was how to make it look BIG eventhough my model of it woudln't be as large. The easiest way to accomplish this was to make strongly contrasted dark vertical stripes - in this case, the inner sides of the gate were orientated in such a way as to emphasize height over width. The major pillars were also made very thick so they visually look very strong as well with the horizontal ledges being thin and slight by comparison. Here are some of the design elements I built into the model to make it easier to cut and build!

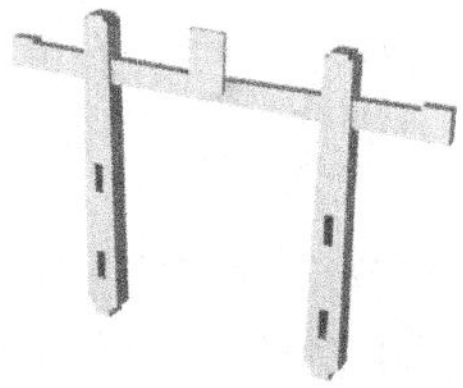

Strong Columns

I knew from the start that I had to get the columns perfect so that this model had the illusion of mass and strength like the original conveys. Out of all the parts of this model I built, the pillars you see on the right were by far the most worked on.

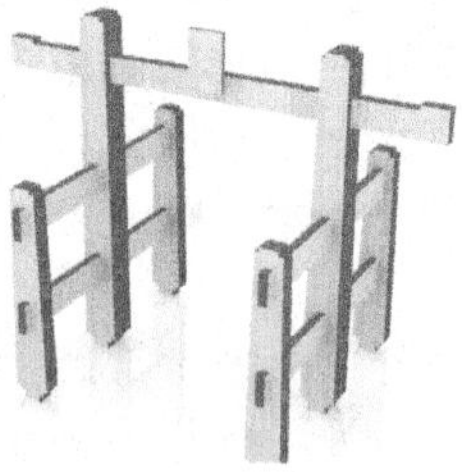

Right Proportions

This is a highly recognizable gate and I wanted it to look as close to the original as possible - the easiest way to accomplish this was by getting all my proportions right. I had to make some changes due to the medium but generally, this all had to be sorted FIRST before I got too involved in the model as otherwise, it's too much work to edit later into the process.

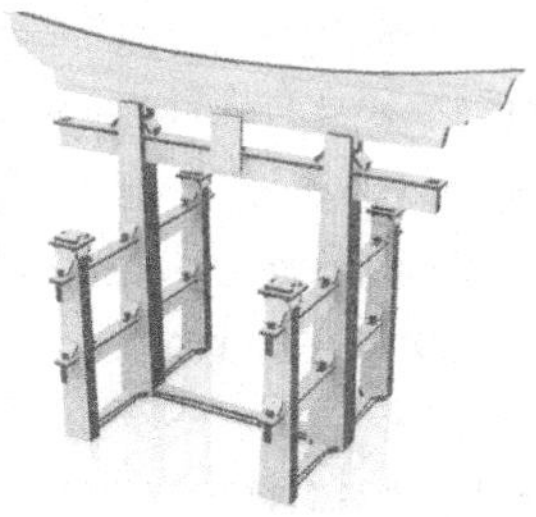

Roof Line

Believe it or not, the roof line of this gate took endless hours to get right. I played with different slopes and builds before I settled on a full-on solid surface. The original has a beautiful roof but the more I tried to build this into the model, the worse and less accurate it became. My resolution was 4 mm, to do a good job, I would have had to use 2 mm.

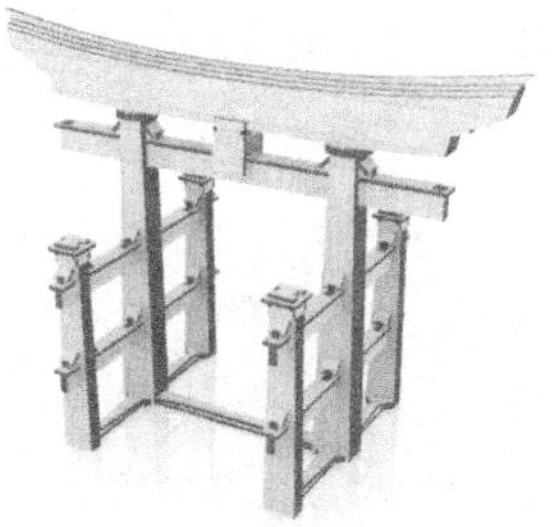

Roof Line 2

After adding lots of little accessories to improve the contrasts throughout this model, I continued to play with the roof line. It just always bothered me so after a few more hours, I decided to build a staggered roof line that somewhat gave the illusion of the original.

Unlike the Eiffel Tower where the holes were causing me problems with the mesh, in this case, it was the roof line that had me in a loop. I've found that making models using 4 mm or 6 mm thick material that require lots of fine details end-up looking bad if you try to overthink them. This Japanese gate's roof line was a prime example.

My goal with CNCKing.com is to make easy projects, though this model has slightly over a hundred pieces, they are all distinct and repetitive so it isn't like assembling a 100 piece jigsaw puzzle but a 25 piece one. Up until this point, I had always designed using 6 mm thick material as my base but with the Eiffel Tower, I knew I'd run out of resolution so I settled with 4 mm - I'm so glad I, out of habbit from my last model, decided to build this gate out of 4 mm as well as it gives me a third more resolution.

4 + 4 + 4 = 12 vs 6 + 6 = 12 ........ gain 1/3 build resolution by going down by 2 mm thick!

As all my models are built from the same material thickness, I've since found 4 mm to be superior design wise, it's quicker to cut with a laser, just as strong materially (if model is built correctly) and actually cheaper overall! The one issue I've found is worldwide availability of 4 mm isn't as good as 6 mm but you can always scale my plans by 50% and everything is fine as my models are based off relative material thickness.

# CNC Table Router Technological Overview

CNC projects found within this section were cut using a ShopBot Desktop, which is manufactured by ShopBot Tools Inc. out of North Carolina (USA). It has a 6 mm (1/4") thick sheet measuring 500 mm (19.8") long by 400 mm (15.8") wide using a standard 1/8" straight double-flute bit. Keep lots of bits around you; they really come in handy.

Some of these CNC router templates show rounded inner edges while others are sharp. The reason for this is that I'm still trying to figure out the most helpful convention to show this. Unlike laser cutters, the curves and dog-bones you see in the templates are dependent on the bit (larger bit equals larger arc). While you can "unfocus" the beam of light coming from a laser, it's nowhere near what you can do with different sized bits. I learned so much from this "little" CNC machine... can't recommend it enough!

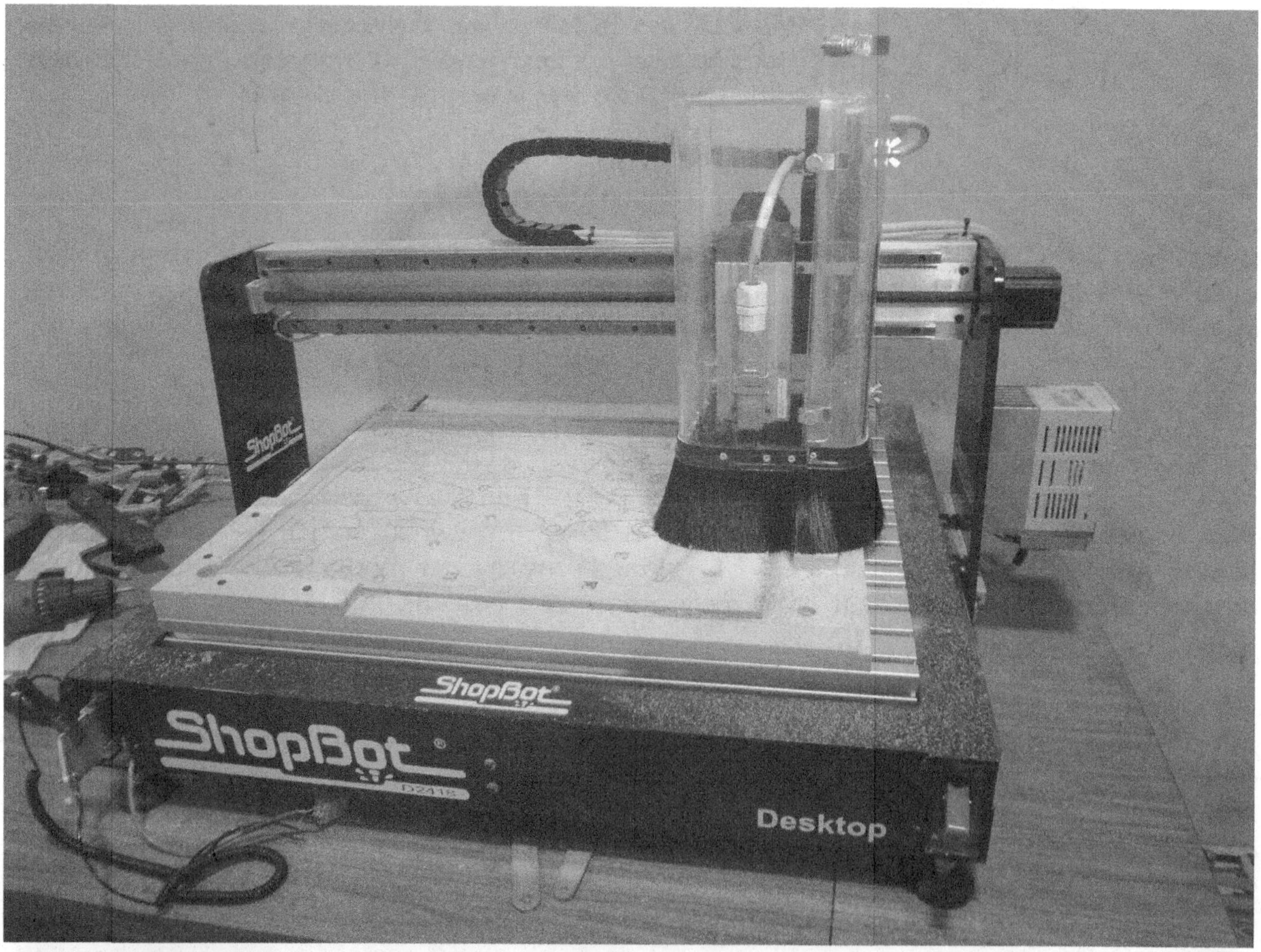

The ShopBot Desktop was my first real foray into the world of CNC cutting; previously, I designed all my models virtually (including assembly) and had them cut every once in a while with a laser. I was only getting half the feedback from the design process! After designing, assembling and playing around with almost 40 of my own CNC table router creations, I realized that the more parts a model had, the more headaches were involved with assembly. I also realized that my models were way too small!

When I first started designing, I paid so close attention to cutting costs that it became second nature to make compromises size-wise to make the bill a little lighter -- this is something that isn't much of a concern when using a table router, though there is more wastage. I forgot about this for YEARS but, thanks to ShopBot, I started building things much bigger across all my CNC designs. My original goal was to build toys, not miniatures!

There are a few things to keep in mind when ShopBotting -- sound so great as a verb! The first is to make sure you have both hearing protection and woodshop glasses. Stuff will fly around when you least suspect it! Luckily, my ShopBot Desktop has a very good acrylic enclosure, so it's difficult for something to fly around. It goes without saying that you should never stick your fingers inside a router when it's spinning -- I always turned the machine off using the safety whenever my fingers were anywhere near the bit.

Things to keep in mind when shopping for a CNC table router:

**Table Router Buying Tip #1: You get what you pay for!**

Maybe I'm a special person, but I like things to work out of the box without me having to adjust things or buy some products locally. As with everything in life, you get what you pay for. Sure, you can build your own CNC table router rather easily if you are so technologically inclined, but I am not -- nor do I have time to deal with all the resultant headaches, such as having to manually program G-code or build my own motherboards.

My very best advice, and it's valid across all CNC machines types, is to buy a little more than you can afford ... don't try to skip out on having to buy xyz accessory because it's over budget. The ShopBot comes with two options, an industrial spindle or a manual router. The cost difference spread out over the life of the machine is pennies a day, but the benefits you get from using an industrial spindle are MASSIVE.

Same with a laser -- get more tube than you need, as the cost difference is equally as trivial over the life of your machine.

**Table Router Buying Tip #2: Get a rigid and overbuilt machine!**

Regardless of the type of table router (or laser) you get, make sure it's overbuilt and incredibly rigid. Having a machine precise to within a fraction of an inch is worthless if the table router you have use a gantry system that's bending under pressure or the desk of your machine is made out of MDF. The rigidity of the machine is paramount for creating projects that are consistent and with as few flaws as possible. A bit spinning at high velocity is the only concern you should have, not if the machine will break apart from the vibrations!

Look for the fine details like rubber foot rests, a t-slotted aluminum deck, and, most importantly, high quality stepper motors for absolute precision.

**Table Router Buying Tip #3: Safety and reliability**

Like buying a car, you want to get something that's both safe and reliable. CNC machines are absolutely no different. The first thing I look for in a CNC machine is "Where's the emergency stop?!" This is vital ... when things go wrong (and they WILL go wrong,) you need to be able to easily reach and cancel a job INSTANTLY. A safety switch that has a delay isn't very useful if you are unlucky enough to have your finger a few seconds from being turned into hamburger meat.

I really can't emphasize how important this is, and it's something that scares the daylights out of me when I see people who made their own CNC machines and forgot to stick an emergency switch on them. CNC machines are incredibly safe by their nature -- you may be smart but always one dumb move away from membership in the missing finger club!

What I suggest to you is to actually try (with the machine not working) and see how easy it actually is to get your fingers where they shouldn't be. The harder you need to try to find a way to hurt yourself, the safer the machine will be in full operation.

**Table Router Buying Tip #4: Use a dedicated computer**

Although this is an advice I don't follow myself, it's recommended to use one dedicated computer for each CNC machine in your shop. There are many reasons for this, from software conflicts to safety (having a Skype call pop up during a cutting process isn't good). I've had my fair share of COM errors due to a trivial background process.

These dedicated computers are generally barebones as well; they don't need to have the latest operating system or even a good CPU or dedicated GPU, or be loaded with RAM. A laptop that costs only a few hundred dollars is usually more than capable, as the design process is "suppposed" to happen on another far more powerful computer anyhow. I've seen people salvage old and discarded computers that for any other purpose would be completely worthless. You have to remind yourself that this computer only sends X, Y and Z coordinates to the CNC's motherboard, so the processing power required is minimal.

If you can, I'd suggest you let the manufacturer of your CNC machine provide you with a computer loaded up with all the software and hardware that is needed from the start. This makes things easier for both parties should issues arise -- you can say "it's your computer and CNC machine, you fix it!"

## CNC Opportunities: Size Matters

A general theme I see in the CNC world is that of bigger bed sizes leading to more profitable operations and growth over the long term. Why? The first reason is efficiency -- if you have 10,000 name tags (I hope you used my earlier tip about print-merge to save you days of typing,) doing them all in one shot instead of across several jobs will make you far more efficient. The most difficult and time consuming work after design is actually the loading and unloading of the table; the less of that you can do, the better. This is doubly true for table routers as you have hold-downs to contend with! It's nice to press the "print" button and do other work for an hour rather than having to look the machine over every five minutes because a job takes 15 minutes to complete.

The other advantage of having a larger bed size is the ability to take on larger, more profitable projects where there is less local competition. Anybody can buy an entry level machine, but not many have the funds or business acumen to buy one that's larger than average within the market you are mining. The bigger and more powerful the CNC machine you end up getting, the higher-end jobs you will be able to take on, which will have a significant effect on your bottomline. Of course, you can still take the smaller jobs to augment your income, but you now have a far wider customer base to generate revenue with -- and economies of scale kick in rather quickly especially for CNC machines.

Although I have many friends who have a combination of CNC machines, from plasma to laser, table routers to 3D printers, the ones that seem to be the most profitable in the market they serve tended to buy the biggest and best they could afford. Their first purchase was rarely a small machine. They instead borrowed money off the equity of their home, got loans with the bank, manufacturer and friends and family if needed, and bought the top of the line instead of settling for the average. When you are competing against every Tom, Dick and Harry, it becomes a race to the bottom. The less competition for your services or the higher hourly rate you can generate from your CNC machine, the more free time you'll have towards business development and growing your enterprise rather than working in it. As a friend used to tell me, you can either be a farmer or a cow -- in the CNC world, you want to be a farmer as fast as possible and a big, powerful machine offers you that opportunity.

Going back to table routers, their bed sizes relative to lasers or 3D printers are far larger relative to money you invested. Take full advantage of this! Other features that will save you a lot of time and money are things like automatic tool changers (do a whole carpentry job by pressing one button instead of having to stop and change the bit yourself,) getting a vacuum bed (no more turning the machine off, making guide holes, screwing and unscrewing,) and building jigs. Here are a few more top secret tips to make your CNC table router more profitable.

**Table Router Top Secret Tip #1: Learn CAD**

There really is no easy way to break this to you, but if you get a table router or any other CNC machine type, you will have to bite the bullet and learn CAD -- but that's only part of the story. Your CNC machine usually has job software of its own and sometimes an intermediate program too! I use Autodesk's 3DS Max to design, export the files to Adobe Illustrator for positioning, and then open the files in PartWorks for tool pathing -- THEN I export those files into ShopBot's job control software. Each software has its benefits and drawbacks regardless of the manufacturer who built your machine.

It is vital that you know this in advance and become comfortable with moving electrons from one program to another. Whether you buy a laser, 3D printer or a table router, make sure it comes with job software that's as powerful as your ambitions for the machine (otherwise, the CNC machines will be useless) and be sure to MASTER IT.

Make sure you don't have any loose wires. You need an easy to find and use emergency switch. You also want a cage around your router bit and spindle!

**Table Router Top Secret Tip #2: Time**

The biggest sources of stress in life usually involve time. You have a project that's due yesterday and you are out of bits AND your partner just called you to tell you they are seeking a divorce. Don't let all these issues affect you when making a purchasing decision on new capital equipment. Time is never going to be on your side, but make sure you always have enough of it to make sound judgements and decisions.

Life is about priorities; if you ignore them, they build up, and the consequences seem to hit you all at once. If you are shopping for a CNC machine, give yourself time to make a decision; don't let anyone pressure you. There is absolutely nothing worse for productivity than having too much to do in not enough time. Visit a few trade shows, get client lists, go on shop tours ... TAKE YOUR TIME. When you make a decision, follow through and don't waste the salesperson's time. You want X for $Y, let's sort the details out.

**Table Router Top Secret Tip #3: Drill holes, pockets, inside, outside**

Want to save yourself lots of problems? Follow this simple rule: have your CNC table router first drill guide holes for your hold-down -- I use screws. This will prevent you from having your bit go over random screws. Next up are pockets; each hole you put THROUGH your board will weaken it, so try to keep any major throughput to as far in the cutting process as you can. Pockets don't go all the way through your material (if you have your z-depth correct) so they won't substantially weaken the board. Afterwards, you want to cut all the inside parts. If you cut outside parts first (even with tabs) followed by inside parts, the shearing action will usually dislodge the larger piece, and it will move around -- rendering your entire part useless.

After you've cut all your inside parts, you can then do the outsides. This process remains true especially if you are working with a vaccum table because each hole through your material will affect its strength. If you have a lot of small parts and a few large ones, I've found it's best to cut the outline of the smaller parts first, followed by the larger ones, as major cuts (even with tabs) tend to really affect the board strength at a far higher rate than smaller cuts. By the way, this rule also holds true for other CNC machines, each cut weakens the material and increase the chance of things moving.

**Table Router Top Secret Tip #4: Use a vacuum**

If at all possible, try to put your vaccum tube as close to the spinning bit as possible to maximize material removal -- and buy the strongest vaccum you can afford! The reason for this is two-fold.

First, it allows you to easily judge if the cutting or engraving process is being done correctly, which you can't see if you have an inch of chips on the cutting surface.

Second, having a clean surface will lessen bit wear and tear and reduce the possibilities of dust getting into rails and other important components, including your lungs! A dusty environment isn't good for your computer, your health or your machine.

One more note about the vacuum tube itself: make sure it's long enough to be able to easily travel the entire distance of your bed. Your CNC machine is easily capable of ripping it out of the socket and dragging your entire vaccum system if it isn't, without breaking a sweat. Have more length than you need in case it gets hung up on something.

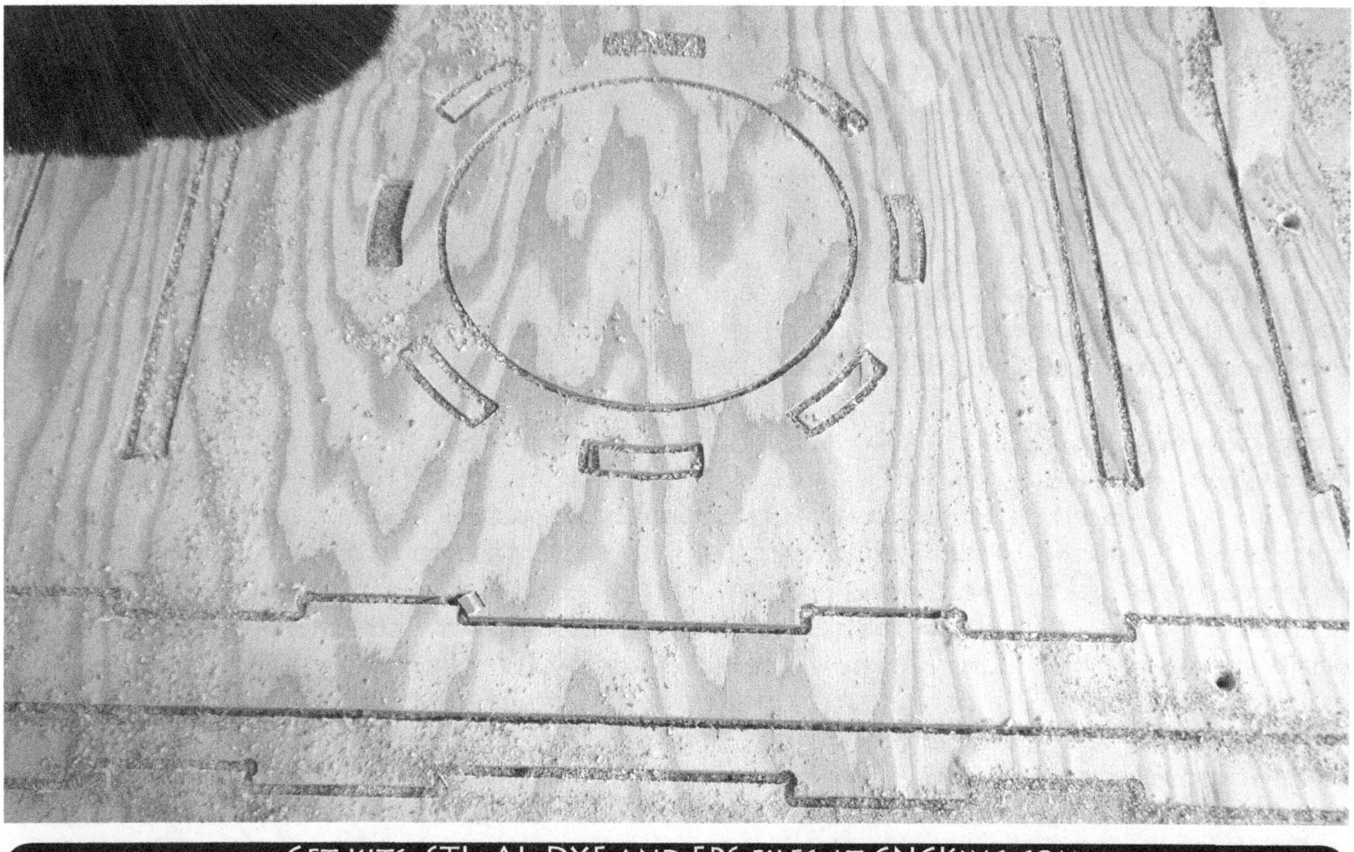

**Table Router Top Secret Tip #5: Never leave your machine running alone!**

This should go without saying but needs repeating. NEVER RUN A CNC MACHINE ALONE. CNC machines have no concept of self-preservation. If your laser has a flameout, it will not stop until it burns out. It will keep right on going until your entire shop is raging in flames and the power cuts out. I have three friends who have experienced this first-hand. Same goes with your CNC table router -- sometimes there is a COM error and all of a sudden, it starts doing some funky stuff like forgeting its X, Y and Z co-ordinates. Not a big deal, you think? What if it suddenly thinks the z-axis surface is actually a meter down? It will go down through your table, causing a nice friction-induced fire in the process. These machines are incredibly powerful yet also very dumb, and a common power surge can screw up your computer and your life if you don't pay attention.

**Table Router Top Secret Tip #6: Material testing**

Unlike a laser, a high quality CNC table router is just as strong at one end of the table as the other. But what most operators fail to realize is that unlike a laser, the bit can wear out. Normally this isn't a big issue, but if you are doing material testing and your bit breaks before you perform your first true cut, don't just put in a new bit using the same settings -- do another quick test. You may be surprised at the difference in chip load that a sharp new bit has over a dull worn one.

**Table Router Top Secret Tip #7: Broken bits**

Sooner or later, your CNC table router will break a bit ... and then proceed to break many more as times goes by. This is a seemingly random occurrence, but the worst thing you can do is simply replace the bit and re-do the job on the same board. Let me tell you from experience: THROW THE ENTIRE BOARD OUT!

What I've found, more often than not, is that small metal shards remain in the cutting path. When your fresh new bit goes over it, guess what ... ANOTHER BROKEN BIT! I once went through four bits in the span of two minutes because I failed to realize what was going on -- I was lazy! When your bit breaks, take some time to investigate what has happened (where along the cutting process did it break -- while doing a curve or a corner, etc.?) and if your bit breaks at exactly the same spot repeatedly, adjust your depth or add another cutting path if needed. You may have built a part that has too much pinching, so your machine simply overpowers through it, breaking the bit in the process. This is why it's important to have lots of extra bits lying around. You don't want to have to delay a job because you ran out and the hardware store is out as well!

**Table Router Top Secret Tip #8: There aren't any clamping rules**

Sometimes you end up making a fantastic project but find that the clamps you have just aren't strong enough or, due to a weird design angle, are simply unable to hold a project down correclty. The easiest solution I've found to this is using weight plates. Gravity is a wonderful thing. It puts pressure on an entire model -- making your clamps more effective in the process.

A general rule of thumb is to have a few kilograms of weight plates hanging around to augment your clamping abilities, along with some string and even some bungee cords. Your set-up might look very unprofessional (like the image of my light table above,) but the results speak for themselves. Whatever works for you, do it -- don't worry about what others may think of it. At the end of the day, once the glue has set, the plates and clamps will all come off  and the customer will be happy with the results.

If you use screws or nails, that's also a great way to keep things together, but I prefer to simply use glue and a good design to accomplish the same task. In my shed, screws are for hold-down, not keeping things together.

## Interview with Ted Hall, CEO of ShopBot Tools Inc.

I first met Ted when I went to ShopBot's HQ in North Carolina to pick-up my ShopBot Desktop with the car; it was a tight fit but we got it in! Ted was a professor of neuroscience at Duke University who enjoyed building boats in his spare time. He was becoming frustrated by the amount of time it was taking to cut the panels for curved hulls with hand power tools so, after unsuccessfully looking for an affordable CNC table router, he founded ShopBot Tools Inc.

While I was at their HQ, the ShopBot Tools staff were awesome; I felt very welcome and spent about a week along with Bill Young (Special Projects Manager) and Angus Hines of Hines Design Labs (angushines.com) learning the ropes of my brand new CNC machine. Coming from the virtual world, it was quite a steep learning curve for me but they were patient and all taught me a lot!

You can find out more about ShopBot Tools by visiting ShopbotTools.com.

**What is ShopBot Tools' unique value proposition in the CNC table router field?**

With the Desktop CNC our focus was on making a production tool that had engraving-level precision and was capable of accurate and reliable machining of plastic, wood, aluminum, and various composites and foams. The Desktop may be a bit overkill for the hobbyist woodworker, but since many home shops now also aspire to do some production work, even as a sideline ... we wanted them to have a full-up machine.

**What features should I look for in a quality CNC table router before purchasing one?**

There is an excellent American Woodworker article (Feb/March 2011) comparing a number of the smaller CNC routers. It describes important features and shows how they lead to smoother and more accurate cutting. These features include a drive system with ample speed, power, resolution, and low-backlash. The article also emphasizes a rigid frame and an overall mass that is substantial enough to support chatter-free cutting.

**ShopBot Tools offers a range of table routers from your small but very capable ShopBot Desktop which I use, to your large industrial PRSalpha, which were the most challenging to build from a R&D perspective?**

Each tool presents a different kind of challenge because one is trying to optimize different factors for each type or range of applications.

For the large tools, power and speed become important because of production demands. For smaller tools, resolution and elimination of backlash become very important because highly detailed work is often done in smaller work areas.

In all cases, ShopBot is oriented to making digital fabrication as affordable as possible. Because we want to put high performance production equipment in the hands of individuals in small shops and make it easy for them to leverage the capabilities of digital processes, optimization requires careful attention to costs and to selecting components that provide the best value.

**Your company is massive in education and has recently partnered with AutoDesk, better known for its 3D and CAD solution - how did that come about?**

Our Autodesk partnership recently supported a project that significantly defrayed the cost of 50 of our Desktop CNC's for schools. The project came about because of our joint interest in getting subtractive digital fabrication equipment into schools.

Low-end 3D printers have become much less expensive in recent years and many schools have been able to purchase them. But it is much harder for schools to afford the subtractive equipment like CNCs and lasers.

Autodesk has a new set of web-based applications, called 123D, that are oriented both to design and making the continuum of digital fabrication. We were able to tie their interest in promoting 123D and supporting digital fab education into this program. Both Autodesk and ShopBot are all convinced that digital fabrication represents the coming revolution in how everything is produced.

Link to Carl Bass (Autodesk CEO) speaking about digital fabrication:
http://www.popsci.com/science/article/2013-03/autodesk-ceo-carl-bass-future-3-d-printing-home?cmpid=mobify

**100KGarages.com and 100KSchools.org are two major initiatives launched by ShopBot Tools, can you tell me a little about the goals you've set for each?**

We are very committed to growing digital fabrication. We see it as an opportunity to establish a new kind of manufacturing and local production in communities everywhere, a type of manufacturing that can provide fulfilling work rather than repetitious drudgery.

100kGarages is our attempt to help support the establishment of small, new, entrepreneurial businesses by funneling work and providing resources and collaborative opportunities. 100kSchools is about getting everyone geared up for the new digital technologies.

Teachers and educators are excited about digital fabrication. It offers them exciting ways to involve today's students in hands-on, project-based learning. It is strongly integrated with STEM topics and programs. But teachers need all the help they can get in implementing digital fabrication technology programs. 100kGarages is a tool for teachers.

It helps them teach their students to use digital fabrication. It helps them to prepare their students for the future. Watch TJ help teachers set up the Desktop CNC: http://www.100kschools.org/desktop_quickstart.php

**What are your views on the exponential growth of hardware and software solutions for digital fabricators? How do you keep-up?**

Chris Anderson, in "Makers:... ", does a pretty good job of explaining why digital fabrication is going to be important to manufacturing and why this is going to be much bigger than the effects of digital revolutions we have already seen in other industries.

Given the pace of technological change, I'm not sure anyone can keep up ... but the emerging enabling technologies digital fabrication continue to impress and excite me.

I just attended a conference at MIT on the Science of Digital Fabrication. Here's a link to the meeting. Click on the names of the major speakers to link to their presentations.

You might start with Hod Lipson to get an overview of some of the main technologies. Lipson and Kurman's new book on 3D printing, "Fabricated:...", is also very informative. Meeting: http://cba.mit.edu/events/13.03.scifab/

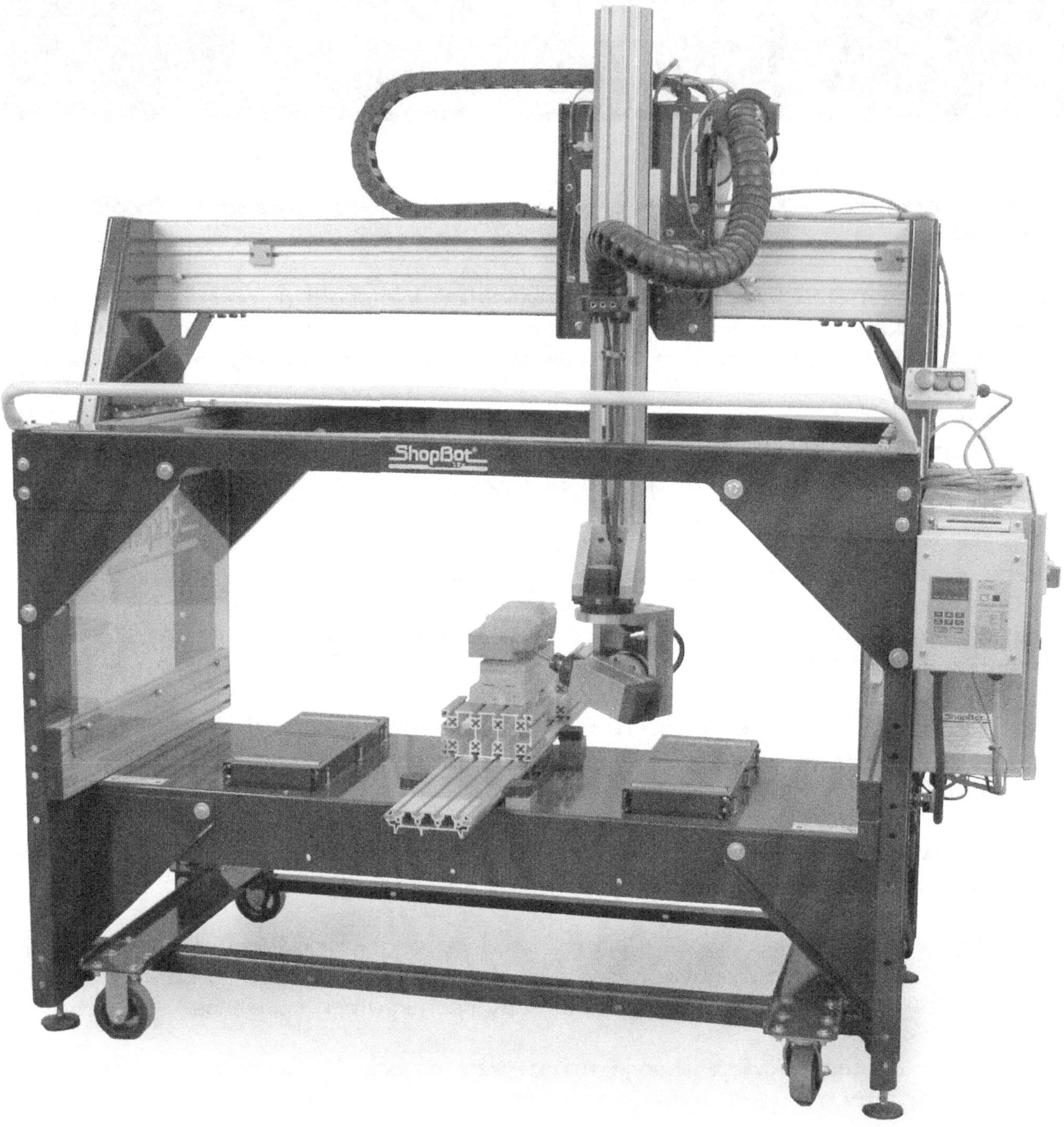


ShopBot Tool's closed-loop 5-Axis PRSalpha CNC is capable of creating 3D objects and shapes out of softwood or urethane foam at a fraction of the price of traditional 5-axis industrial machines.

Right out of the box, it's ready for production cutting, drilling, trimming, as well as prototype-and fixture-machining in a large work envelope, accessible from all angles.

**What are some technological limitations shared by all CNC table routers that represent the next major hurdles for the industry to innovate over? What innovations do you see coming from the table router field over the next 10-20 years?**

You will appreciate from reading some of the material in the last question that the days of subtractive digital fabrication are limited ... probably to something around 30 years.

Over that time, we will develop increasingly sophisticated materials that will allow us to embed much of the design and assembly process for producing things into the fine structure of materials (from molecular to microscopic levels).

This will eventually allow materials to do something like self-assemble into the items we need, and then to un-assemble itself so that the material can be used again. This material-based digital fabrication will eventually replace both the additive and subtractive methods of today -- which will someday seem crude in comparison.

**I see a lot of potential with the Handibot across a variety of fields where the repetition of CNC and the incredible portability of a power tool would come in very handy. What was the reasoning behind having it crowd funded and open sourced vs the traditional internally funded walled garden approach?**

The Handibot could be just a great little CNC tool - there's nothing wrong with that, and at the very least, it's what will happen. But we see a much broader potential for digitally controlled "Smart Tools" to transform how the building and repairing of practically everything is done. That potential depends on the development of a wide-ranging library of job-related and task-related "apps" that make using a Smart Tool essentially push-button. For example, you would download an app for a job to your smartphone, enter a setting or two, hit the "load" button and your Handibot is sitting waiting for you to position it and squeeze the "Start" trigger.

We don't feel like ShopBot, or even a larger company, has the resources and breadth of insights and creativity required to develop these wide-ranging apps, accessories, and refinements of the tool that will so broadly expand its potential. That's why we've turned to the collaborative powers and resources of the crowd.

We see Handibot as an exciting opportunity for growing an innovation and platform and extensive resources for it, based on open and public community involvement. If it takes off, we also hope to produce the tool using a network of distributed manufacturers ... making the tools where they are needed, using locally available materials, and creating local jobs.

You can find out more about the Handibot by visiting http://www.handibot.com.

**When you are ShopBotting at trade shows and Maker Faires across the United States, what material and bit are your favorites to use?**

At trade shows that are oriented to woodworking, plastic, or aluminum ... we go with the flow and use material appropriate to the show. With wood, we use a range of cutters depending on what sort of project we are illustrating: carving = 1/8" tapered ball nose cutter; cabinet making = 3/8" compression cutter; signs = 60deg or 90deg v-bit and 1/4" up spiral. With plastic and aluminum (unless we are doing something small) we use 1/4" "O" flute cutters.

Maker Faires are a whole different thing. We like to cut projects, both big and small. Large projects out of wood we cut with a 1/4" down-spiral to make material hold-down easier at shows; with plastic we would use "O" flute. For small projects we would use a 1/8" straight bit and cut anything 1/4" and thicker with multiple passes. We also do carving at Maker Faires to illustrate the Vectric software and show the detail one can get in carving. For this we use a tapered ball nose, either 1/8" or 1/16".

For work at shows and any sort of standard production, we are likely to be using Onsrud Cutters, though there is other good tooling for CNC.

**ShopBot Tools all came about when you were looking for a machine that could help you finish your home-built boat and couldn't find what you were looking for, it has been several years since then and ShopBot Tools has grown into a big company - when do you plan on finishing your boat?**

By the end of the first year or two of ShopBot development, we were making tools that were capable enough to finish my boat. Yet, 18 years after ShopBot got started, the boat still sits unfinished in my barn. It does not seem like I am going to stop working on digital fabrication equipment anytime soon ... but I would like to get back to that boat someday.

# 1911 Ford Model-T Torpedo Runabout

Completed 39 piece model dimensions @ 6 mm L: 25 cm, W: 8 cm, H: 9 cm
(1/4 inch L: 9-27/32, W: 3-5/32, H: 3-35/64)
Laser version found in WoodMarvels.com Volume 3: Evolution of Wooden Designs (page 15)

Product Renders

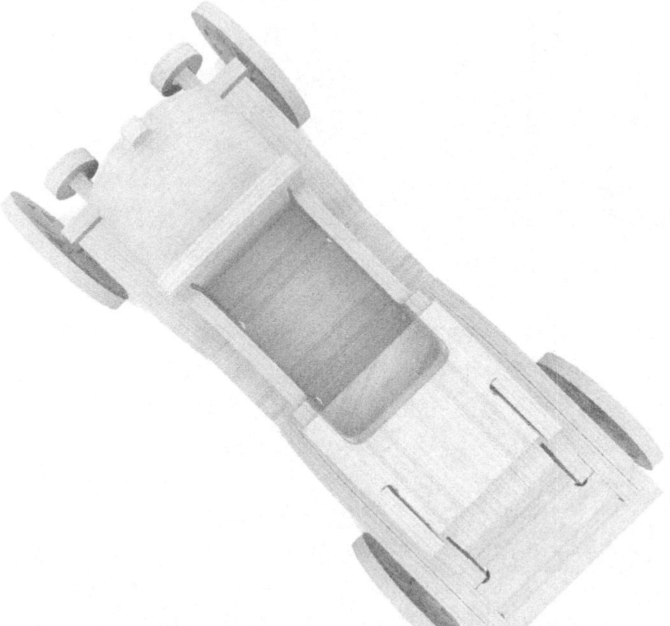

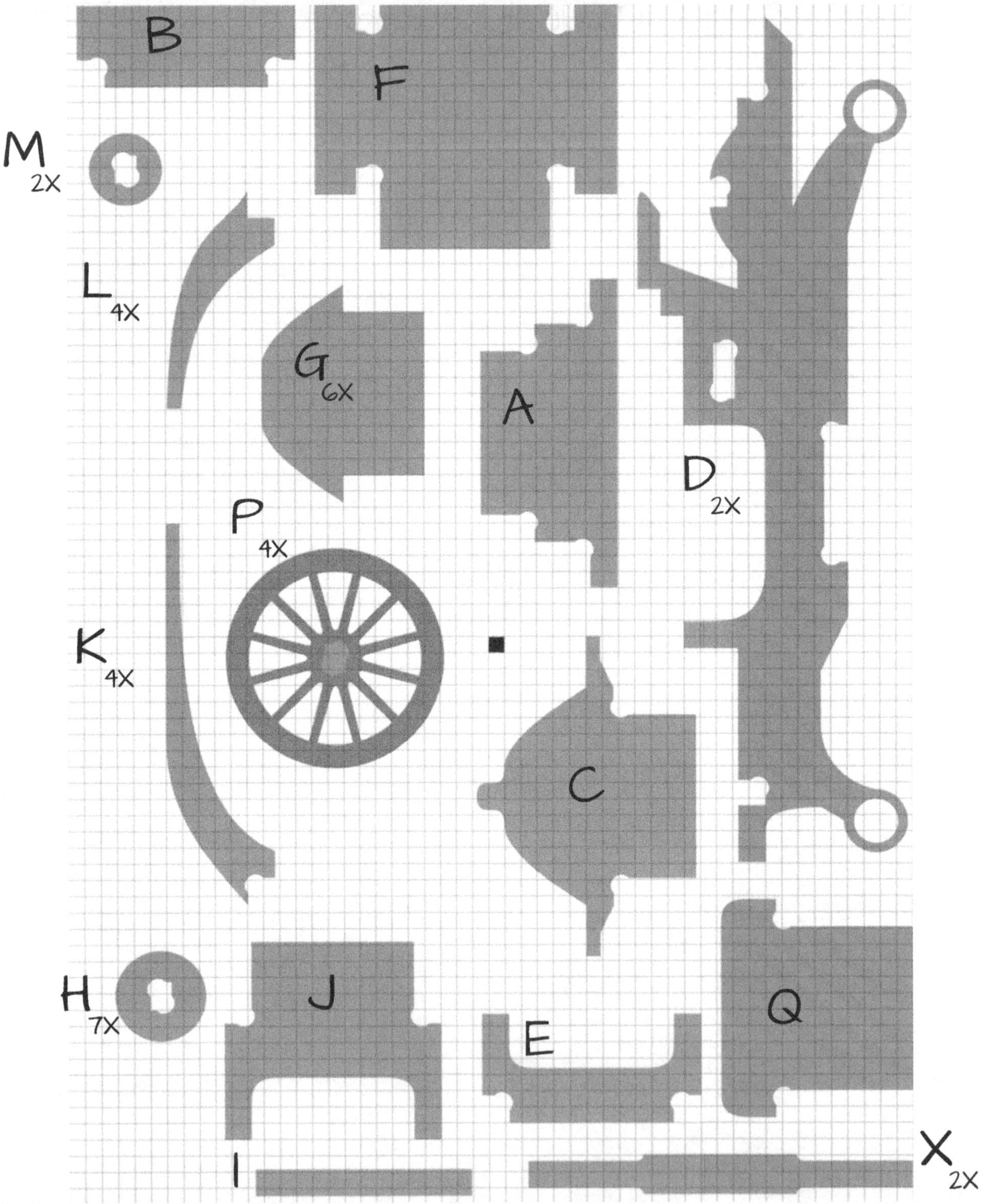
B
F
M
2X
L
4X
G
6X
A
D
2X
P
4X
K
4X
C
H
7X
J
E
Q
I
X
2X

# Step-by-Step Assembly Instructions

**Step 1: Label and check inventory**

Once you have all the pieces in front of you, the process of putting them together is no different from assembling pieces of Lego.

Label, using a pencil, the various parts and double-check your inventory. Take your time and enjoy the building process! Before jumping to the glue, be sure to do a dry run.

**Step 2: Start the cab's foundation**

Slide and glue together the back support (part A,) front seat (part B,) and front grill (part C) between two car side panels (part Ds).

Be sure to insert the axles (part Xs) but not to glue them in place, as you want them to freely rotate.

**Step 3: Add cab supports**

Glue and slide the back hood (part E) and the foot rest (part F) under the car followed by the front hood pieces (part Gs).

Allow everything to dry overnight at this point in the building process.

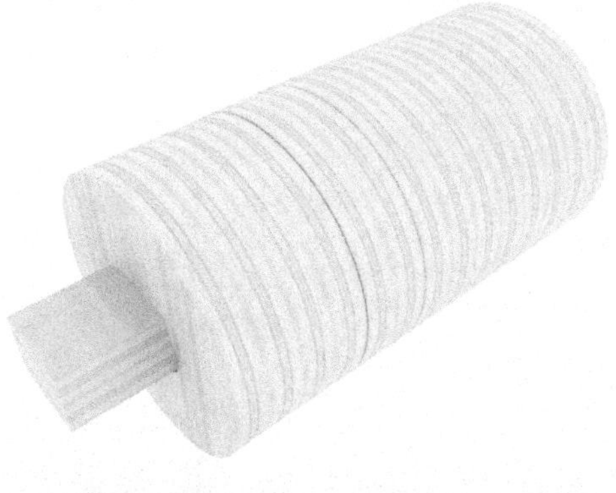

**Step 4: Make the gas tank**

Glue and slide into place all seven rounds (part Hs) in the middle of the wooden rod (part I).

Step 5: Add the accessories
You can now glue the rear gas tank (the assembly you built in step 4) into the notches provided in the back of the side panels (part Ds).

Afterwards, you can glue the back panel (part J) and two pairs of fenders in the back and front (parts K and L,) followed by the front lights (part Ms).

Step 6: Insert the axles
Glue into place the four wheels (part Ps) and then drop the windshield (part Q) into place.

Step 7: Perform some finishing touches
Allow everything to dry overnight and then slide the trailer onto the cab.

There is no need to apply any preservatives or paint; feel free to sand any sharp edges down. ENJOY!

# Army Transport Truck

Completed 63 piece model dimensions @ 6 mm L: 44 cm, W: 15 cm, H: 16 cm
(1/4 inch L: 17-21/64, W: 5-29/32, H: 6-19/64)
Laser version found in WoodMarvels.com Volume 3: Evolution of Wooden Designs (page 82)

## Product Renders

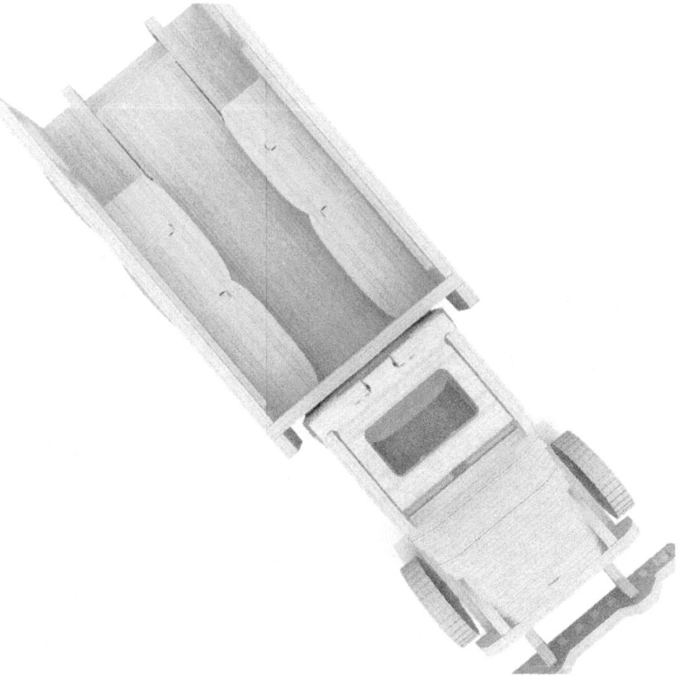

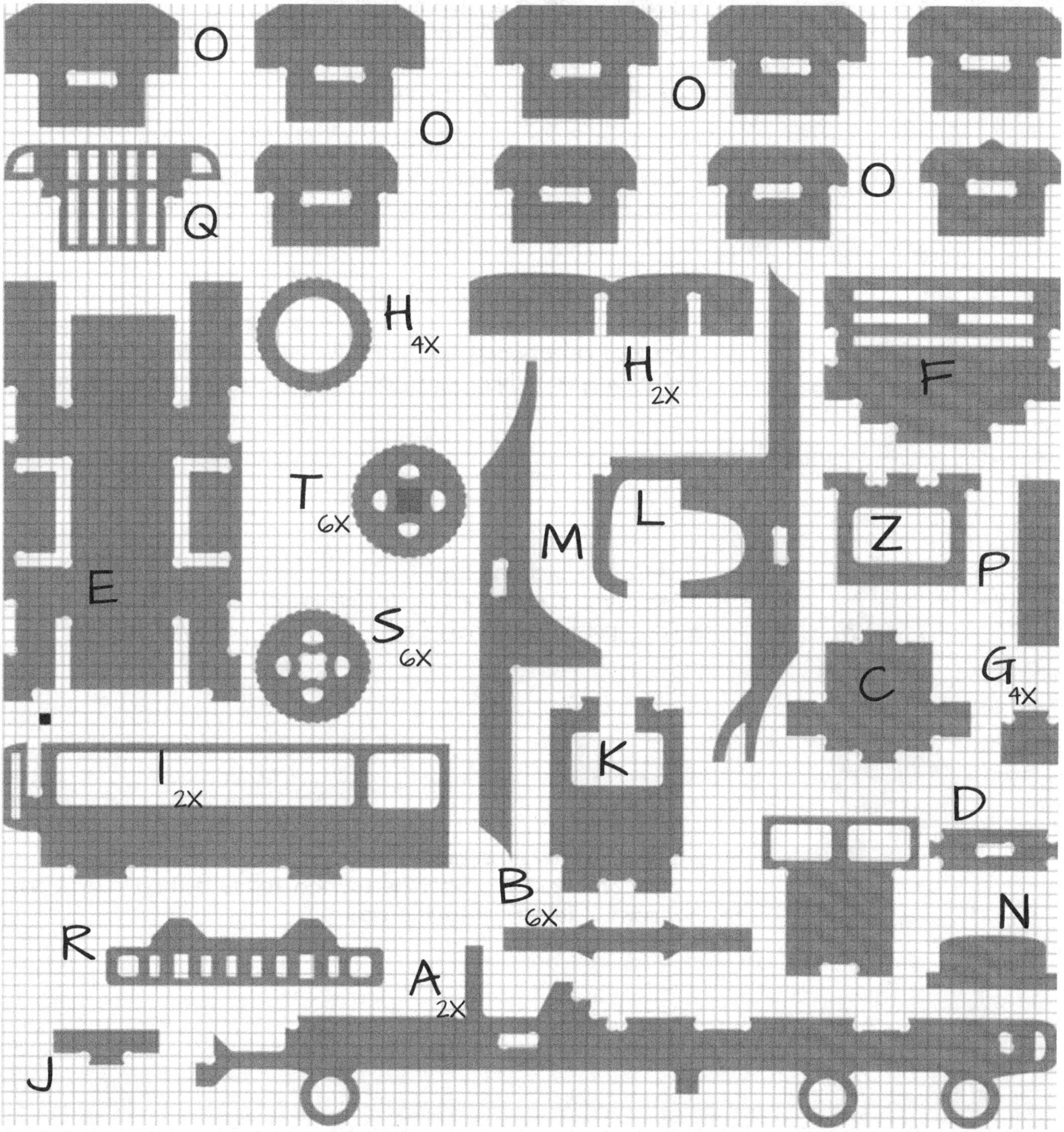
O
O
O
O
Q
H
4X
H
2X
F
T
6X
M
L
Z
P
E
S
6X
G
4X
C
I
2X
K
D
B
6X
N
R
A
2X
J

# Step-by-Step Assembly Instructions

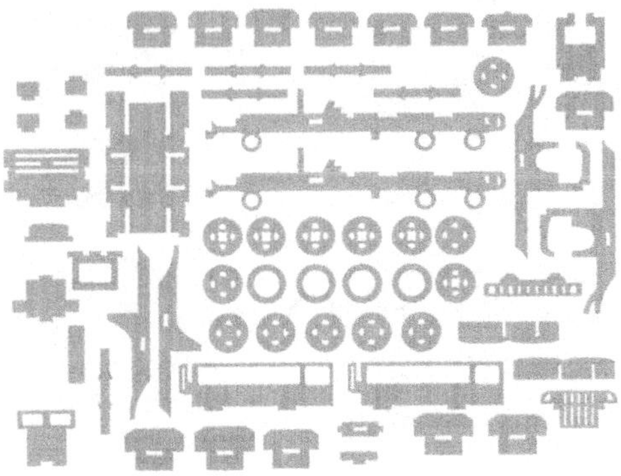

Step 1: Label and check inventory

Once you have all the pieces in front of you, the process of putting them together is no different from assembling pieces of Lego.

Label, using a pencil, the various parts and double-check your inventory. Take your time and enjoy the building process! Before jumping to the glue, be sure to do a dry run.

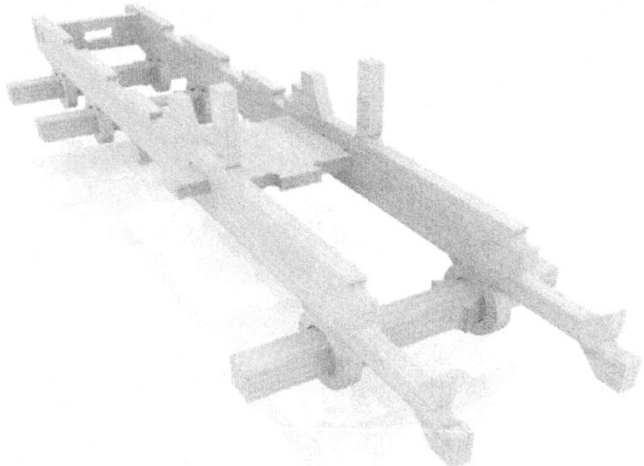

Step 2: Start the foundation

Slide and glue a part C (cab floor) and D (rear bumper holder) in between two part As (sides). Without applying glue, prepare three sets of doubled up part Bs (axles).

Step 3: Start the bed

Slide and glue a part E (trailer base,) F (bed head,) and all four part Gs (bench feet) onto the part As.

Step 4: Finish the bed

Slide and glue the part Hs (bench tops,) Is (bed sides,) and a part J (step) onto the bed.

Step 5: Start the cab

Starting in the cab's back, slide and glue into place part K (vertical back cab). On each side, glue into place a part L (side with roof) and M (side without roof).

Inside the cab, glue into place part N (seat).

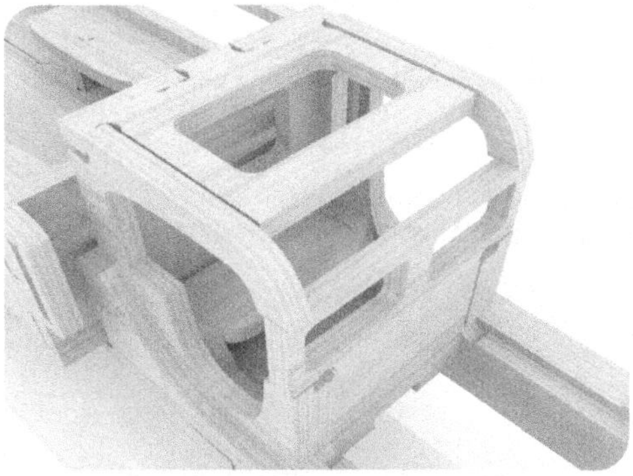

Step 6: Finish the cab

Slide and glue into place part Z (sun roof) and, on the front, part P (windshield).

Allow everything to dry overnight.

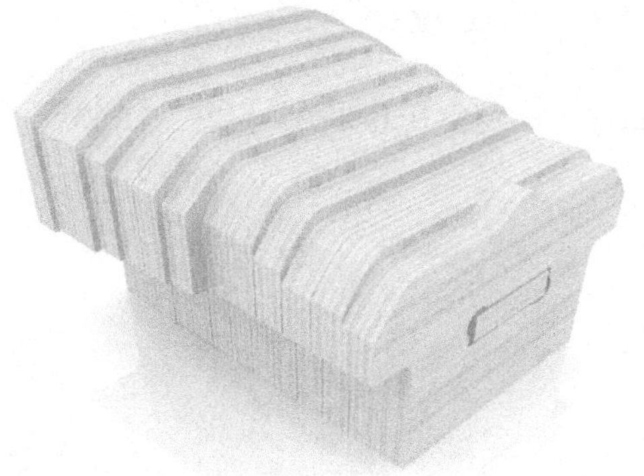

Step 7: Create engine mount

Slide and glue the different sized part Os (engine mounts) into a part P (engine guide). Depending on the REAL wood thickness, you may need fewer or more part Os to fill part P completely from end to end.

Step 8: Drop the engine

Slide and glue into place the engine mount you created in the previous step. In the front, glue into place part Q (radiator) and part R (front bumper).

Step 9: Add rear wheels

On each side, glue and slide into place a part S (wheel, full hole,) T (wheel, half holed,) and U (wheel rim) on both back axles.

Step 10: Add other wheels

Slide and glue into place (on the front axle) a part S (wheel, full hole) and T (wheel, half holed).

Apply clamps to keep pressure on the wheels while it dries overnight.

Step 11: Perform some finishing touches

There is no need to apply any preservatives or paint; feel free to sand any sharp edges down. ENJOY!

# Brontosaur

Completed 17 piece model dimensions @ 6 mm L: 36 cm, W: 5.5 cm, H: 25 cm
(1/4 inch L: 14-11/64, W: 2-11/64, H: 9-27/32)
Laser version found in WoodMarvels.com Volume 2: 50 Laser Cutting Plans (page 5)
Dowel version found in WoodMarvels.com Volume 1: Top Sellers (page 23)

## Product Renders

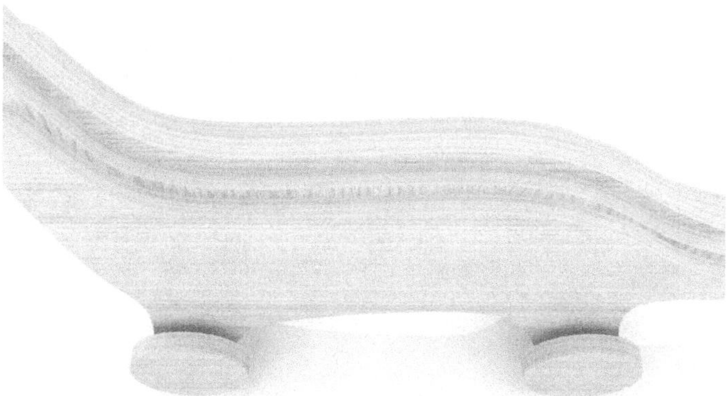

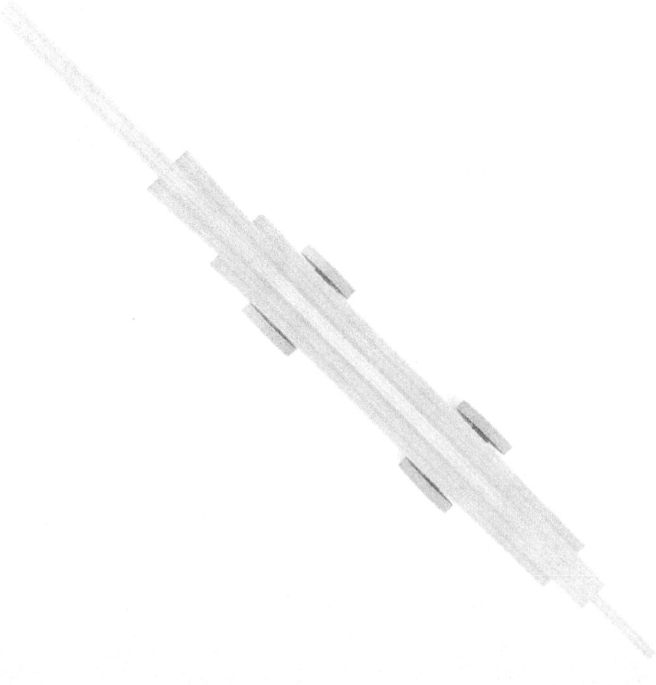

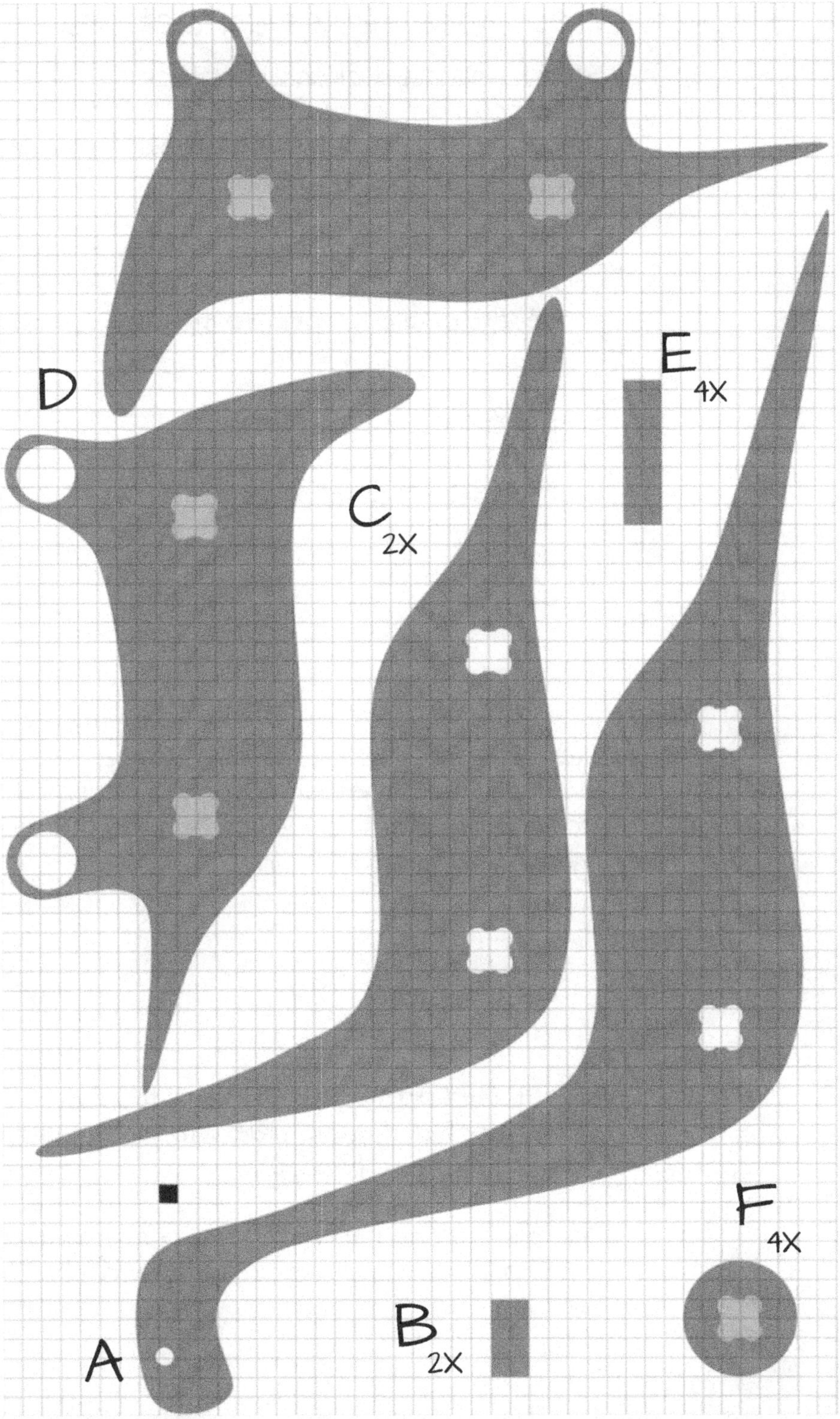
D
C
2X
E
4X
F
4X
A
B
2X

# Step-by-Step Assembly Instructions

Step 1: Label and check inventory

Once you have all the pieces in front of you, the process of putting them together is no different from assembling pieces of Lego.

Label, using a pencil, the various parts and double-check your inventory. Take your time and enjoy the building process! Before jumping to the glue, be sure to do a dry run.

Step 2: Start the body

Slide and glue into place part A (body with head) with all four part Bs (guides).

Step 3: Complete the body

Slide and glue together a part C (body sides) and D (body with legs). Use clamps to hold the model together tightly until its body is dry.

Step 4: Perform some finishing touches

In the front and back, slide into place part Es (axle) in between part Fs (wheel).

There is no need to apply any preservatives or paint; feel free to sand any sharp edges down. ENJOY!

# Centrosaur

Completed 17 piece model dimensions @ 6 mm L: 35 cm, W: 5 cm, H: 15 cm
(1/4 inch L: 13-25/32, W: 1-31/32, H: 5-29/32)
Laser version found in WoodMarvels.com Volume 2: 50 Laser Cutting Plans (page 8)

Product Renders

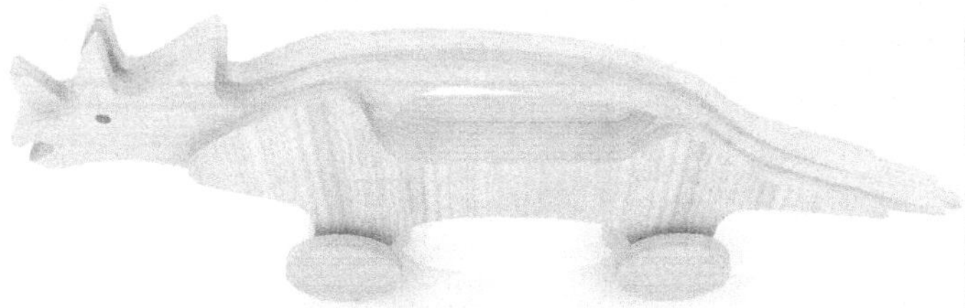

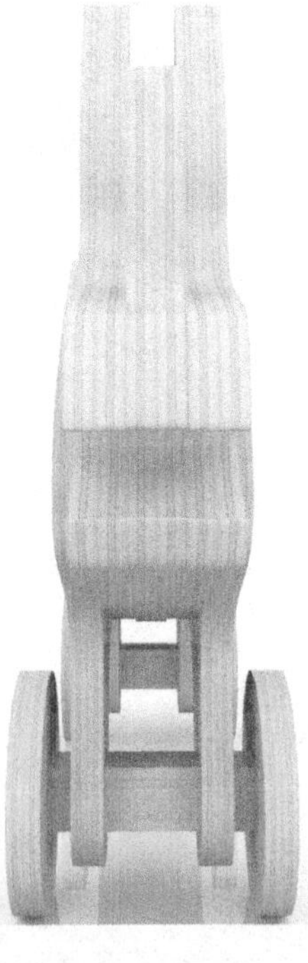

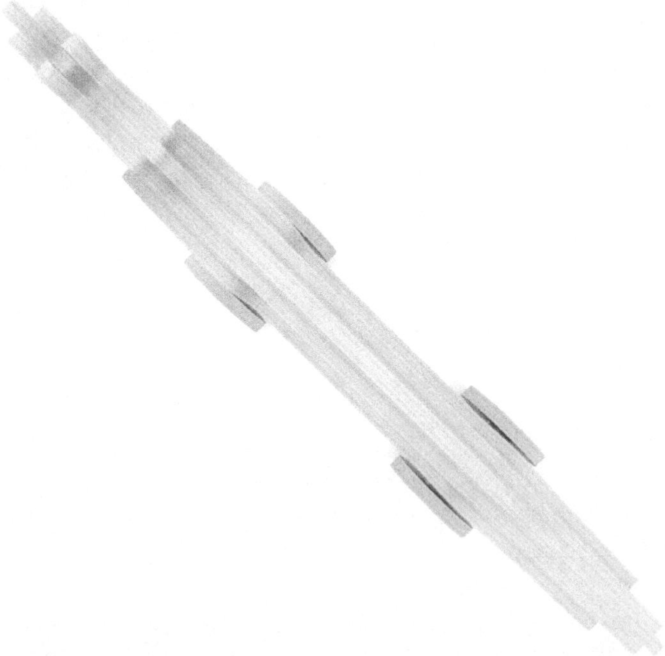

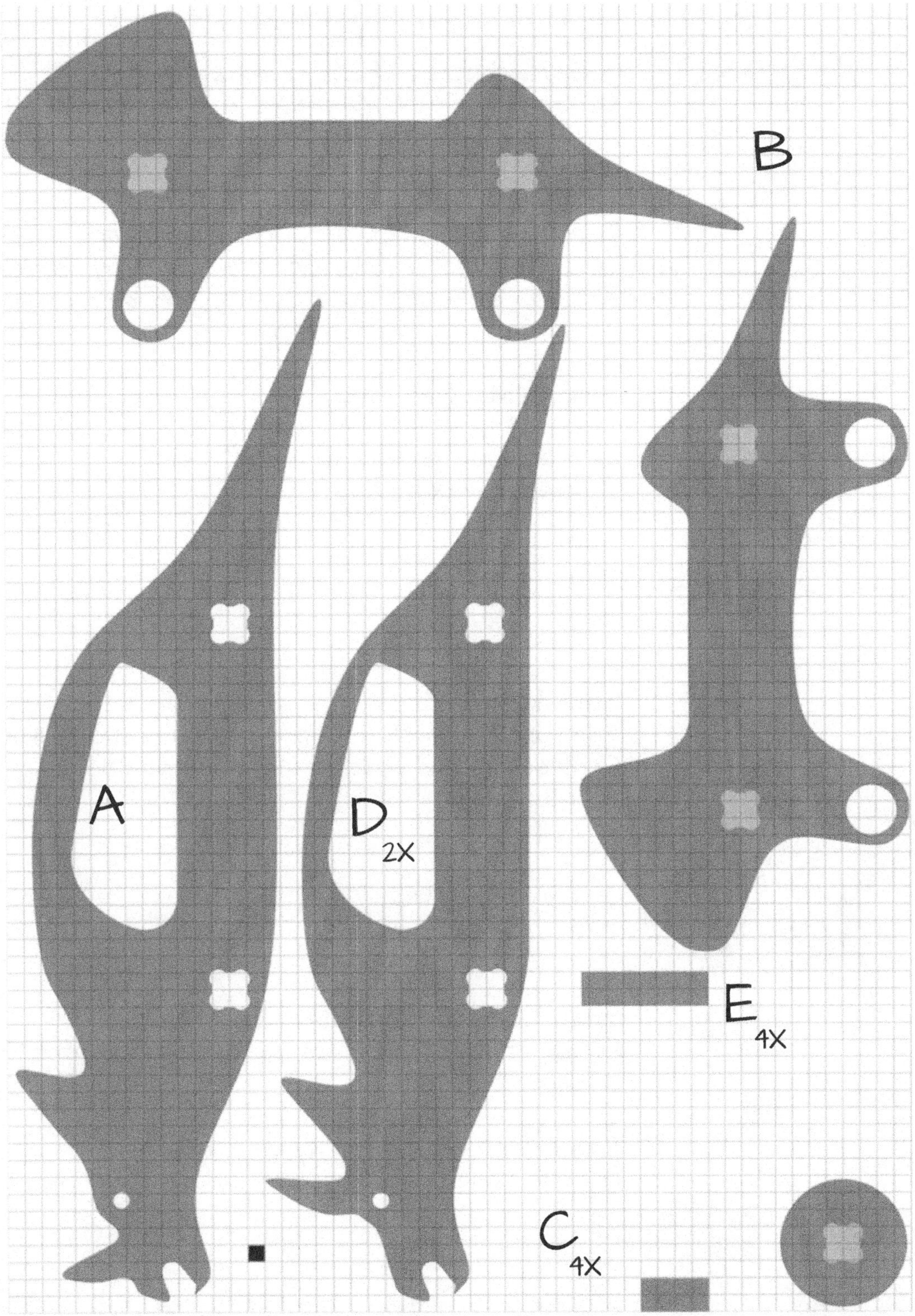
B
A
D
2X
E
4X
C
4X
F
4X

# Step-by-Step Assembly Instructions

Step 1: Label and check inventory

Once you have all the pieces in front of you, the process of putting them together is no different from assembling pieces of Lego.

Label, using a pencil, the various parts and double-check your inventory. Take your time and enjoy the building process! Before jumping to the glue, be sure to do a dry run.

Step 2: Start the body

Slide and glue into place part A (large body) with both part Bs (side panel with horns,) using part Cs (guides) to help align everything.

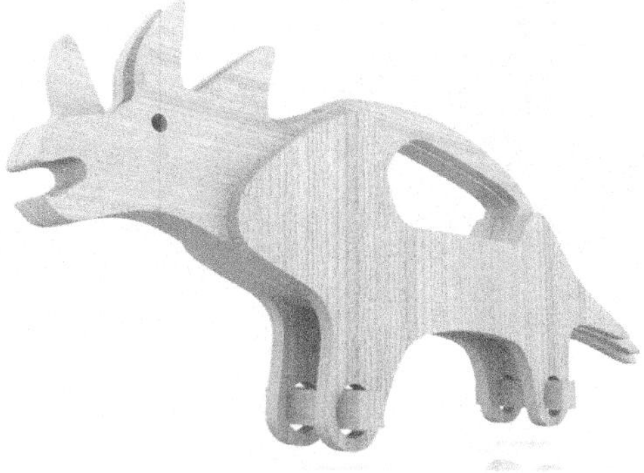

Step 3: Add sides

Add part Ds (pocketed side panels) on both sides of the Centrosaur, and underneath, glue together two sets of part Es (axles).

Step 4: Perform some finishing touches

At the end of both axles, glue into place part Fs (pocketed wheels).

There is no need to apply any preservatives or paint; feel free to sand any sharp edges down. ENJOY!

# CH47 Chinook Helicopter

Completed 49 piece model dimensions @ 6 mm L: 76 cm, W: 35 cm, H: 20 cm
(1/4 inch L: 29-59/64, W: 13-25/32, H: 7-7/8)
Laser version found in WoodMarvels.com Volume 3: Evolution of Wooden Designs (page 40)

## Product Renders

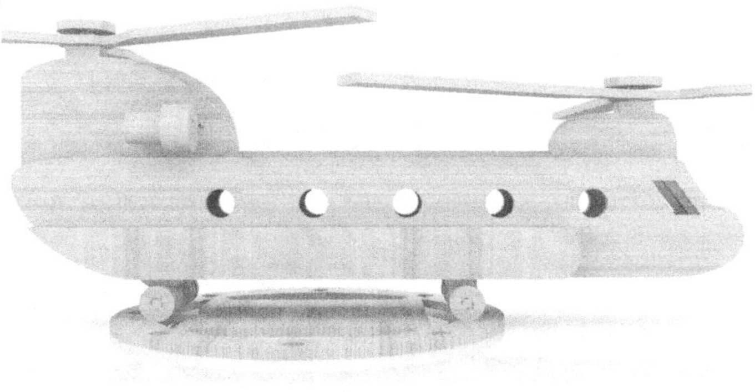

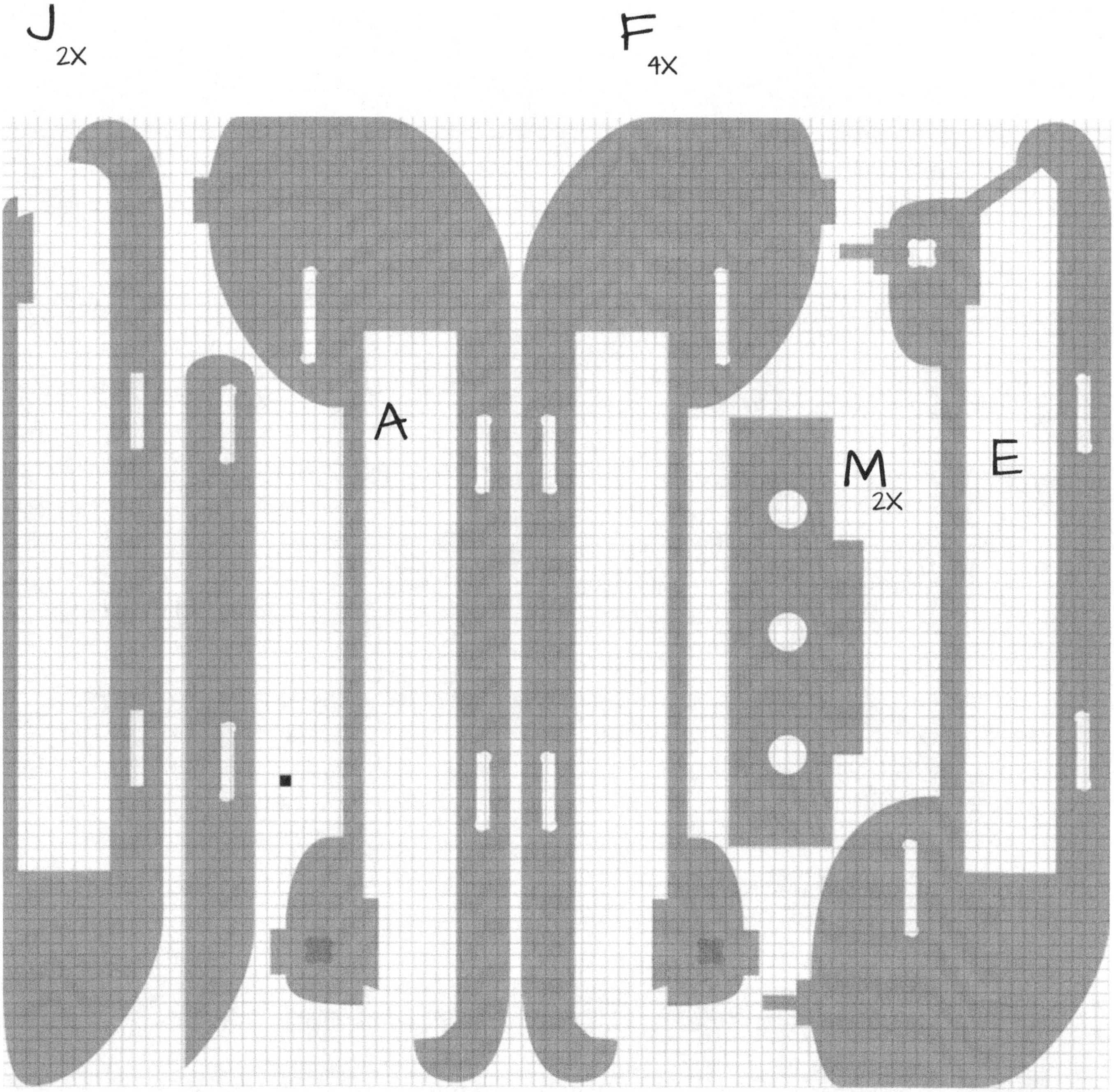
J 2X
F 4X
A
M 2X
E

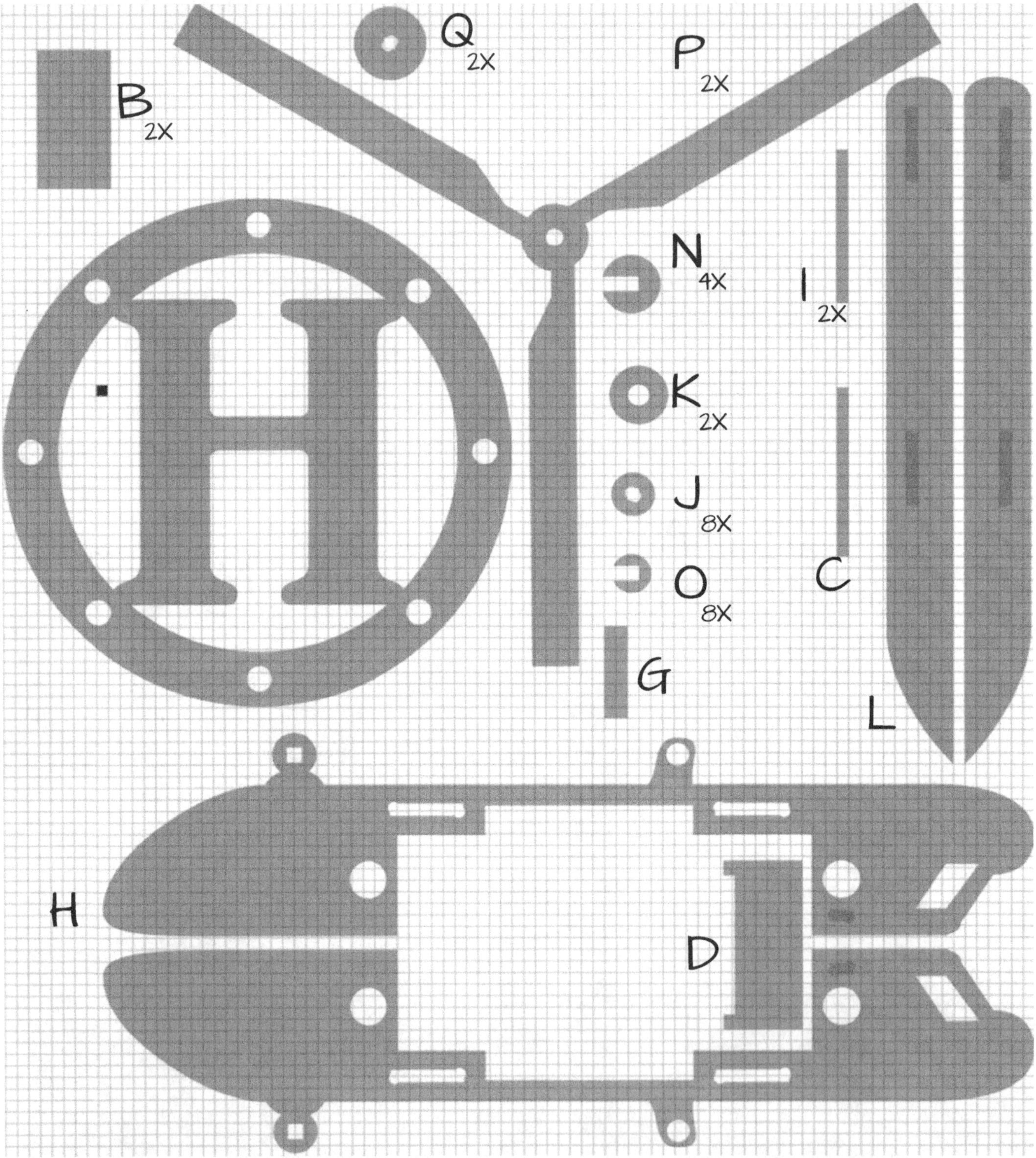
Q
2X
P
2X
B
2X
N
4X
I
2X
K
2X
J
8X
O
8X
C
G
L
H
D

# Step-by-Step Assembly Instructions

Step 1: Label and check inventory

Once you have all the pieces in front of you, the process of putting them together is no different from assembling pieces of Lego.

Label, using a pencil, the various parts and double-check your inventory. Take your time and enjoy the building process! Before jumping to the glue, be sure to do a dry run.

Step 2: Start the body

Slide and glue into place part A (body with propeller pegs,) both part Bs (floor guides,) C (engine lock peg,) and D (engine holders).

Step 3: Build onto the body

Slide and glue together a part E (body with elevated pins) and two part Fs on both sides of the Chinook's body.

You also want to glue into place a part G (cockpit support peg) to help keep the roof from dropping down with use. The ends of part G will slide into the holes found in the part Hs.

Step 4: Complete the cockpits

Once you have slid and glued part Hs (closed cockpit) to the rest of the Chinook, slide into place two part Is (axles,) inserting part Js (wheels) on the inside and outsides of the part Hs. You want to glue them all together, as the wheels are not meant to rotate.

Step 5: Add gas tanks

Add a part J (tank with holes) on the sides, using the pegs as guides with glue on both sides.

On the top engine parts, glue and slide into place, within the extended knob, a part K on both sides.

Step 6: Finish the gas tanks

Slide and glue into place part Ls on both sides (gas end caps,) followed by sliding into place (without glue) part Ms.

On the engine, glue and slide into place two part Ns on both sides. Allow everything to dry overnight.

Step 7: Finish the Chinook

After you have allowed everything to dry, glue and slide into place the final engine components by putting four part Os at the end of the part Ns.

On the top of the helicopter, slide (without glue) both part Ps into place.

Step 8: Perform some finishing touches

On the top, glue into place a part Q (rotor blade cap) onto the end of the pins.

There is no need to apply any preservatives or paint; feel free to sand any sharp edges down. ENJOY!

# Desktop Organizer

Completed 9 piece model dimensions @ 6 mm L: 23 cm, W: 23 cm, H: 23 cm
(1/4 inch L: 9-1/16, W: 9-1/16, H: 9-1/16)
Laser version found in WoodMarvels.com Volume 2: 50 Laser Cutting Plans (page 8)

Product Renders

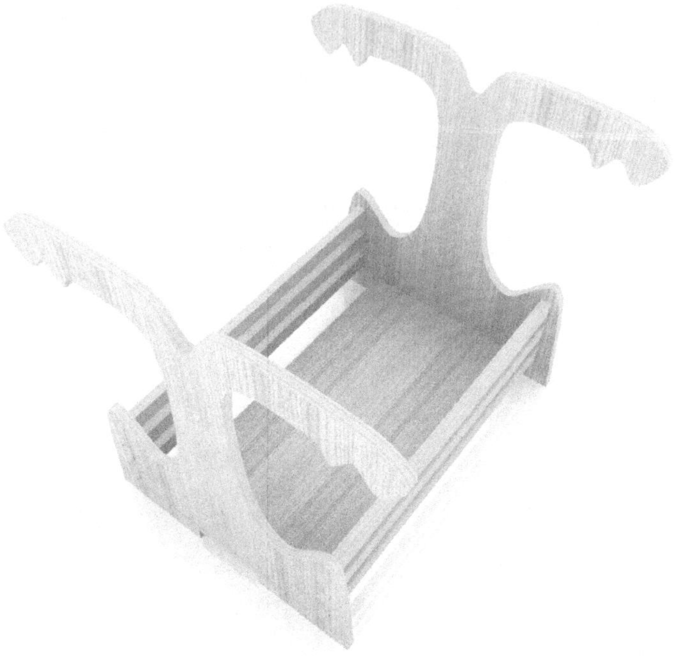

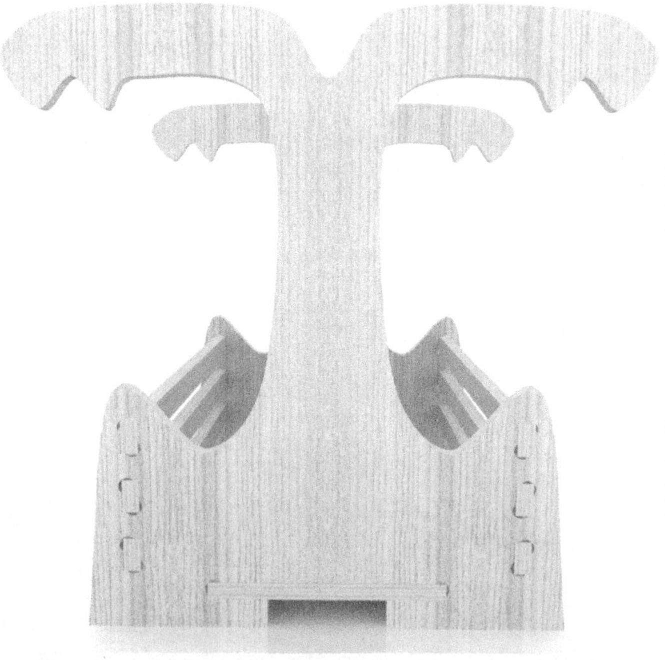

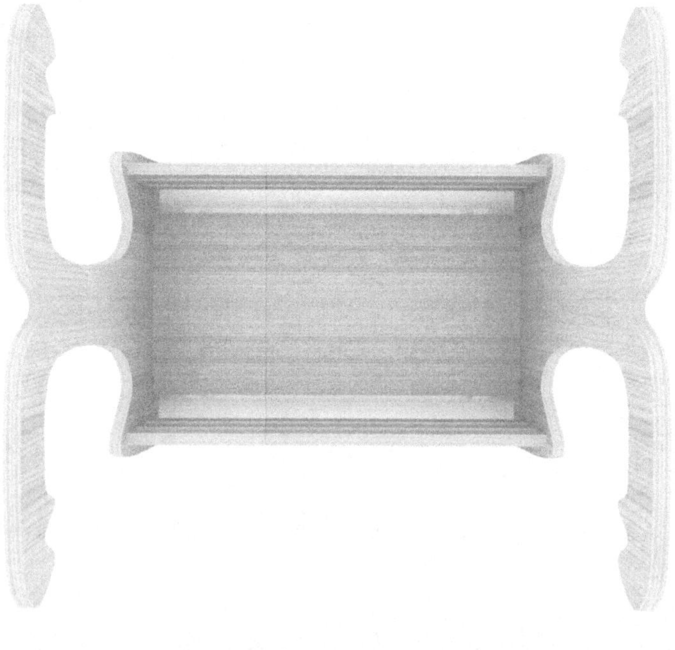

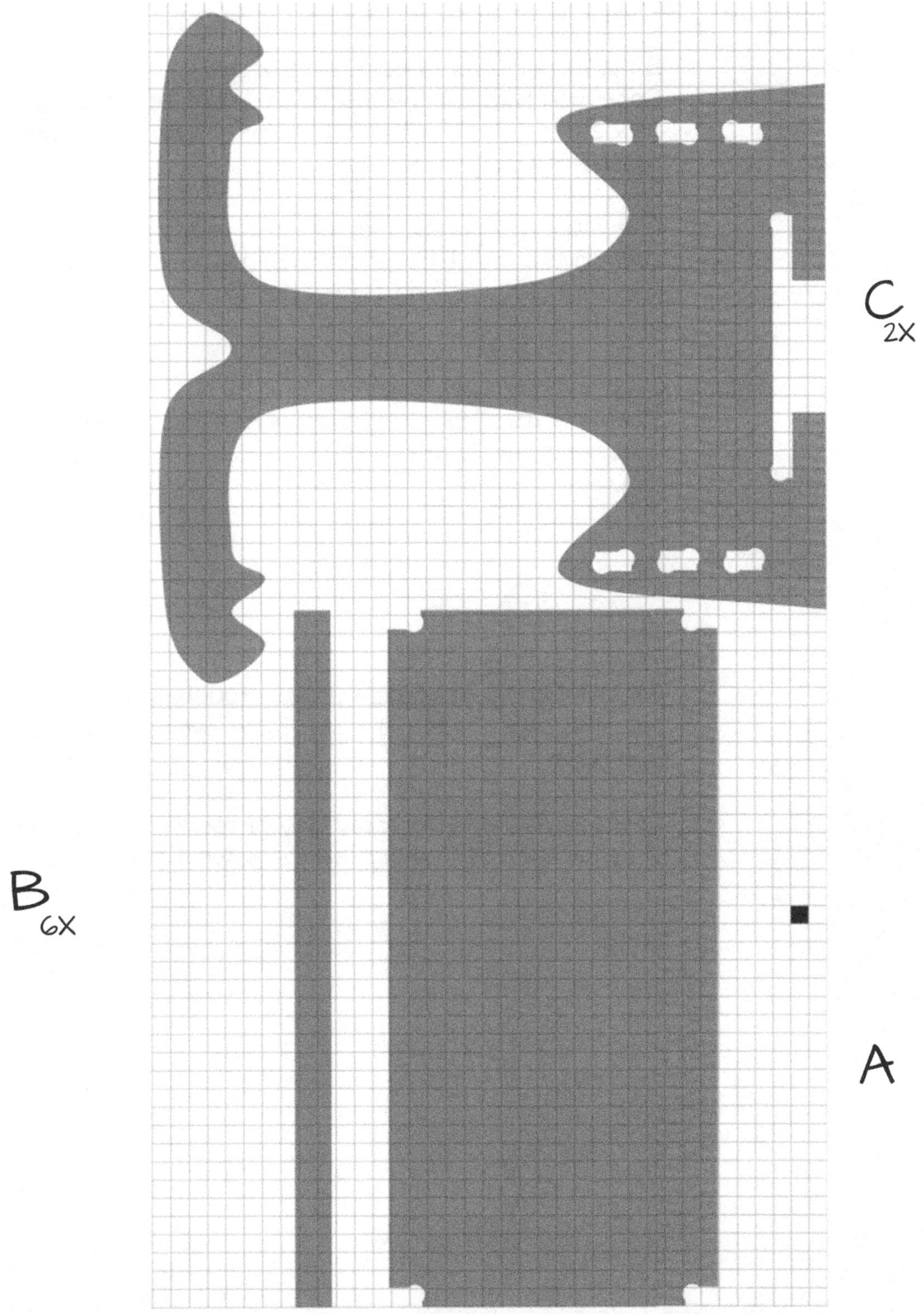
C
2X
B
6X
A

# Step-by-Step Assembly Instructions

Step 1: Label and check inventory

Once you have all the pieces in front of you, the process of putting them together is no different from assembling pieces of Lego.

Label, using a pencil, the various parts and double-check your inventory. Take your time and enjoy the building process! Before jumping to the glue, be sure to do a dry run.

Step 2: Start the base

Slide and glue together onto part A (base) all six part Bs (wall supports) within a part C (side) to keep everything together.

Step 3: Perform some finishing touches

Add the other side part and allow everything to dry overnight.

There is no need to apply any preservatives or paint; feel free to sand any sharp edges down. ENJOY!

# Desktop Organizer B

Completed 10 piece model dimensions @ 6 mm L: 23 cm, W: 20 cm, H: 31.6 cm
(1/4 inch L: 9-1/16 W: 7-7/8 H: 12-7/16)

# Product Renders

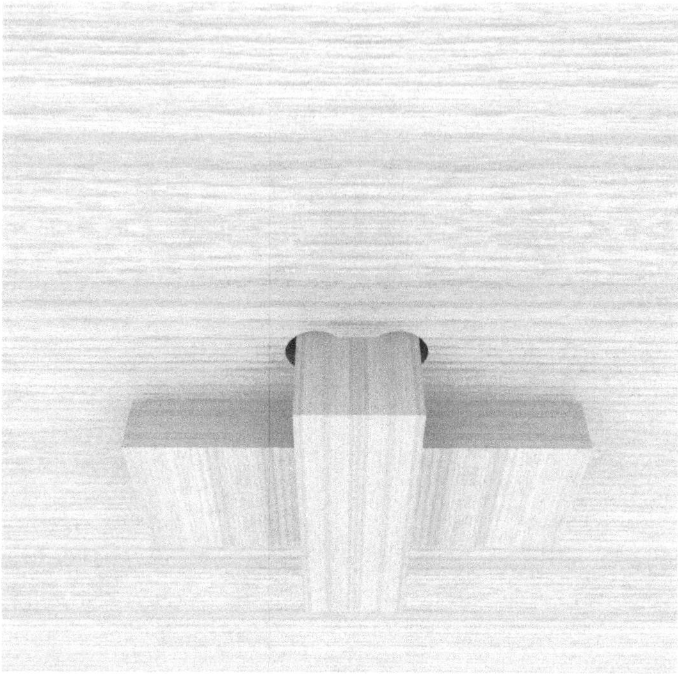

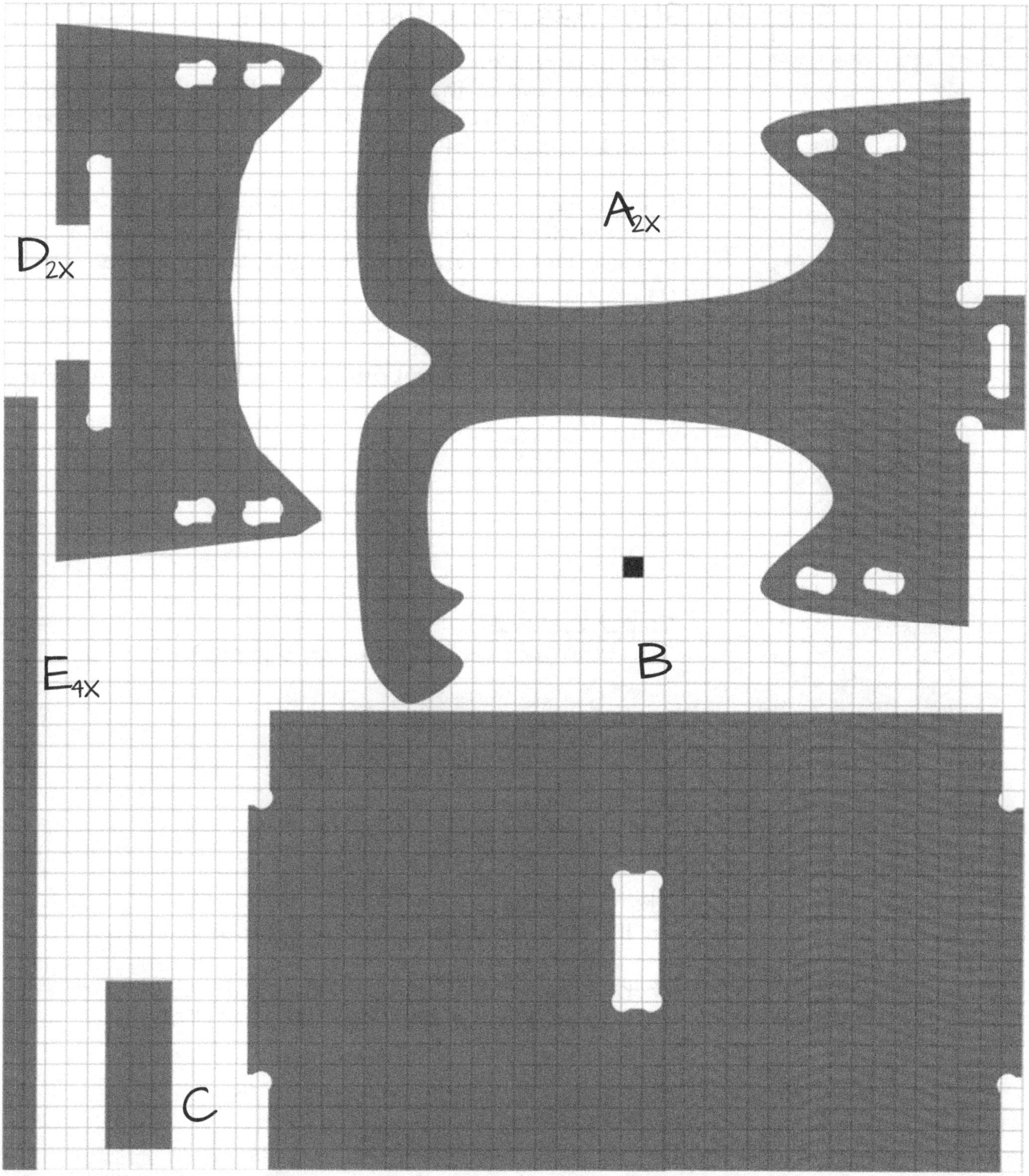
D2X
A2X
B
E4X
C

# Step-by-Step Assembly Instructions

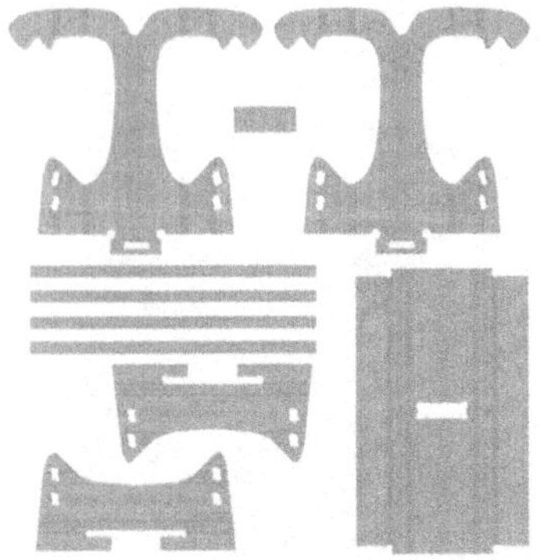

Step 1: Label and check inventory
Once you have all the pieces in front of you, the process of putting them together is no different from assembling pieces of Lego.

Label, using a pencil, the various parts and double-check your inventory. Take your time and enjoy the building process! Before jumping to the glue, be sure to do a dry run.

Step 2: Start the base
Slide and glue together both part Bs (wall supports) onto part A (base).

Step 3: Add walls
Slide both part Cs (wall tops) onto part Bs.

Step 4: Add sides
On the outside edges, add a part D (sides).

There is no need to apply any preservatives or paint; feel free to sand any sharp edges down. ENJOY!

# Desktop Organizer C

Completed 7 piece model dimensions @ 6 mm L: 24 cm, W: 24 cm, H: 8 cm
(1/4 inch L: 9-29/64, W: 9-29/64, H: 3-5/32)
Laser version found in WoodMarvels.com Volume 2: 50 Laser Cutting Plans (page 72)

## Product Renders

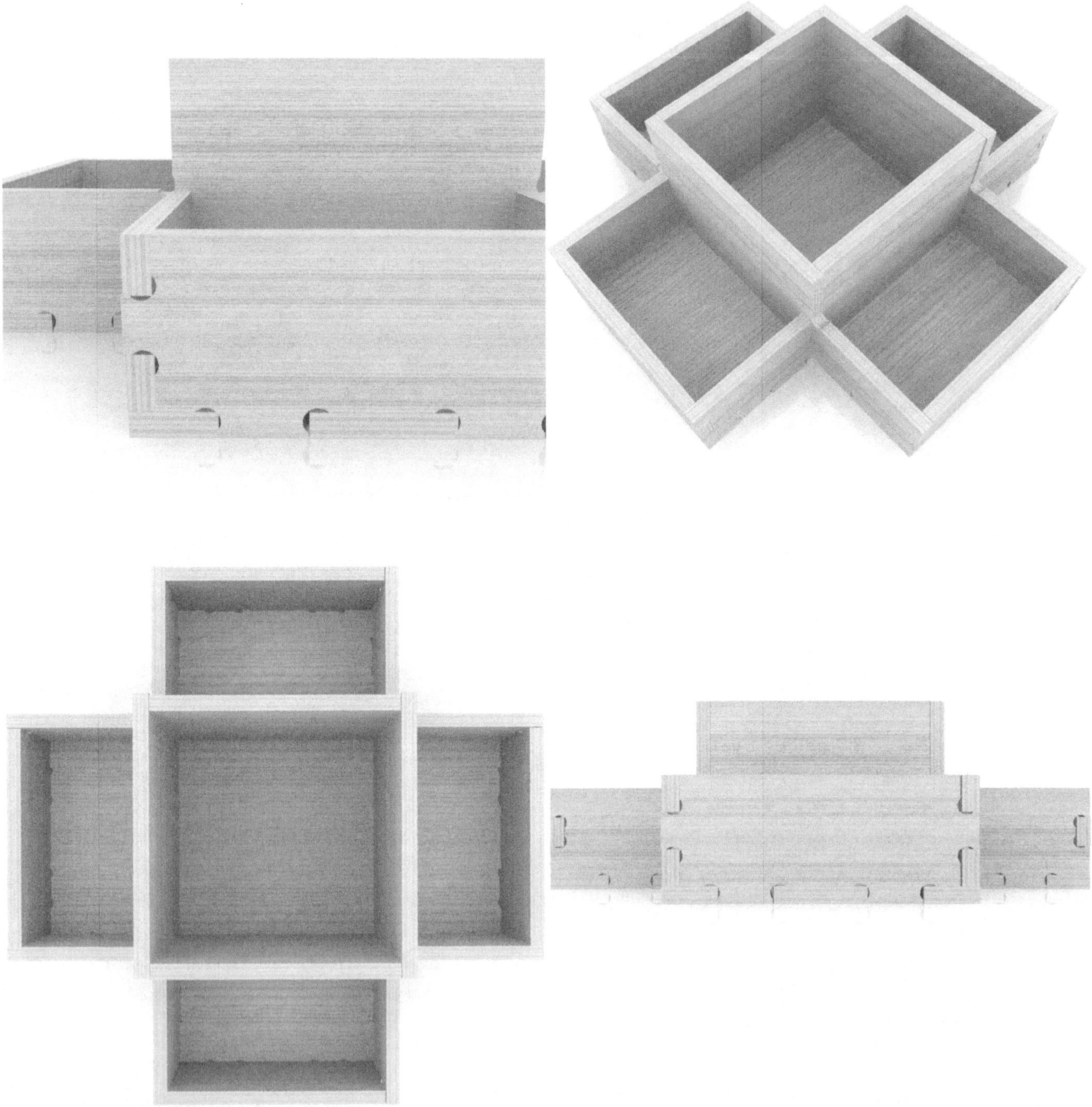

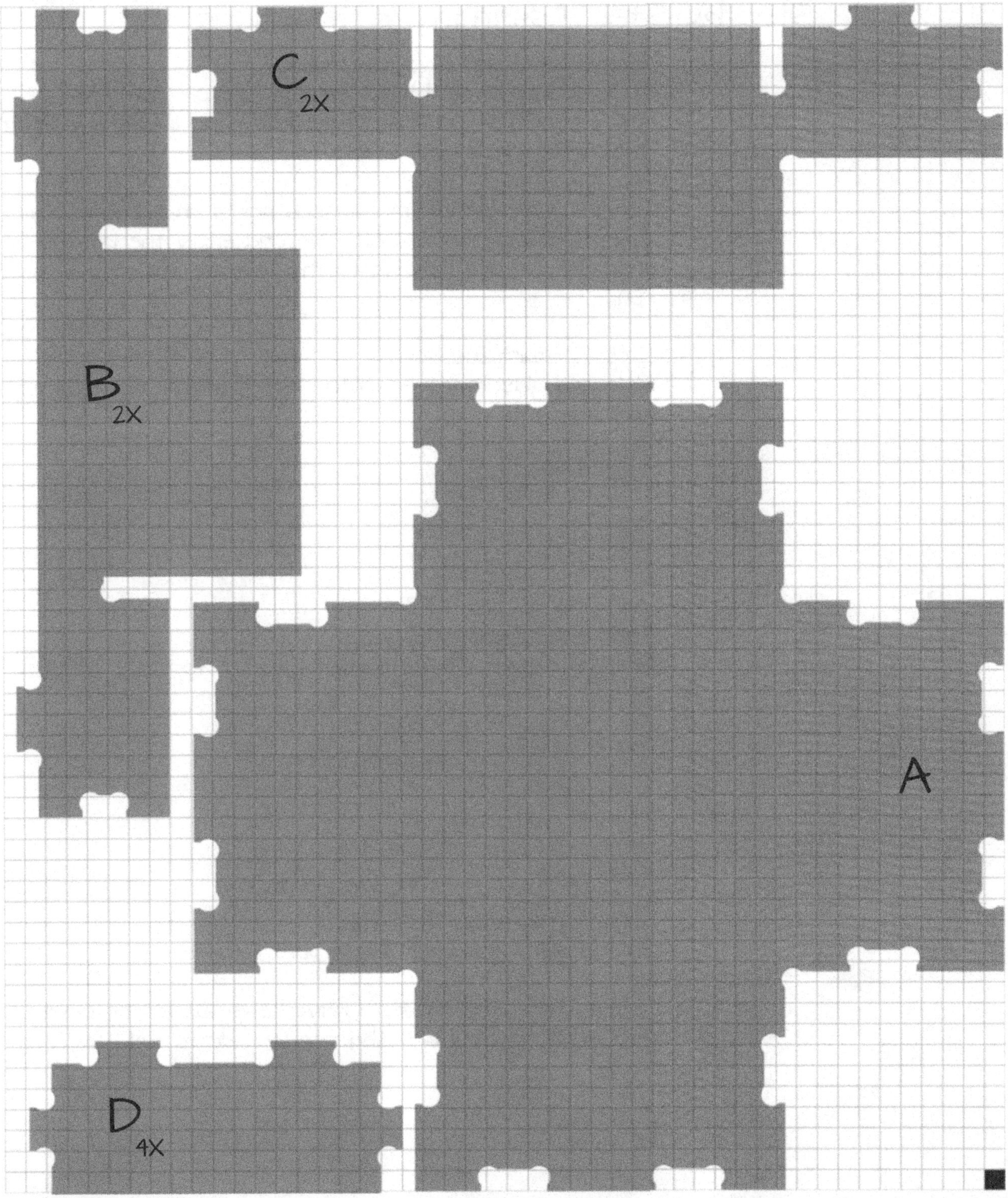
C
2X
B
2X
A
D
4X

## Step-by-Step Assembly Instructions

Step 1: Label and check inventory

Once you have all the pieces in front of you, the process of putting them together is no different from assembling pieces of Lego.

Label, using a pencil, the various parts and double-check your inventory. Take your time and enjoy the building process! Before jumping to the glue, be sure to do a dry run.

Step 2: Start the base

Slide and glue together onto part A (base) both part Bs (wall supports).

Step 3: Add walls

Slide onto part Bs, both part Cs (wall tops).

Step 4: Add sides

On the outside edges, add a part D (sides).

There is no need to apply any preservatives or paint, feel free to sand any sharp edges down. ENJOY!

# Dolphin

Completed 16 piece model dimensions @ 6 mm L: 39 cm, W: 13 cm, H: 26 cm
(1/4 inch L: 15-23/64, W: 5-1/8, H: 10-15/64)
Laser version found in WoodMarvels.com Volume 2: 50 Laser Cutting Plans (page 11)
Dowel version found in WoodMarvels.com Volume 1: Top Sellers (page 26)

Product Renders

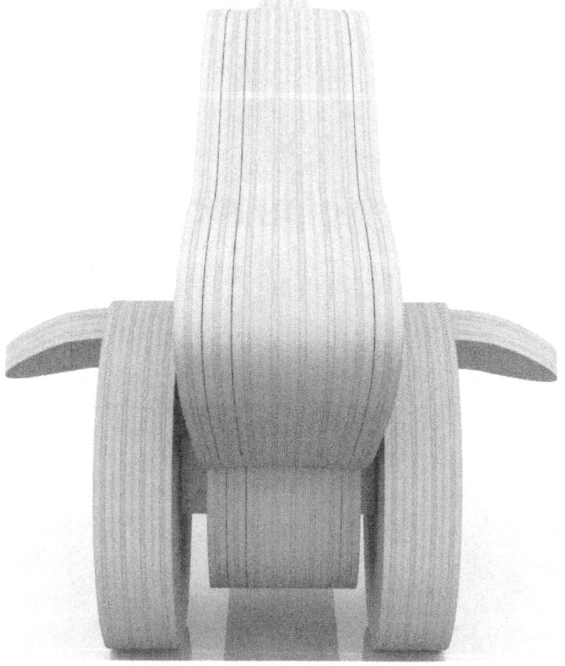

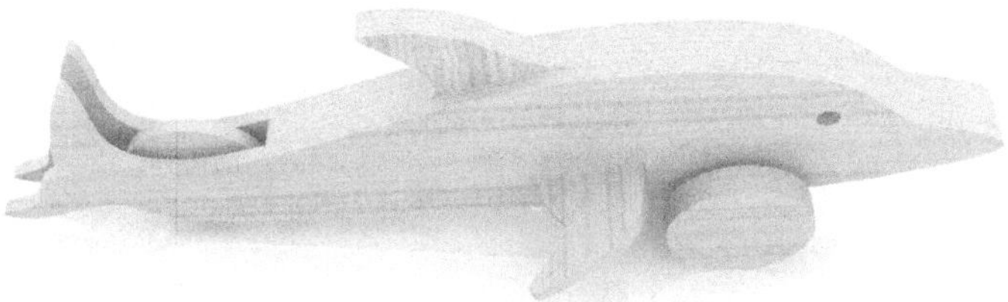

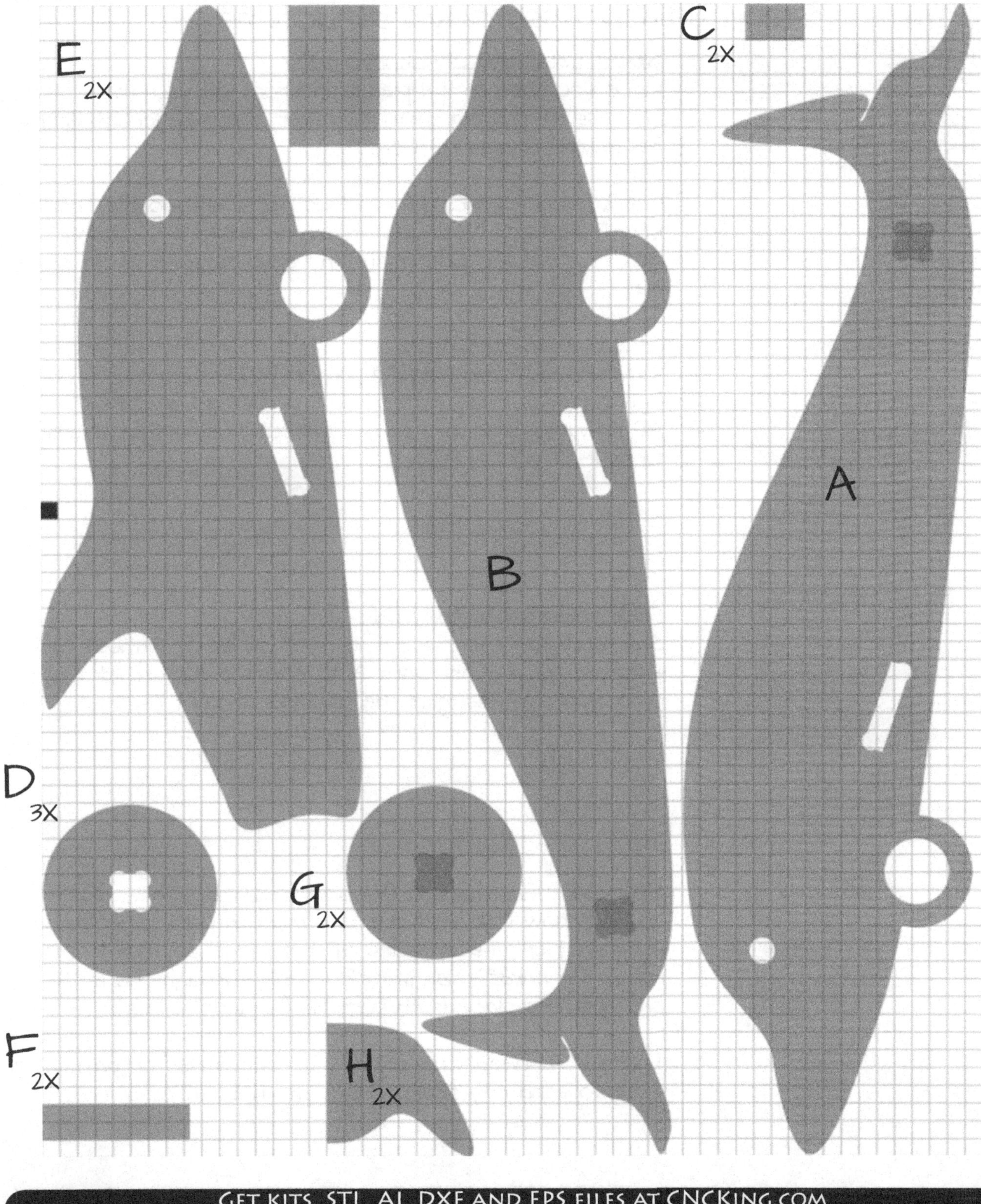
E 2X
C 2X
A
B
D 3X
G 2X
F 2X
H 2X

# Step-by-Step Assembly Instructions

Step 1: Label and check inventory
Once you have all the pieces in front of you, the process of putting them together is no different from assembling pieces of Lego.

Label, using a pencil, the various parts and double-check your inventory. Take your time and enjoy the building process! Before jumping to the glue, be sure to do a dry run.

Step 2: Start the body
Slide and glue into place a part A (body with tail,) a part B (fin holder,) and, in the back, a doubled up part C. Afterwards, slide into place, without glue, a part D (back wheel).

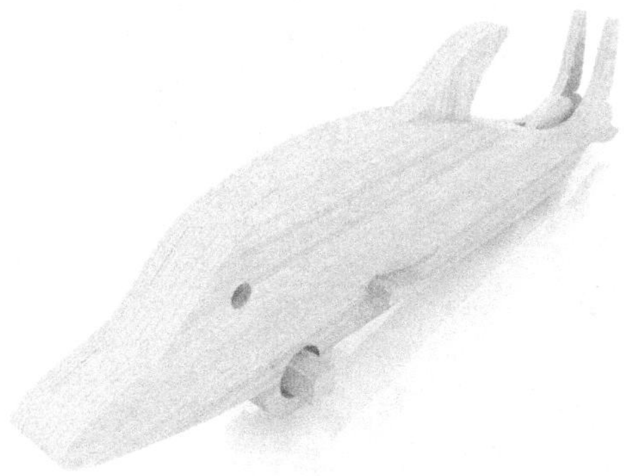

Step 3: Complete the body
Slide and glue into place two part Es (body without tail) followed by another part A to cap the dolphin. At this time, you can also glue the doubled up part Fs.

Step 4: Perform some finishing touches
You can now slide and glue into place a part D, followed by a part G on both sides of the model so that it can roll. Afterwards, push part B evenly through the body and glue into place both part Hs (horizontal fins).

There is no need to apply any preservatives or paint; feel free to sand any sharp edges down. ENJOY!

# Easter Island Head Storage

Completed 31 piece model dimensions @ 6 mm L: 12 cm, W: 19 cm, H: 34 cm
(1/4 inch L: 4-23/32, W: 7-31/64, H: 13-25/64)
Laser version found in WoodMarvels.com Volume 3: Evolution of Wooden Designs (page 44)

# Product Renders

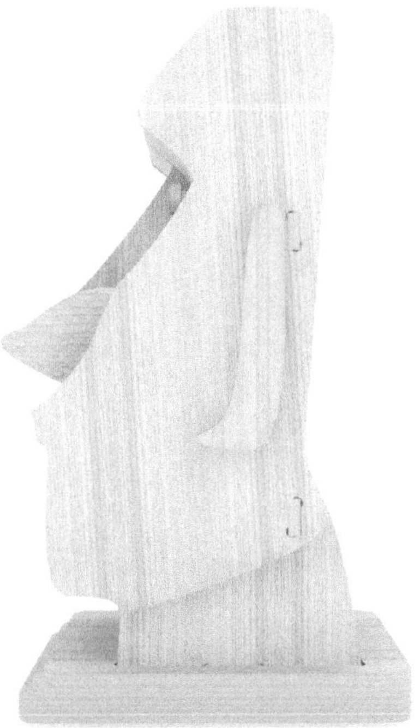

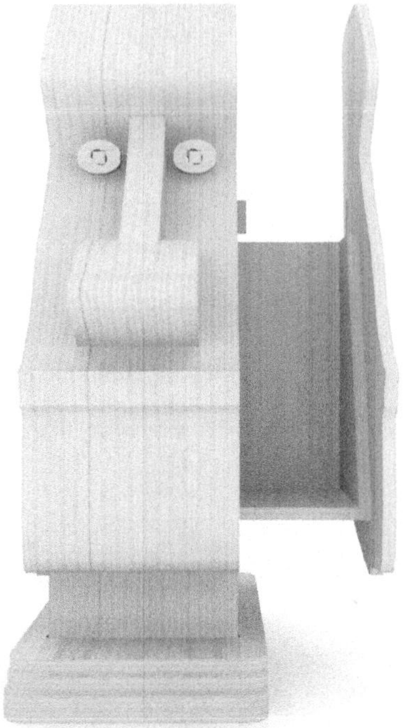

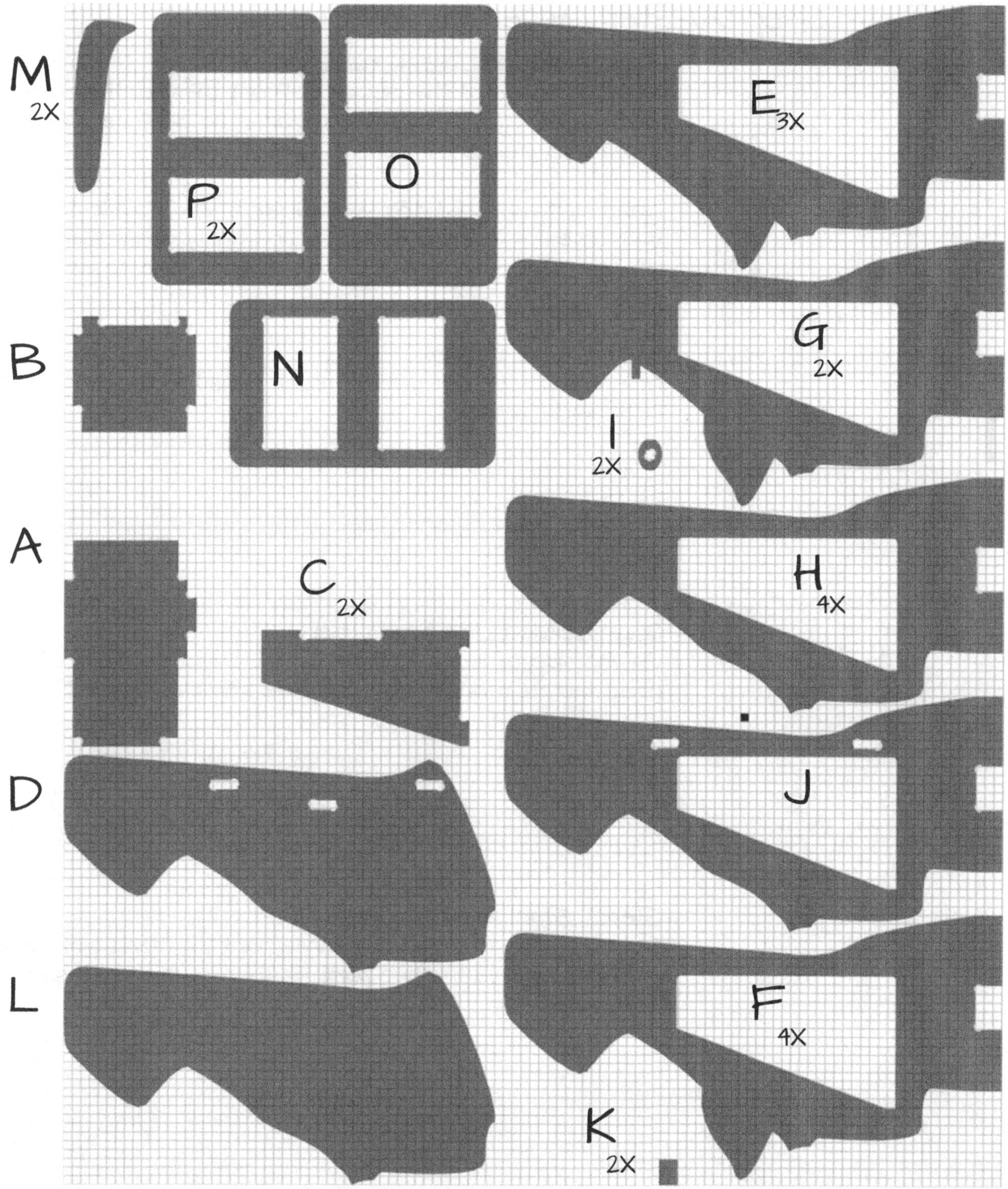
M
2X
P
2X
O
E
3X
B
N
G
2X
I
2X
A
C
2X
H
4X
D
J
L
F
4X
K
2X

# Step-by-Step Assembly Instructions

Step 1: Label and check inventory

Once you have all the pieces in front of you, the process of putting them together is no different from assembling pieces of Lego.

Label, using a pencil, the various parts and double-check your inventory. Take your time and enjoy the building process! Before jumping to the glue, be sure to do a dry run.

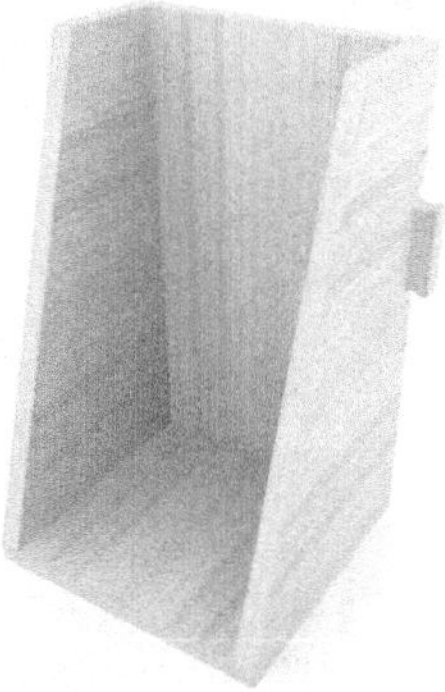

Step 2: Build the box

Glue into place both part A (back panel), B (bottom) and both part Cs (sides).

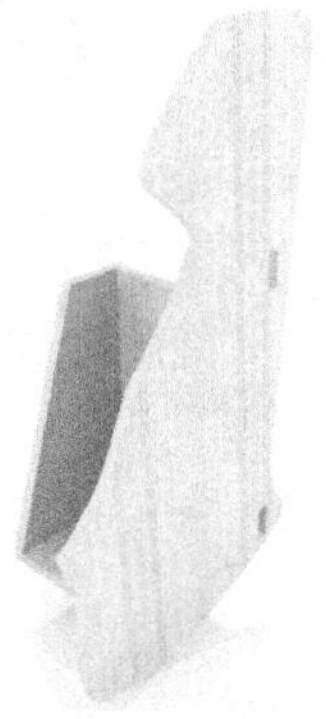

Step 3: Slide in the head panel

On the side of part A, there is a tap that comes out, using this as a guide, glue and slide into place part D.

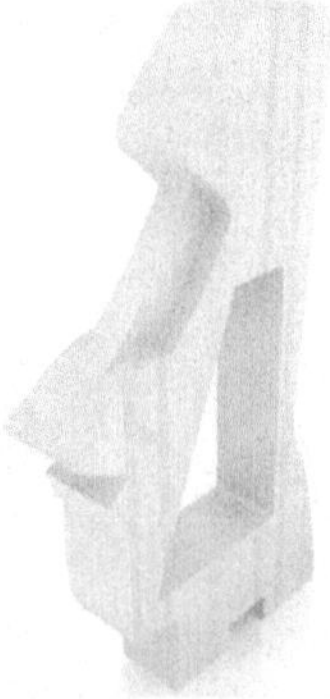

Step 4: Start from the nose

Slide and glue into place all three part Es (head with nose bridge) followed by two part Fs (head with only nose) on both sides.

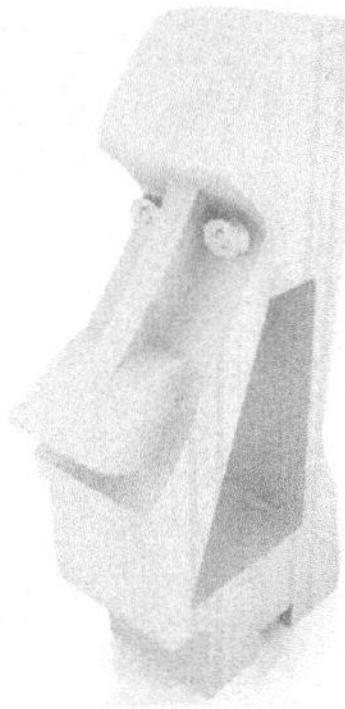

Step 5: Add eyes

Glue and slide into place a part G (head with eye guides) on both sides followed by two part Hs (head without nose) on both sides.

Glue and slide into the eye guides, a part I.

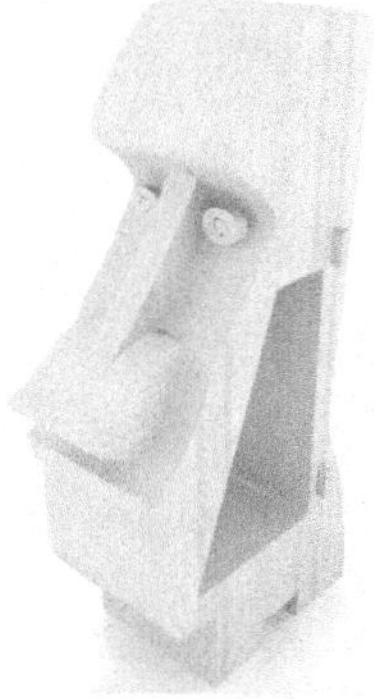

Step 6: Finish the holed head

Glue into place a part H on the left side and J (head without nose with two holes) on the right. Glue and slide into part J both part Ks (guides).

If you wish to have the opening on the other side of the head, mirror step 2 and 3 and put a part H on the right and J on the left.

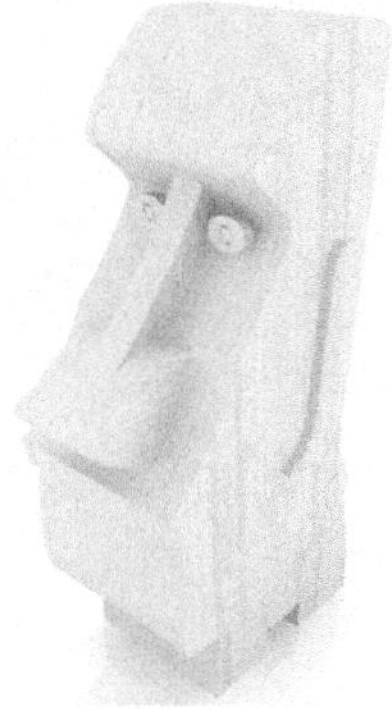

Step 7: Slide in the head panel

On the left hand side, glue into place a part L (solid face panel) and the entire assembly of step 3 without glue.

On both sides, glue into place a part M (ear).

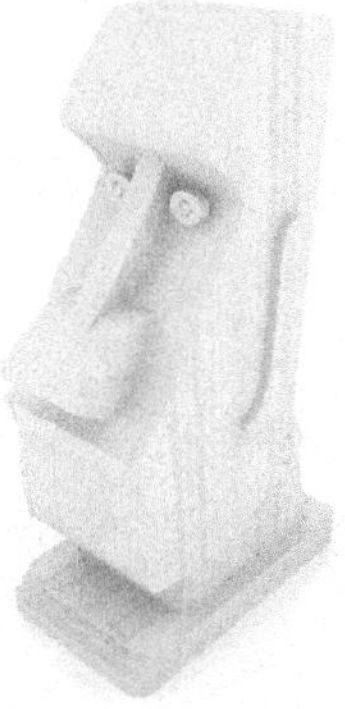

Step 8: Finishing touches

Slide and glue into place from the bottom part N, O and both part Ps and allow everything to dry overnight with clamps to hold things together.

There is no need to apply any preservatives or paint, feel free to sand any sharp edges down. ENJOY!

# Elephant

Completed 25 piece model dimensions @ 6 mm L: 22 cm, W: 12 cm, H: 16.5 cm
(1/4 inch L: 8-21/32, W: 4-23/32, H: 6-1/2)
Laser version found in WoodMarvels.com Volume 2: 50 Laser Cutting Plans (page 15)

Product Renders

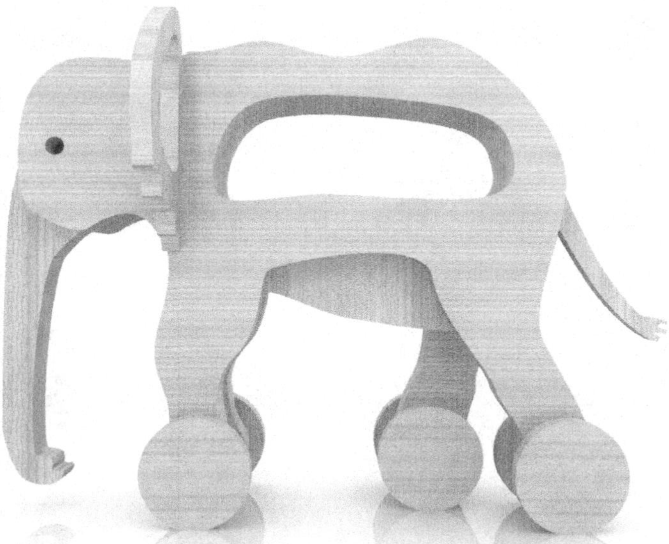

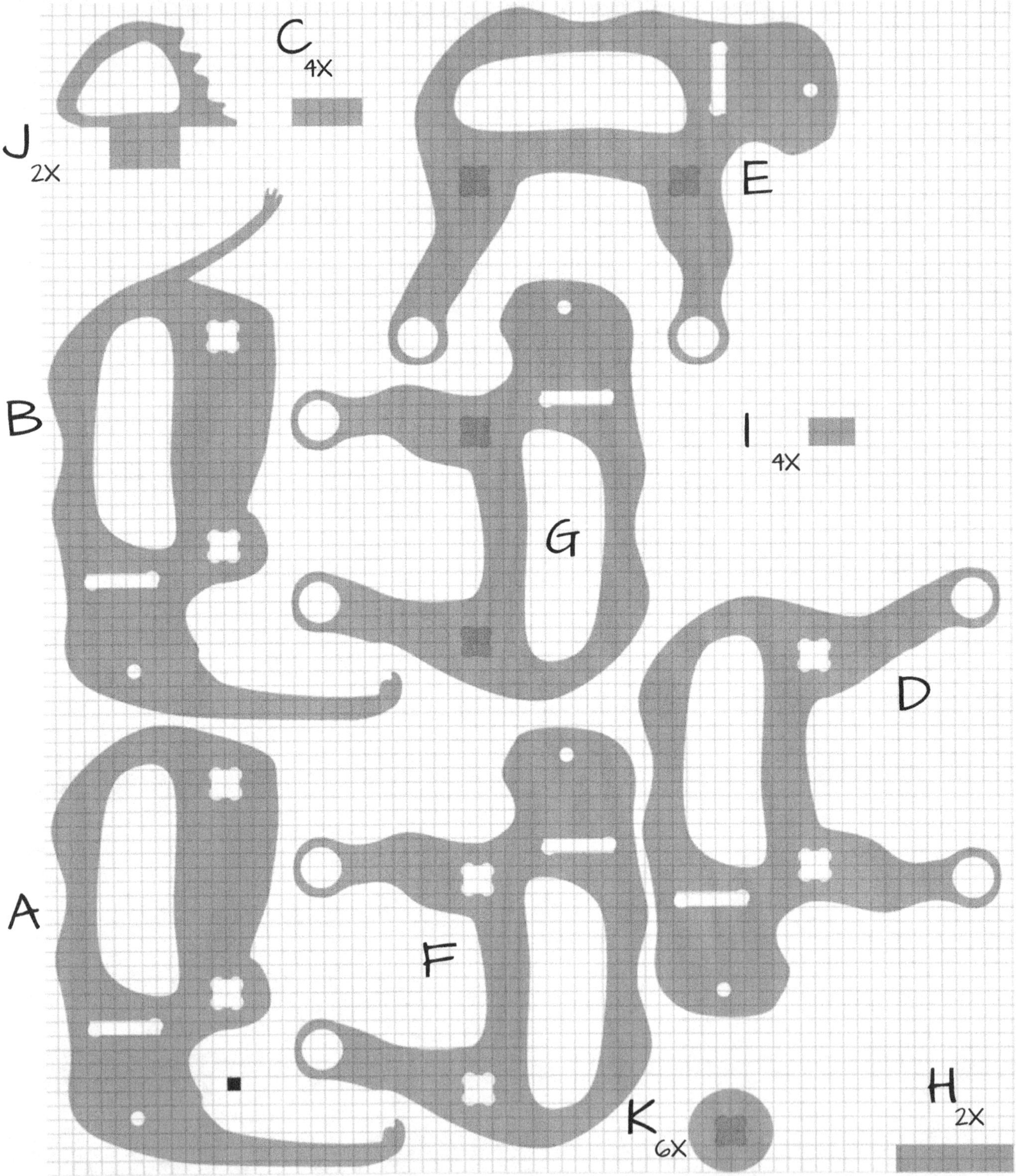
C
4X
J
2X
E
B
I
4X
G
D
A
F
K
6X
H
2X

# Step-by-Step Assembly Instructions

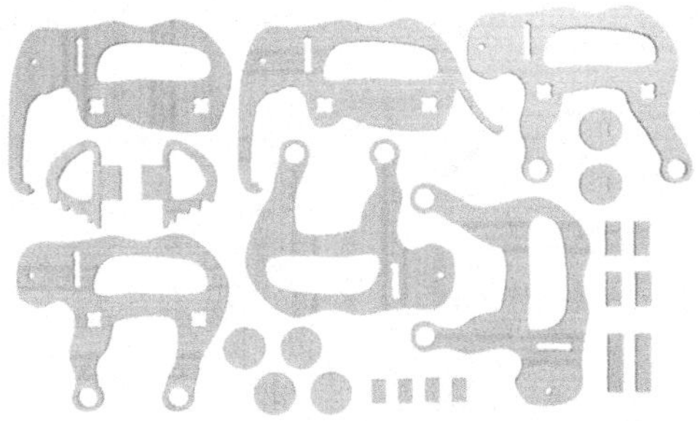

Step 1: Label and check inventory

Once you have all the pieces in front of you, the process of putting them together is no different from assembling pieces of Lego.

Label, using a pencil, the various parts and double-check your inventory. Take your time and enjoy the building process! Before jumping to the glue, be sure to do a dry run.

Step 2: Start the body

Slide and glue into place part A (body with trunk and tail) and B (body with trunk but no tail) using all four part Cs (inner guides).

Step 3: Finish the body

Slide and glue together a part D and E on one side, followed by F and G onto the other.

On the bottom, glue together doubled up part Hs (long axles) and two sets of doubled up part Is.

Step 4: Perform some finishing touches

You can now glue and slide into place a part J (ears) followed by part Ks (wheels).

There is no need to apply any preservatives or paint; feel free to sand any sharp edges down. ENJOY!

# Future Car

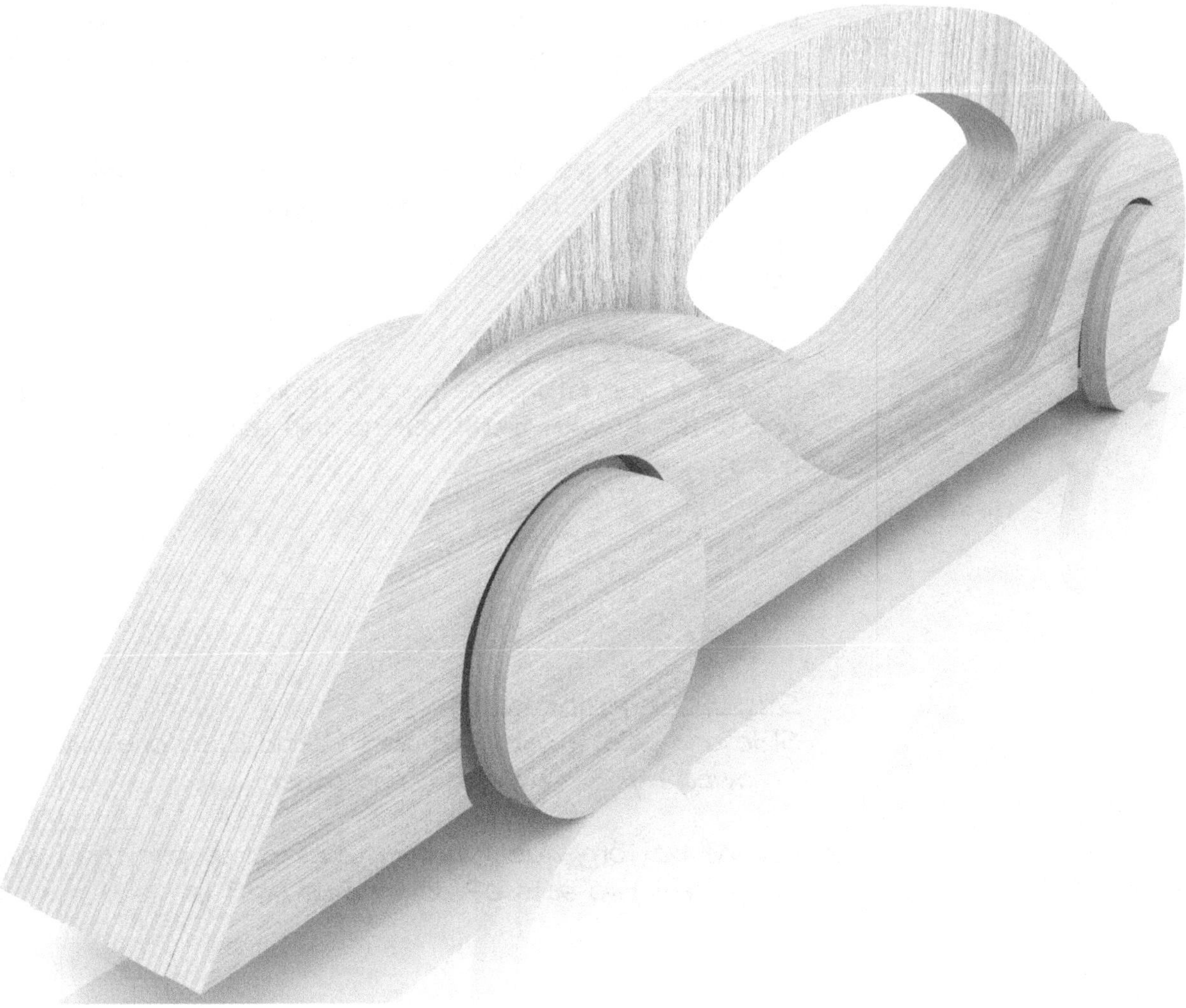

Completed 18 piece model dimensions @ 6 mm L: 35 cm, W: 4.5 cm, H: 10 cm
(1/4 inch L: 13-25/32, W: 1-49/64, H: 3-15/16)
Laser version found in WoodMarvels.com Volume 2: 50 Laser Cutting Plans (page 18)

Product Renders

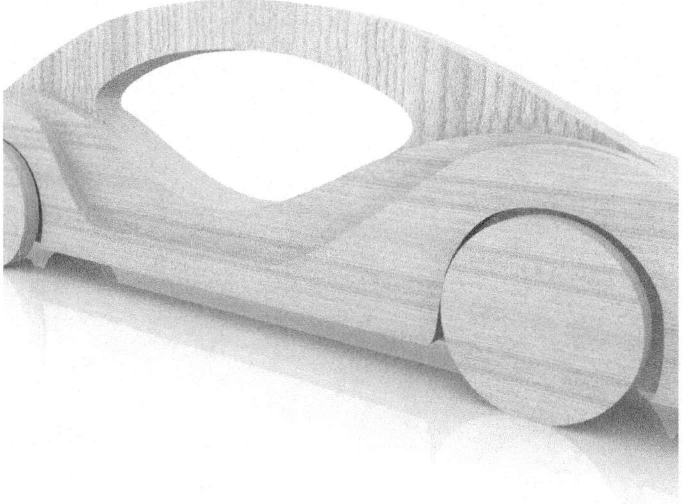

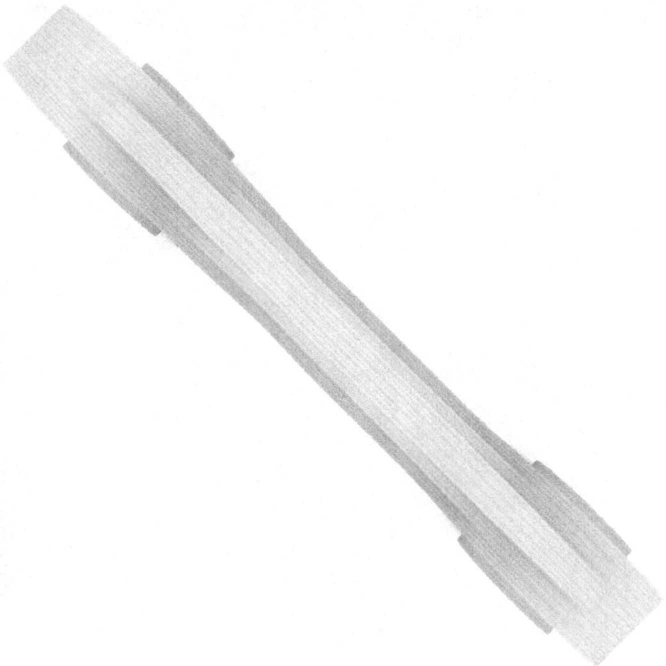

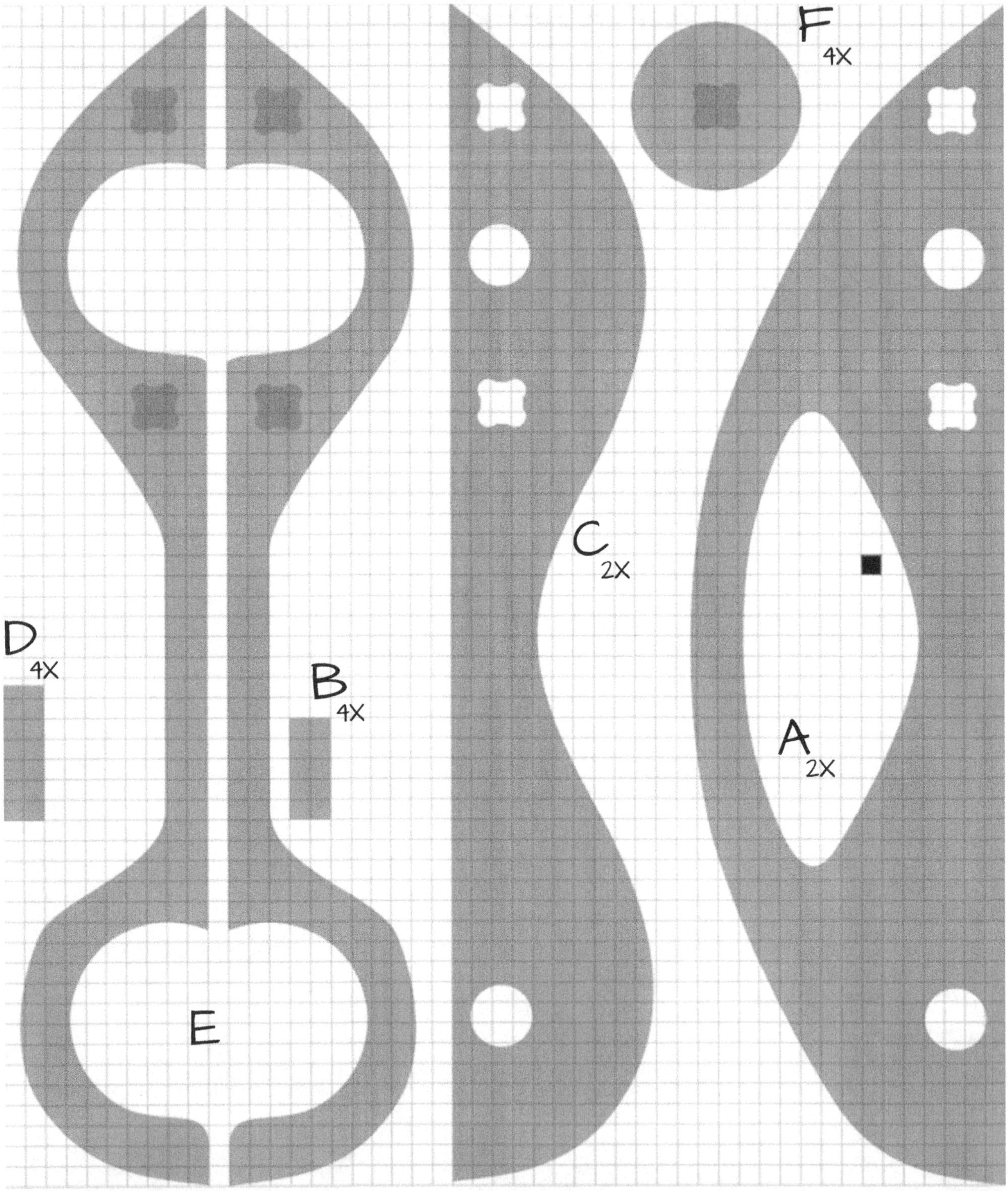
F
4X
C
2X
D
4X
B
4X
A
2X
E

# Step-by-Step Assembly Instructions

Step 1: Label and check inventory

Once you have all the pieces in front of you, the process of putting them together is no different from assembling pieces of Lego.

Label, using a pencil, the various parts and double-check your inventory. Take your time and enjoy the building process! Before jumping to the glue, be sure to do a dry run.

Step 2: Start the body

Glue and slide together both part As (raise roof panels) followed by inserting all four part Bs (short guides) into the square holes.

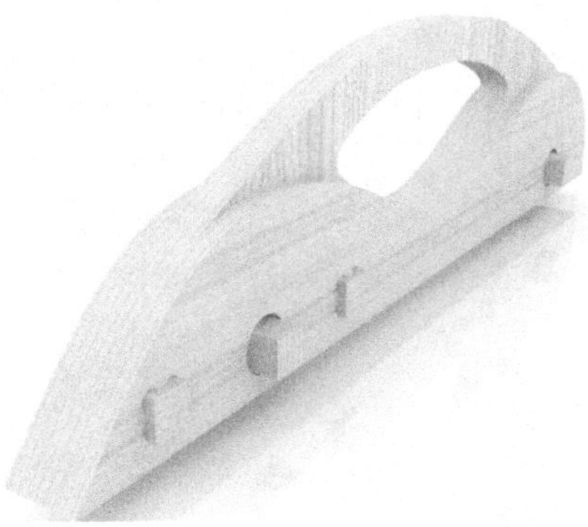

Step 3: Add sides

On both sides, glue and slide into place a part C (panel without roof). Insert into the round holes a set of glued and dried part Ds (axles).

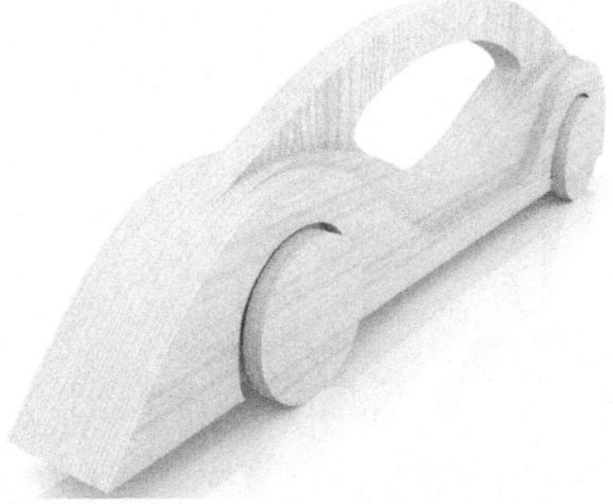

Step 4: Perform some finishing touches

At the end of both axles, glue into place part Fs (pocketed wheels) after you have glued into place part Es (pocked sides) on both sides.

There is no need to apply any preservatives or paint; feel free to sand any sharp edges down. ENJOY!

# Future Car B

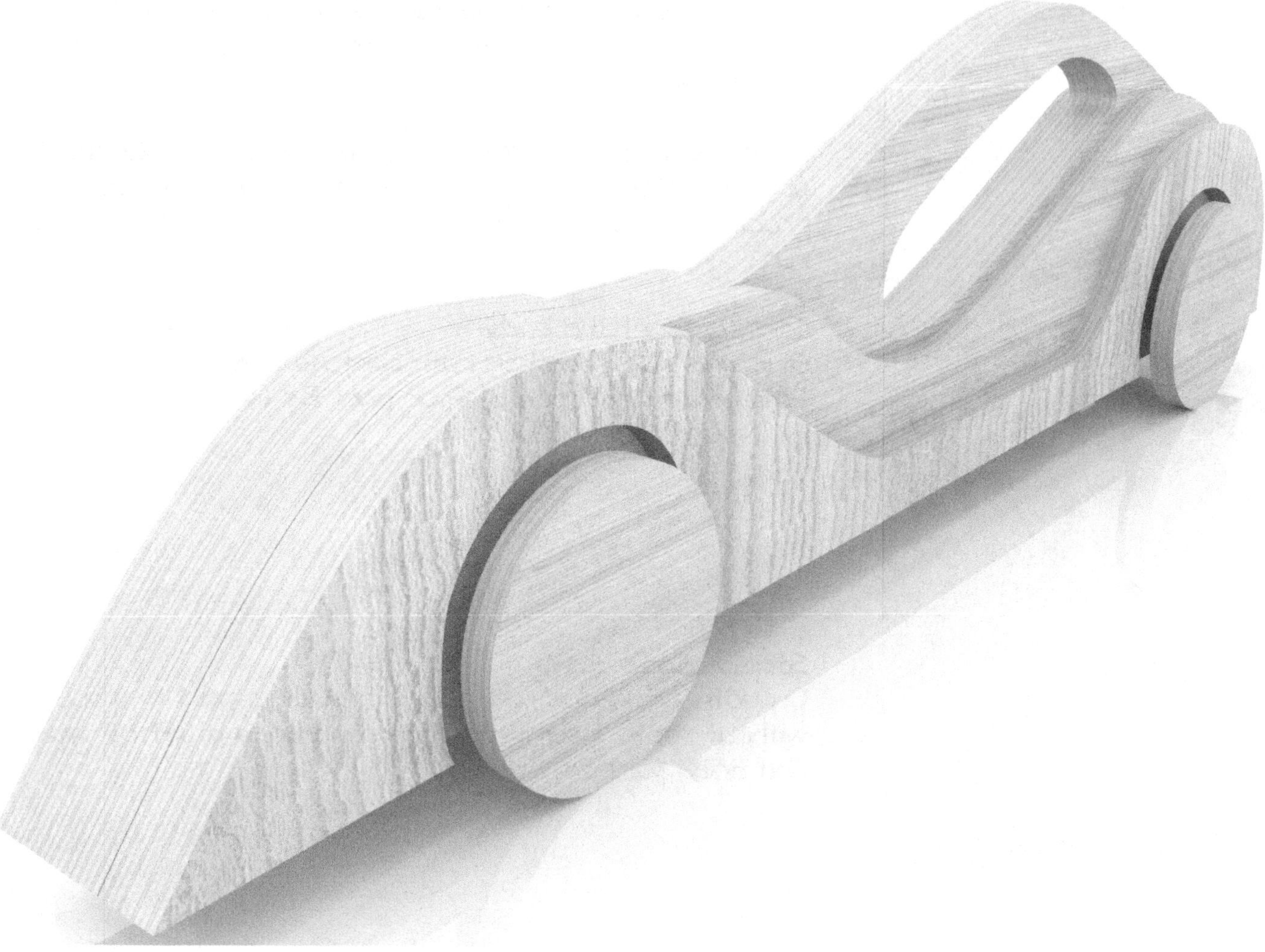

Completed 18 piece model dimensions @ 6 mm L: 37 cm, W: 4.5 cm, H: 10 cm
(1/4 inch L: 14-9/16, W: 1-49/64, H: 3-15/16)
Laser version found in WoodMarvels.com Volume 2: 50 Laser Cutting Plans (page 21)
Dowel version found in WoodMarvels.com Volume 1: Top Sellers (page 30)

## Product Renders

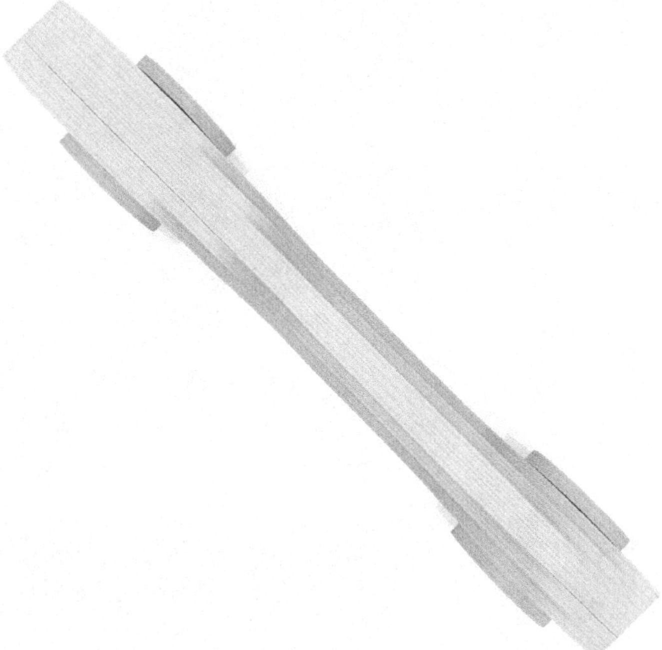

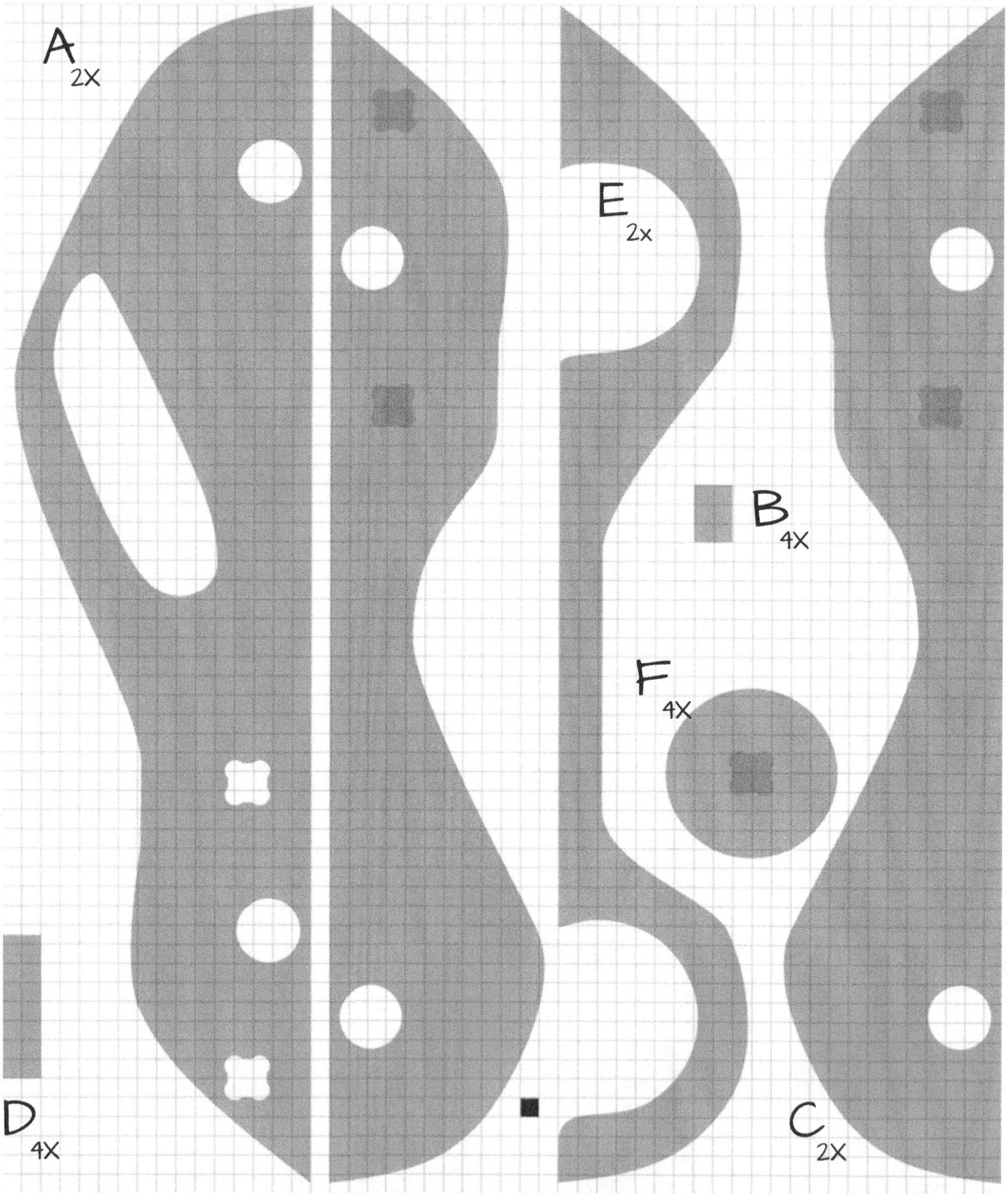
A
2X
E
2x
B
4X
F
4X
D
4X
C
2X

# Step-by-Step Assembly Instructions

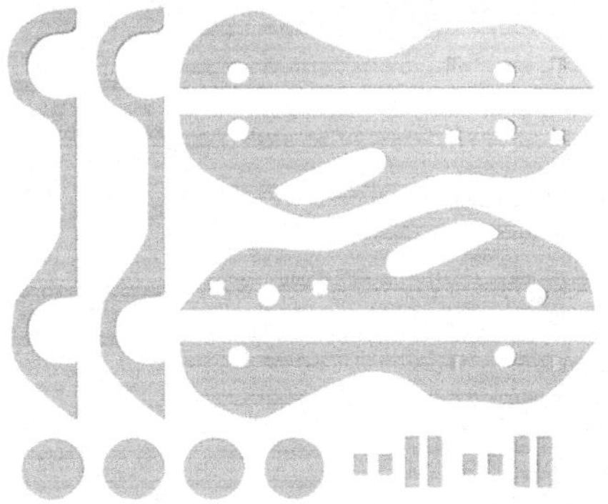

Step 1: Label and check inventory

Once you have all the pieces in front of you, the process of putting them together is no different from assembling pieces of Lego.

Label, using a pencil, the various parts and double-check your inventory. Take your time and enjoy the building process! Before jumping to the glue, be sure to do a dry run.

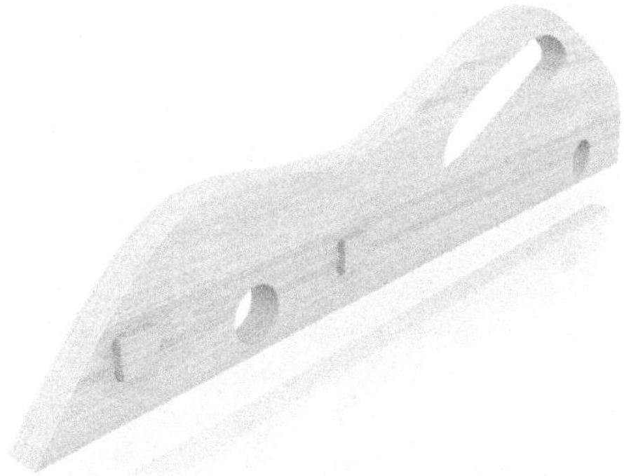

Step 2: Start the body

Glue and slide together both part As (raise roof panels) followed by inserting all four part Bs (short guides) into the square holes.

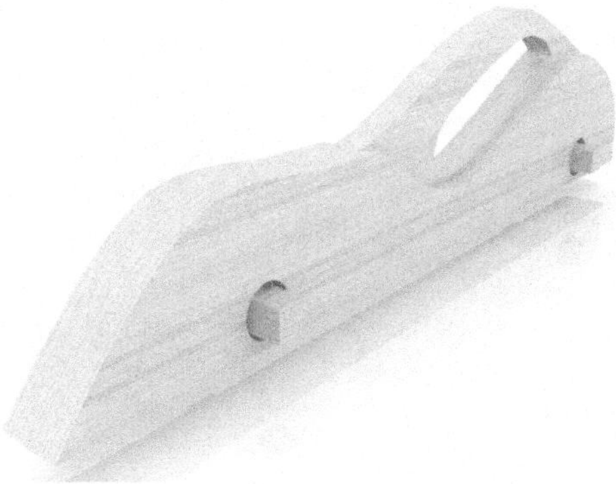

Step 3: Add sides

On both sides, glue and slide into place a part C (panel without roof). Insert into the round holes a set of glued and dried part Ds (axles).

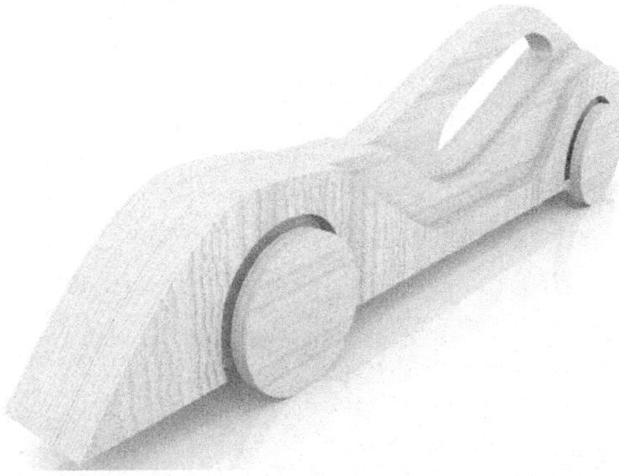

Step 4: Perform some finishing touches

At the end of both axles, glue into place the part Fs (pocketed wheels) after you have glued into place part Es (small side panels) on both sides.

There is no need to apply any preservatives or paint; feel free to sand any sharp edges down. ENJOY!

# Light Table

Completed 10 piece model dimensions @ 6 mm L: 46 cm, W: 30 cm, H: 28 cm
(1/4 inch L: 18-7/64, W: 11-3/16, H: 11-1/32)
Laser version found in WoodMarvels.com Volume 2: 50 Laser Cutting Plans (page 112)

Product Renders

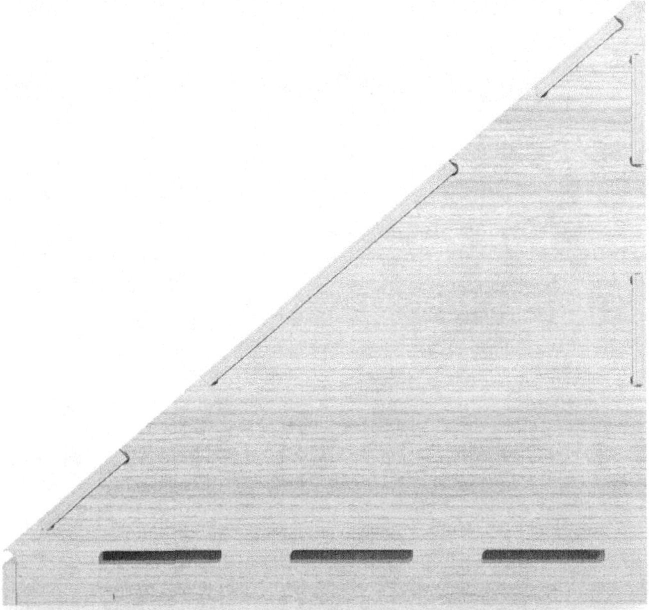

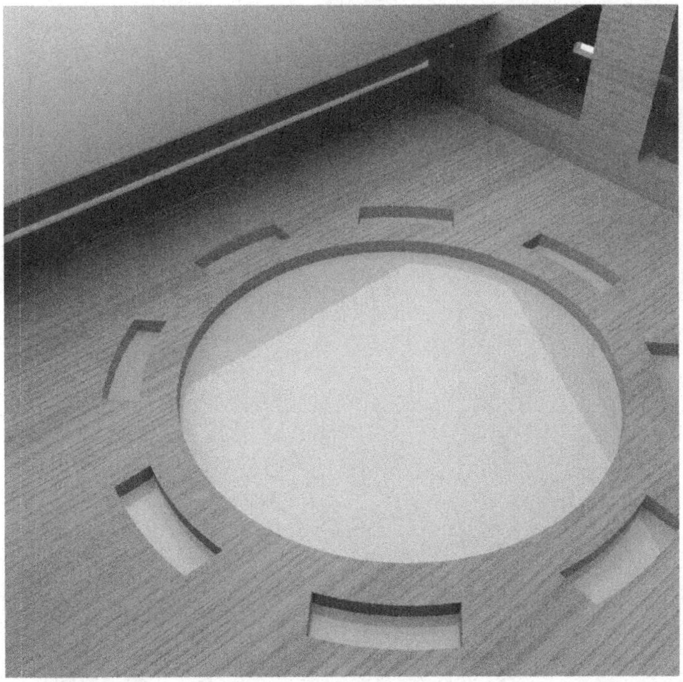

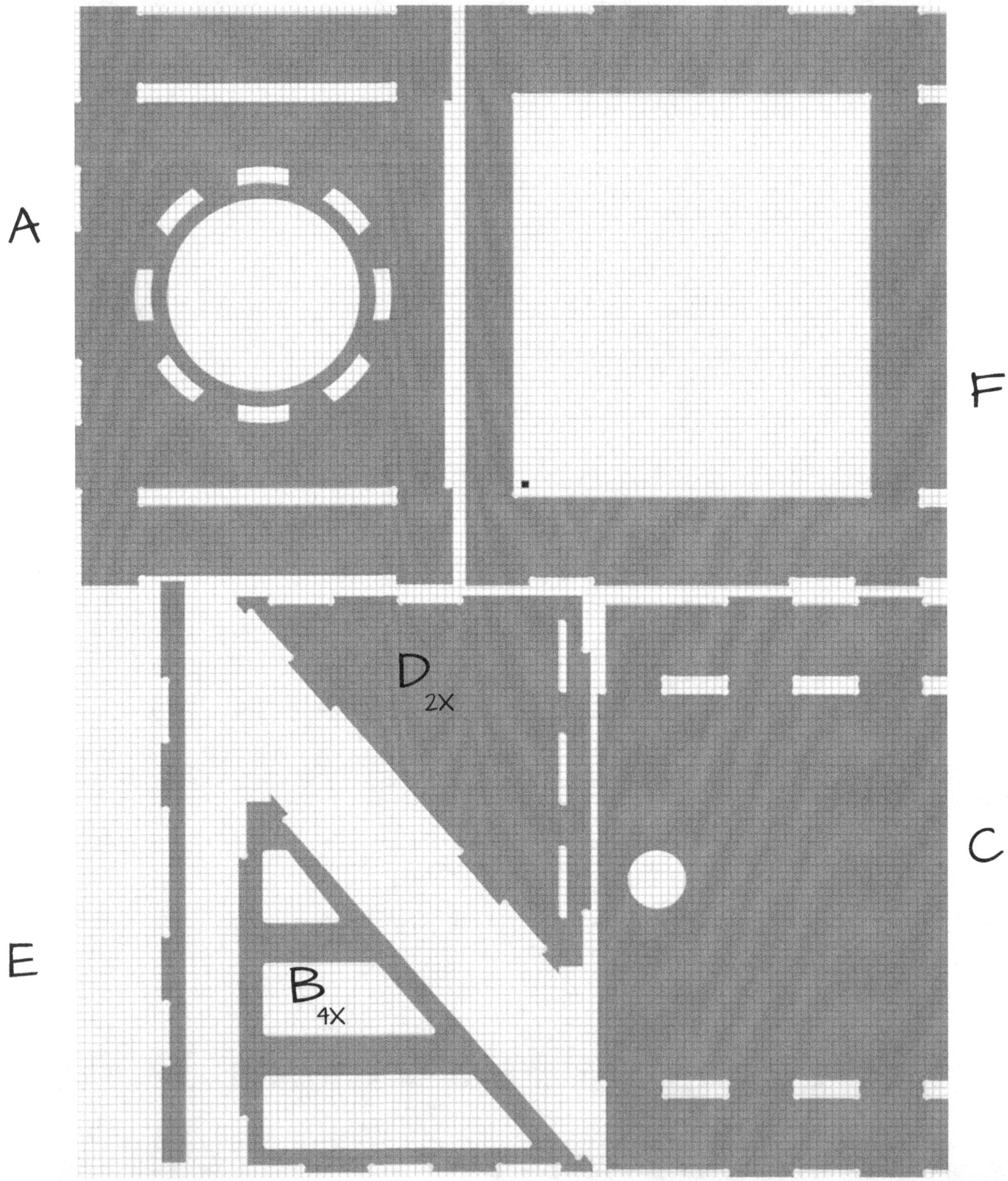
A
F
D
2X
C
E
B
4X

# Step-by-Step Assembly Instructions

Step 1: Label and check inventory

Once you have all the pieces in front of you, the process of putting them together is no different from assembling pieces of Lego.

Label, using a pencil, the various parts and double-check your inventory. Take your time and enjoy the building process! Before jumping to the glue, be sure to do a dry run.

Step 2: Start the foundation

Slide and glue into place two doubled up part Bs (internal supports) into part A (base).

Step 3: Glue supports

Slide and glue part C (back support) onto the back; on each side, glue a part D (edges) and in front, part E (front support).

Step 4: Finish the bed

Slide and glue into place part F on the top and then, without any glue, the glass or acrylic sheet fitting into the hole.

There is no need to apply any preservatives or paint; feel free to sand any sharp edges down. ENJOY!

# Mangonel Siege Weapon

Completed 50 piece model dimensions @ 6 mm L: 30 cm, W: 16 cm, H: 12 cm
(1/4 inch L: 11-3/16, W: 6-19/64, H: 4-23/32)
Laser version found in WoodMarvels.com Volume 3: Evolution of Wooden Designs (page 52)

## Product Renders

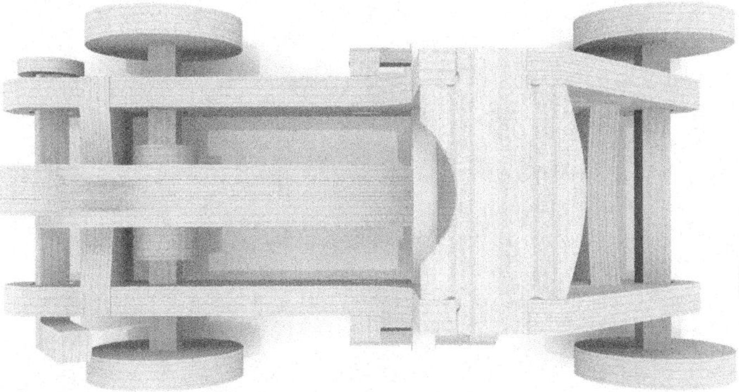

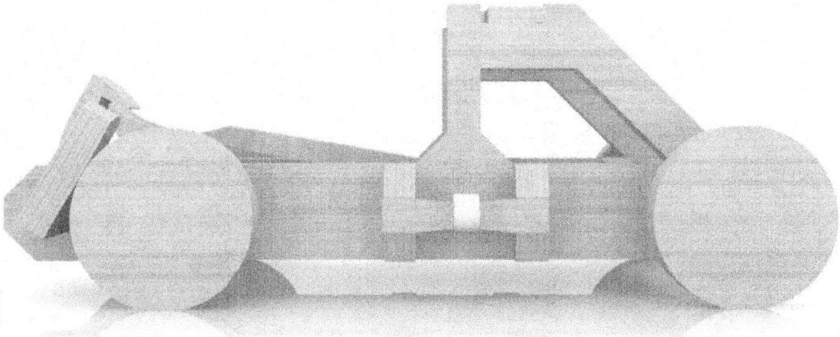

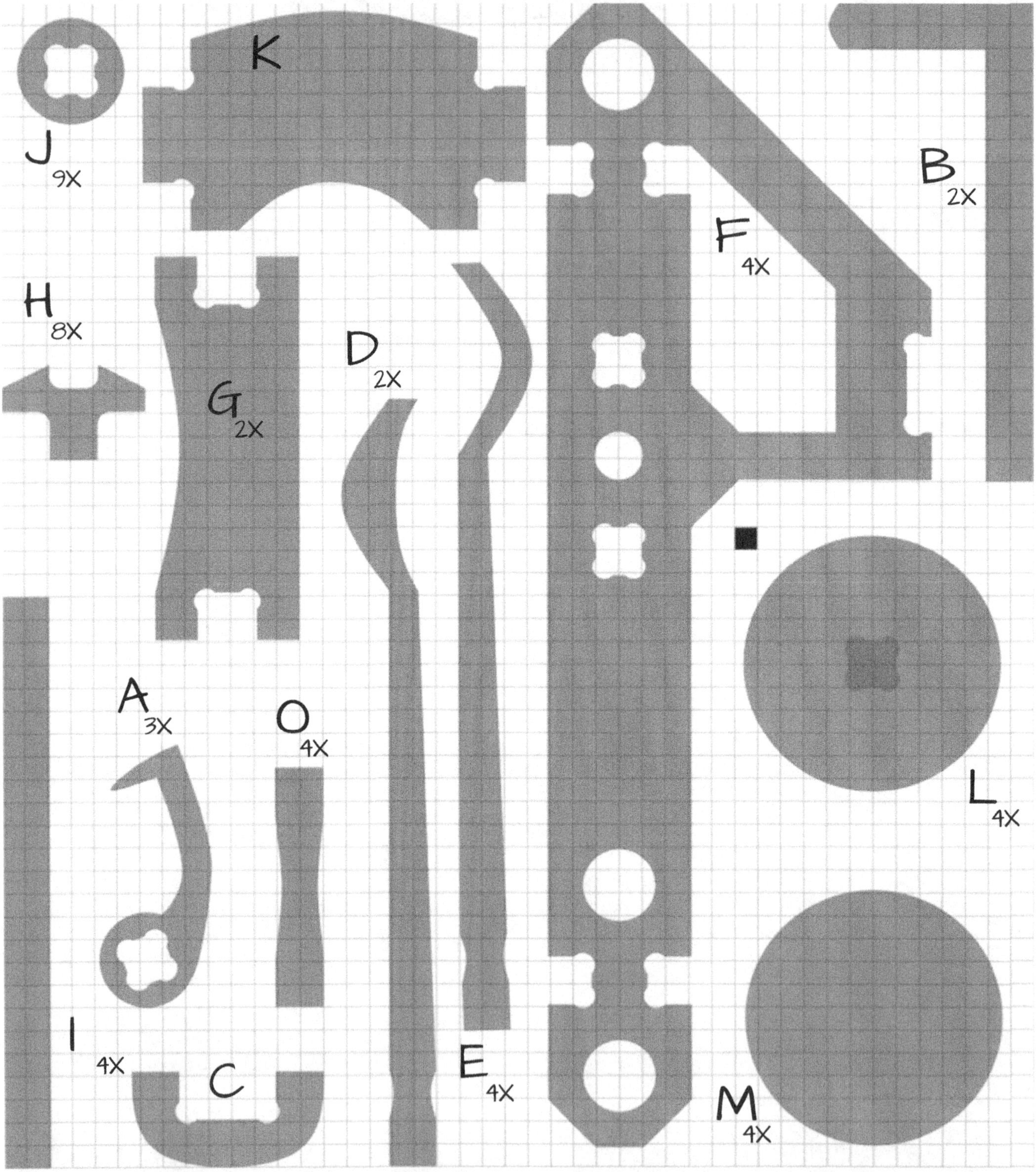
K
J
9X
B
2X
F
4X
H
8X
D
2X
G
2X
A
3X
O
4X
L
4X
I
4X
C
E
4X
M
4X

# Step-by-Step Assembly Instructions

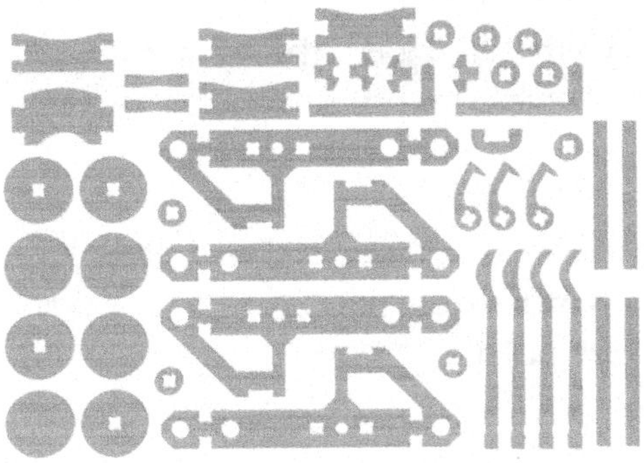

**Step 1: Label and check inventory**
Once you have all the pieces in front of you, the process of putting them together is no different from assembling pieces of Lego.

Label, using a pencil, the various parts and double-check your inventory. Take your time and enjoy the building process! Before jumping to the glue, be sure to do a dry run.

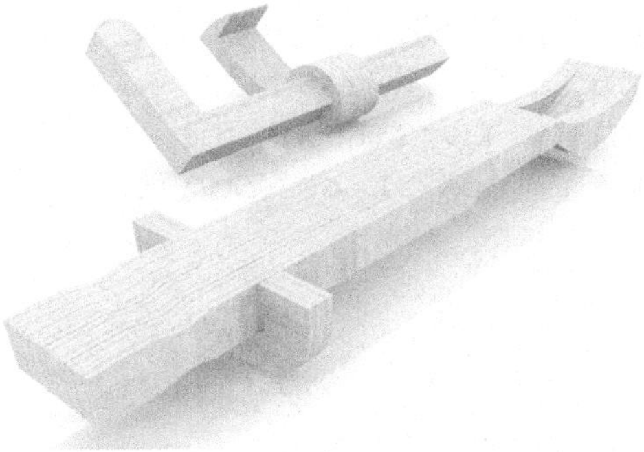

**Step 2: Glue together moving parts**
Slide and glue together both part As (hooks) onto doubled up part Bs (triggers).

In the trough of part C (catapult holder,) slide and glue into place two part Ds (outside edge) with two part Es (inside) between the part Ds.

**Step 3: Doulbe up major components**
Glue together doubled up part Fs (two sets of sides,) Gs (four sets of T-slots,) and Hs (two sets of cross-members).

**Step 4: Start assembling**
In between both doubled up part Fs, slide and glue into place both doubled up part Hs (front and back) and a doubled up part I (axle) in the back with eight glued part Js through it.

On the top, slide and glue into place a part K (catapult stop) and, on the sides, all four doubled up part Gs.

Step 5: Add accessories

In the back, slide into place (without glue) the trigger assembly with a glued stop (part J) at the other end to keep it there.

On the front, slide into place doubled up part Is (axles,) and then slide into place on both sides (back as well,) part L (wheel, holed) and M (wheel, solid).

Step 6: Add the elastic

Using both part Os on either side to hold the elastic in place, slide into the elastic the catapult assembly you glued earlier. To increase tension, just pull out one of the part Os and twist it as many revolutions as you wish.

Step 7: Perform some finishing touches

Allow everything to dry overnight.

There is no need to apply any preservatives or paint; feel free to sand any sharp edges down. ENJOY!

# Mangonel B Siege Weapon

Completed 104 piece model dimensions @ 6 mm L: 52 cm, W: 25 cm, H: 17 cm
(1/4 inch L: 20-15/32, W: 9-27/32, H: 6-11/16)
Laser version found in WoodMarvels.com Volume 3: Evolution of Wooden Designs (page 101)

Product Renders

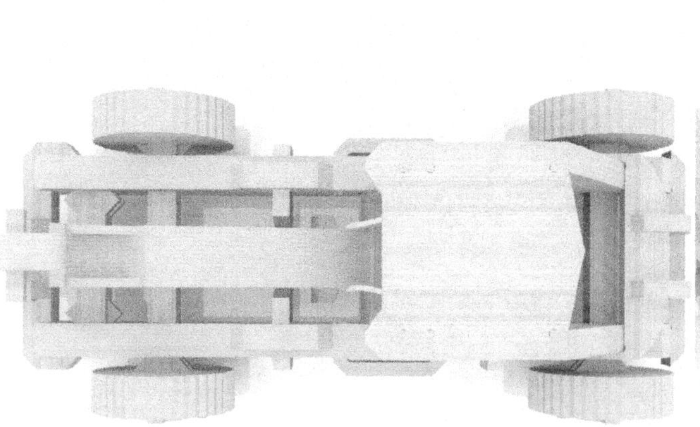

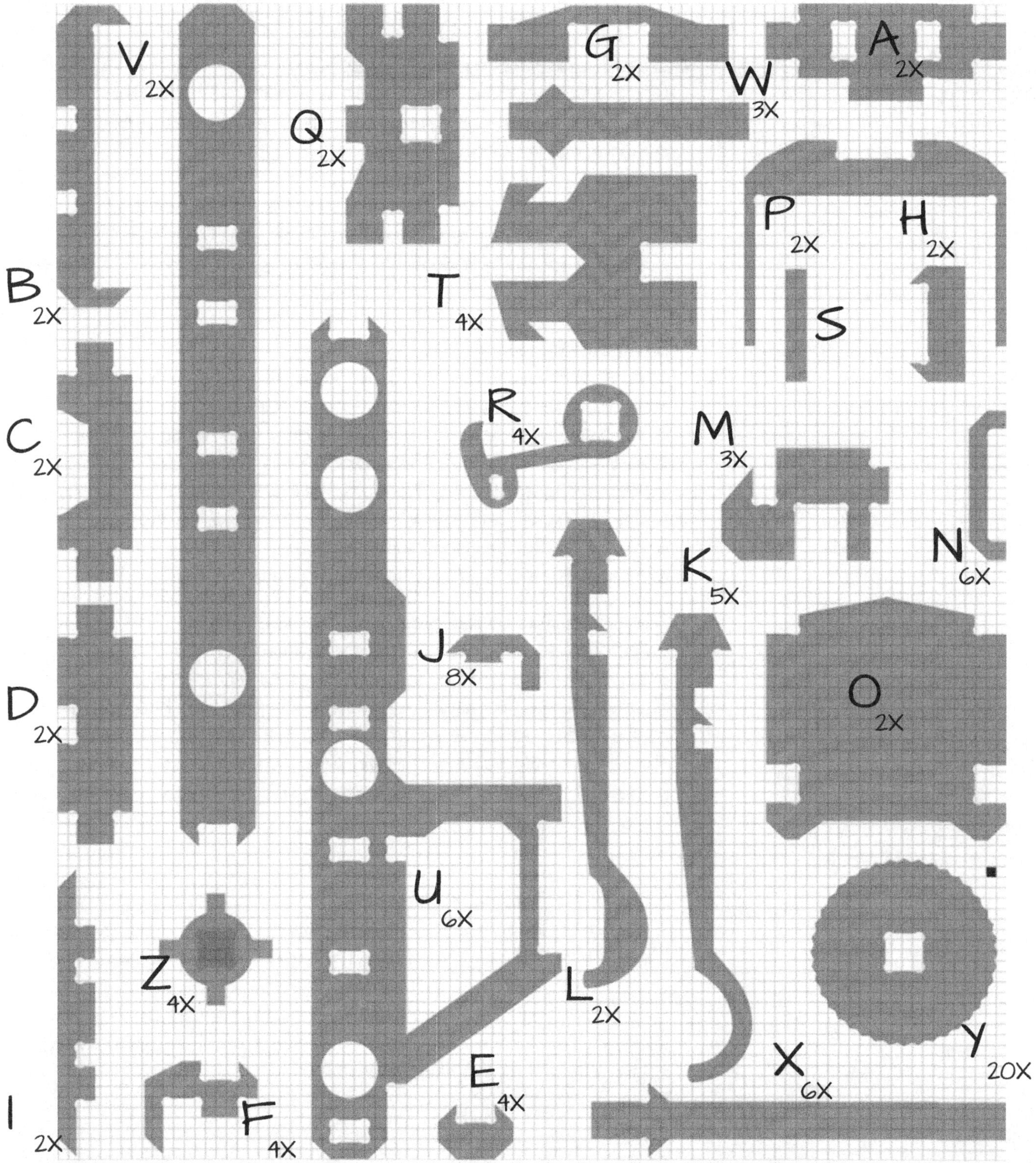

V 2X
G 2X
W 3X
A 2X
Q 2X
P 2X
H 2X
B 2X
T 4X
S
R 4X
M 3X
C 2X
N 6X
K 5X
J 8X
O 2X
D 2X
U 6X
Z 4X
L 2X
Y 20X
X 6X
E 4X
I 2X
F 4X

# Step-by-Step Assembly Instructions

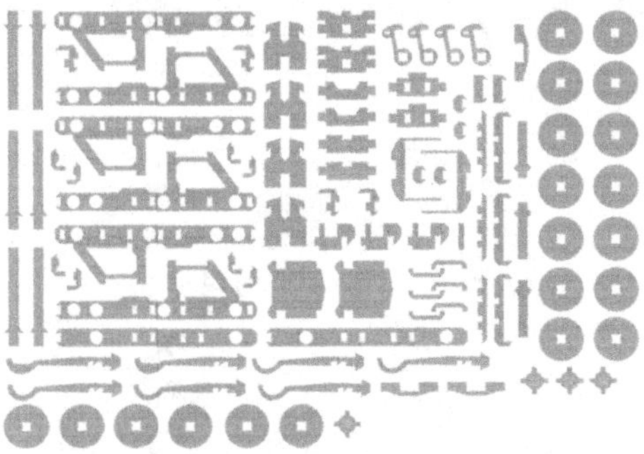

Step 1: Label and check inventory

Once you have all the pieces in front of you, the process of putting them together is no different from assembling pieces of Lego.

Label, using a pencil, the various parts and double-check your inventory. Take your time and enjoy the building process! Before jumping to the glue, be sure to do a dry run.

Step 2: Double up parts

There are many parts that need to be doubled up, including parts A (two-hole front support,) B (back horizontal bumper,) C (pole holder,) D (middle support,) two Es (back vertical bumper,) two Fs (front vertical bumper,) G (pole axis,) H (pole support,) I (front bumper,) eight Js (side bumpers,) five Ks in between two part Ls (catapult,) three Ms (cross-member support,) and two sets of three Ns (long side bumpers).

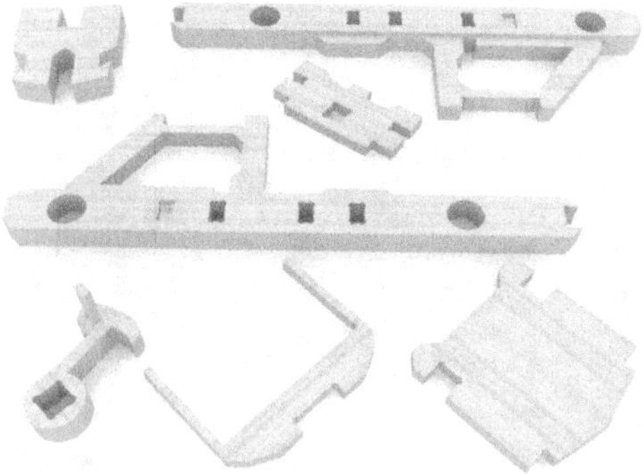

Step 3: Make more doubled up parts

Continue doubling up parts including O (top,) P (front top bumper,) Q (one-hole back support,) four Rs with a part S as a guide (trigger,) and four Ts (elastic holder). To make the two large sides of the mangonel, glue together three part Us (side with window) with a part V (side bumper). Make sure you mirror this so the bumpers will be on the outside.

Step 4: Form the catapult

The catapult is the most important part of the mangonel. It's built by gluing together the dried parts above.

Put the K and L assembly you made earlier on a flat surface and slide and glue into place part G from the top and part H on the underside.

Step 5: Form the elastic holder

Starting with the front, slide and glue into place the following assemblies: Part D (middle support) then into it, M (cross-member support) with T (elastic holder) sliding in from above with Q (one-hole back support) sliding into M's back.

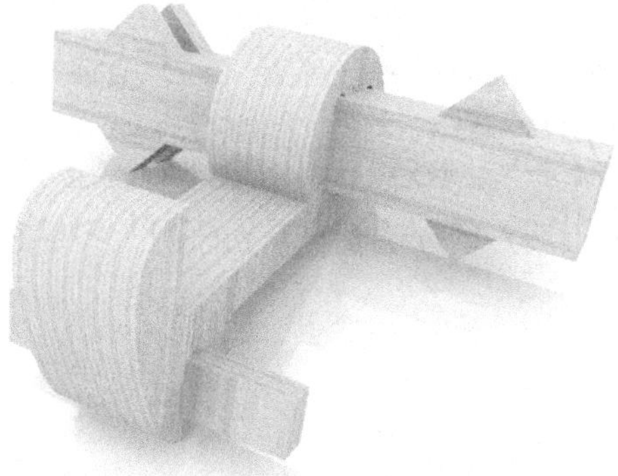

Step 6: Build the trigger

Starting with the trigger assembly you built in step 3, slide and glue alternating part Ws into the tigger hole.

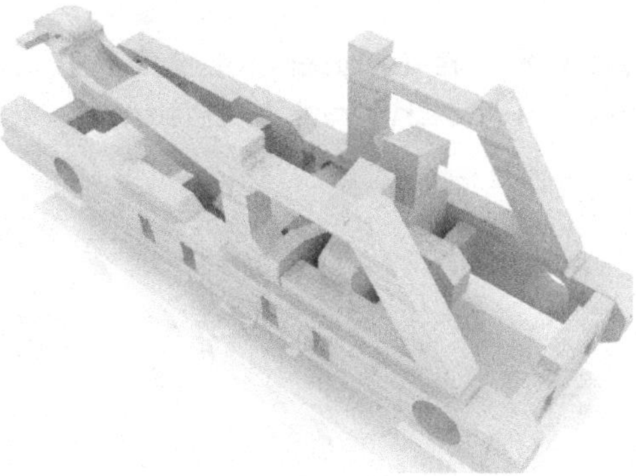

Step 7: Build the foundation

Starting from the front, slide and glue (in between both mirrored side assemblies) part A, the assembly built in step 5, the assembly built in step 4, and part C, followed by step 6's assembly.

For the trigger, do not apply any glue, as you want this to freely rotate.

Step 8: Build the front bumper

Starting on the top, slide and glue into place the part O assembly; then on the front, drop the part P assembly. Then slide and glue into place both doubled up part F assemblies followed by the part I assembly.

Step 9: Build the back bumper

Starting in the middle, slide and glue into place both part N assemblies you created in step 2. On the back, slide and glue into place the part B assembly followed by the doubled up part E assemblies.

Step 10: Add axles

On both sides of the part N assemblies, slide and glue into place a part J assembly. As for the axles, slide and glue into place alternating part Xs together, but make sure to rotate the axles as they dry so they don't stick in the holes.

Step 11: Add wheels

Slide and glue into place five part Ys (wheels,) followed by a part Z (cap) at the end of each part X assembly.

Step 8: Perform some finishing touches

Allow everything to dry overnight and then slide into place the elastic bands you wish to use to power the Mangonel.

There is no need to apply any preservatives or paint; feel free to sand any sharp edges down. ENJOY!

# Medieval Castle Bank

Completed 27 piece model dimensions @ 6 mm L: 19 cm, W: 15 cm, H: 14 cm
(1/4 inch L: 7-31/64, W: 5-29/32, H: 5-33/64)
Laser version found in WoodMarvels.com Volume 3: Evolution of Wooden Designs (page 56)

## Product Renders

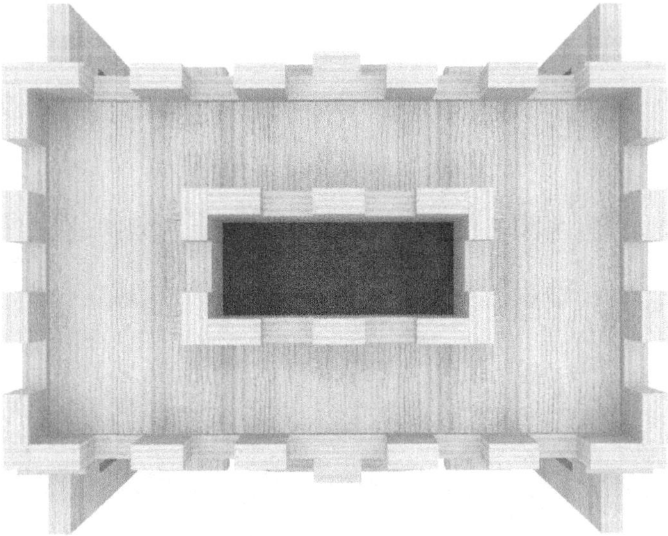

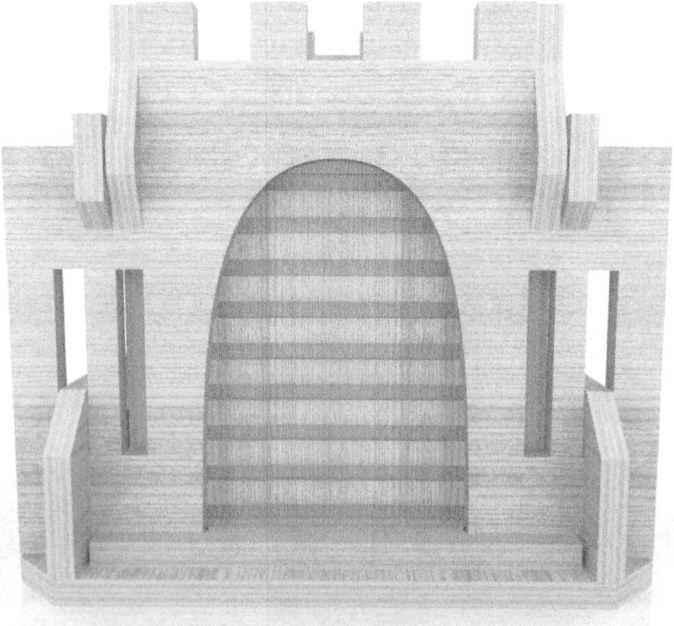

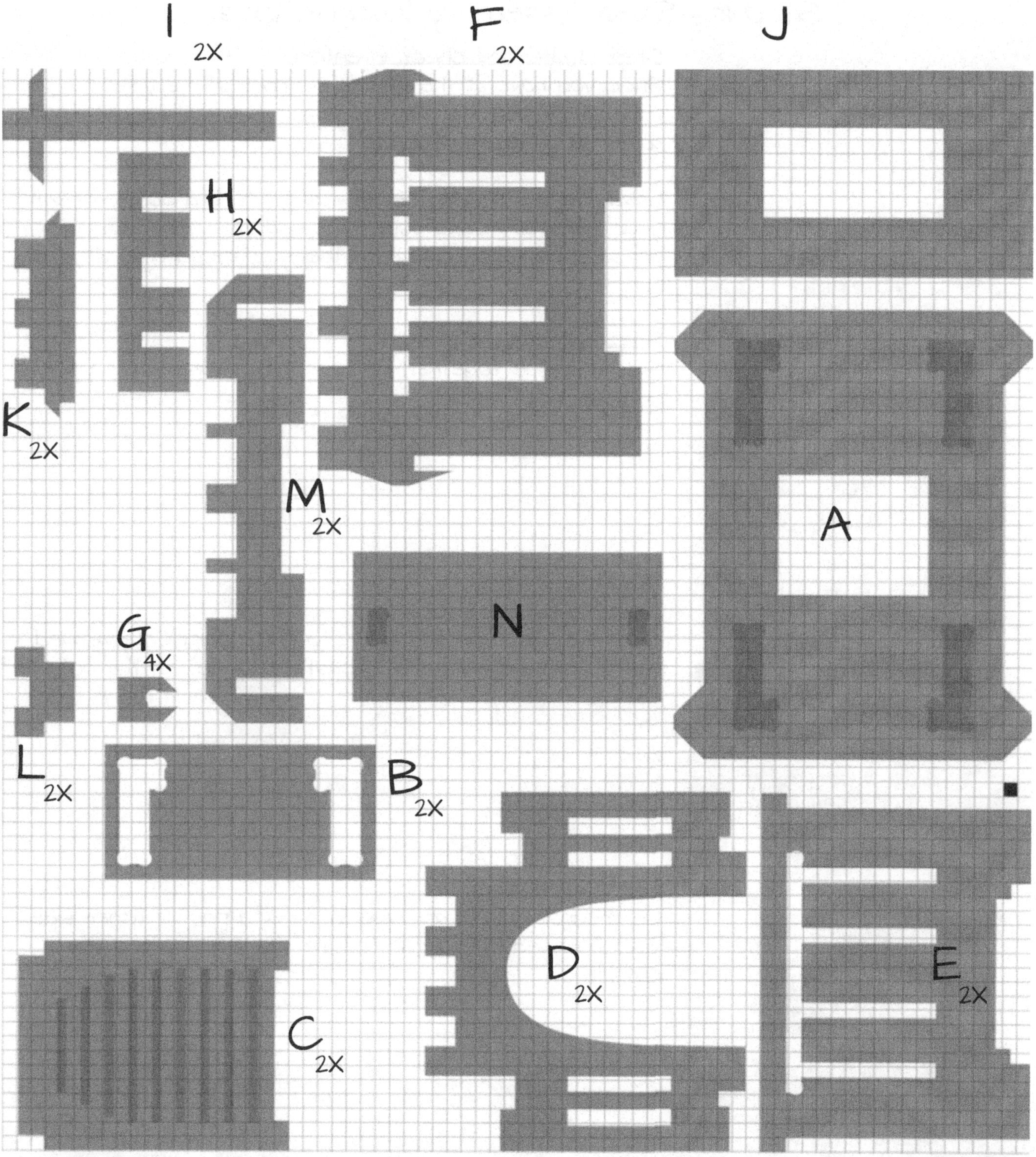
I
2X
F
2X
J
H
2X
K
2X
M
2X
A
N
G
4X
L
2X
B
2X
D
2X
E
2X
C
2X

# Step-by-Step Assembly Instructions

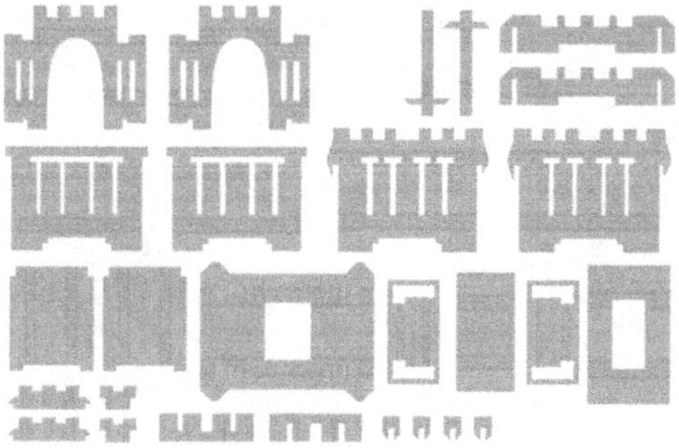

Step 1: Label and check inventory

Once you have all the pieces in front of you, the process of putting them together is no different from assembling pieces of Lego.

Label, using a pencil, the various parts and double-check your inventory. Take your time and enjoy the building process! Before jumping to the glue, be sure to do a dry run.

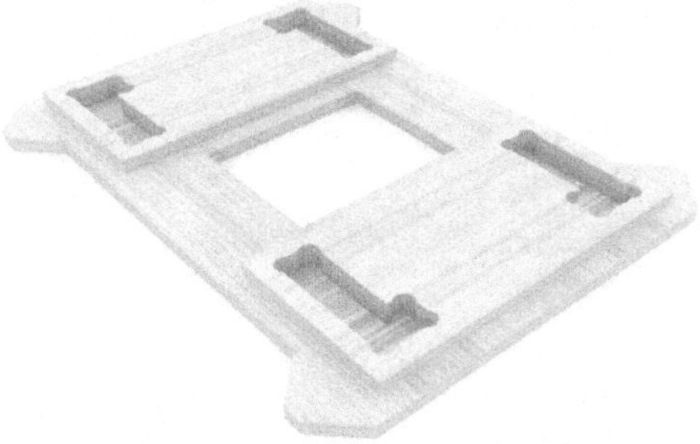

Step 2: Start the base

Slide and glue together both part Bs (base wall supports) onto part A (base).

Step 3: Add the doors

On the inside, slide and glue into place part C (engraved door) followed by part D (castle facade) on the outside.

Step 4: Add sides

On the inside, slide and glue into place part E (inner side wall) followed by part F (castle side facade) on the outside.

Step 5: Add accessories

On each of the corners to lock into place the side walls, glue and slide into place a part G. From the inside, glue and slide into place part H (sword support) while on the outside, slide into place part I (swords) without glue.

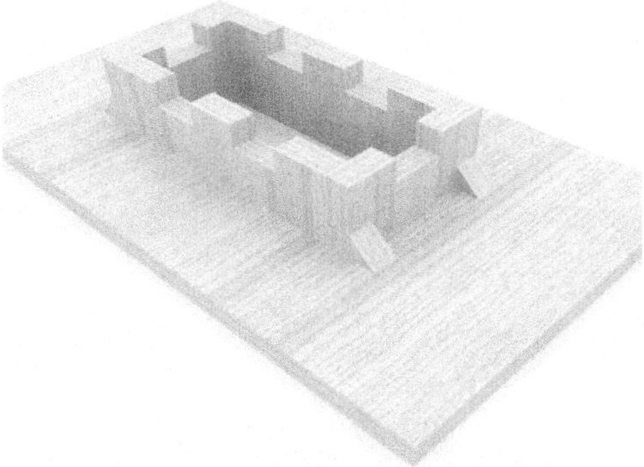

Step 6: Create the cover

To create the top, glue and slide parts K (side) and L (front/back) onto part J (base).

Step 7: Add fences

Slide and glue into place part M (fence) on both sides, but be sure not to glue it to the sword.

Step 8: Perform some finishing touches

Allow everything to dry overnight.

There is no need to apply any preservatives or paint; feel free to sand any sharp edges down. ENJOY!

# P38 Lightning Aircraft

Completed 38 piece model dimensions @ 6 mm L: 37 cm, W: 48 cm, H: 10 cm
(1/4 inch L: 14-9/16, W: 18-57/64, H: 3-15/16)
Laser version found in WoodMarvels.com Volume 3: Evolution of Wooden Designs (page 60)

Product Renders

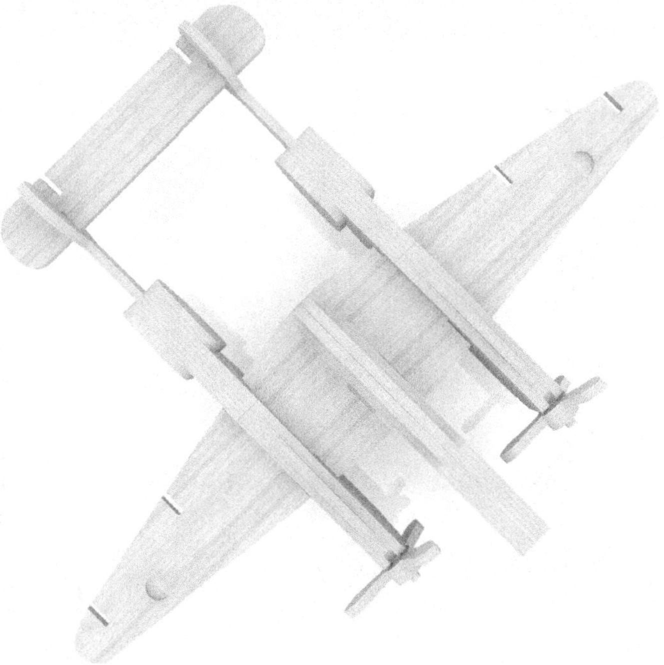

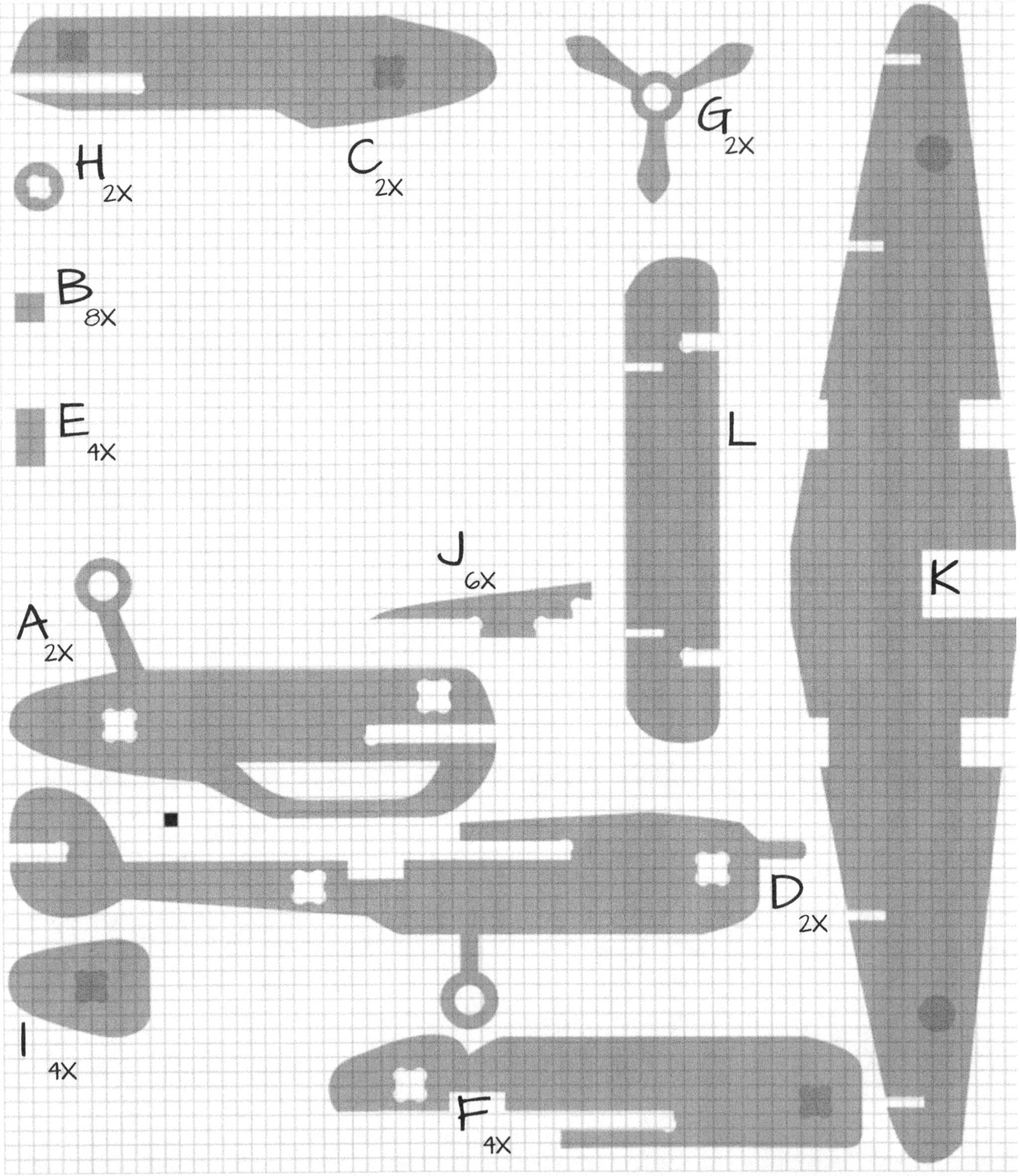
C 2X
G 2X
H 2X
B 8X
E 4X
L
J 6X
K
A 2X
D 2X
I 4X
F 4X

# Step-by-Step Assembly Instructions

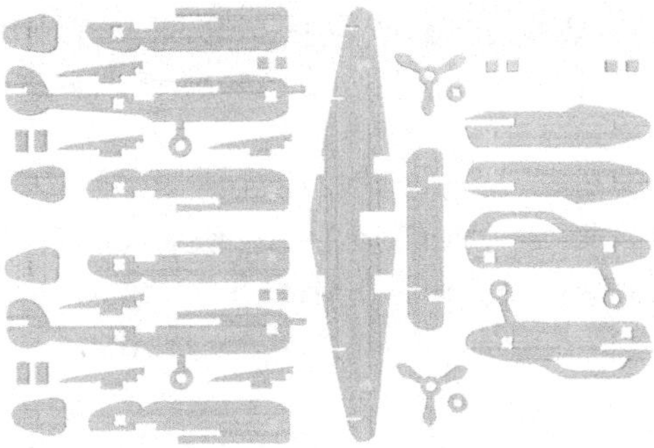

Step 1: Label and check inventory

Once you have all the pieces in front of you, the process of putting them together is no different from assembling pieces of Lego.

Label, using a pencil, the various parts and double-check your inventory. Take your time and enjoy the building process! Before jumping to the glue, be sure to do a dry run.

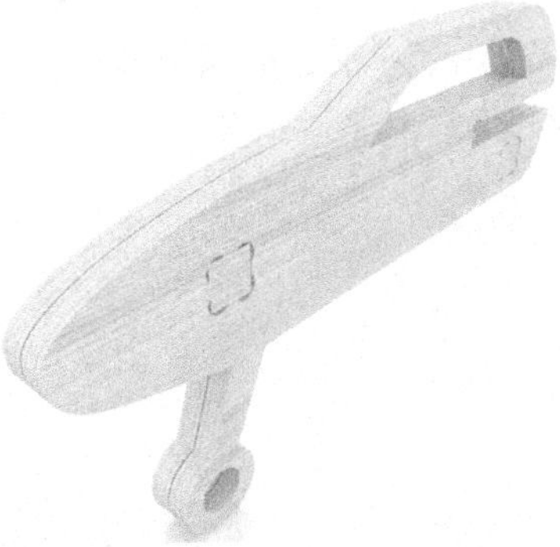

Step 2: Start the cockpit

Slide and glue to one another both part As (cockpit with wheel) and, inside, the two holes, a total of four part Bs (small guides).

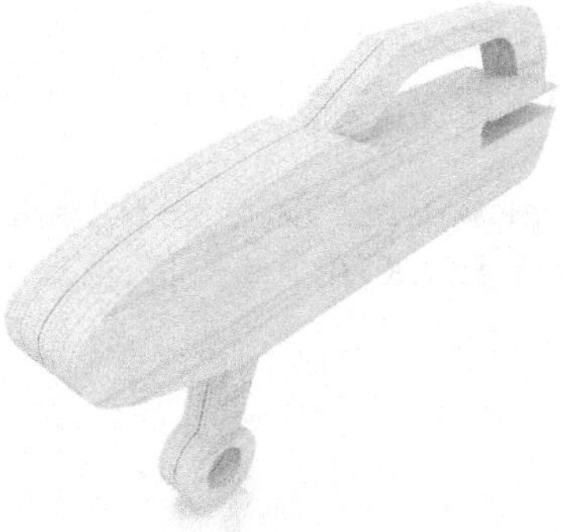

Step 3: Finish the cockpit

Glue into place part Cs (cockpit without wheels) on both sides to complete the cockpit.

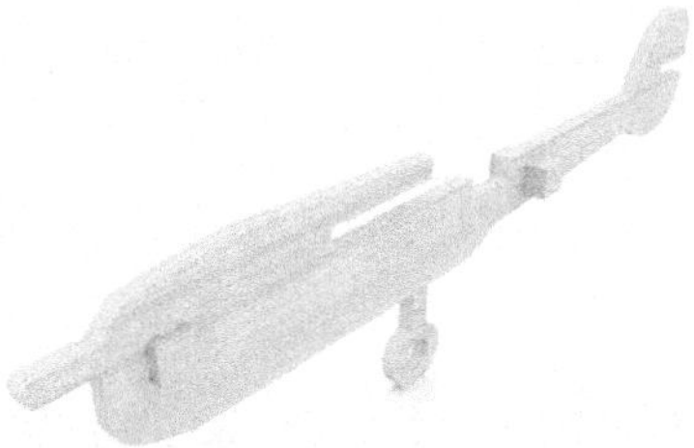

Step 4: Start the engine assembly

Using a part D, in the front of the engine where the propeller is, glue and slide into place two part Bs (small guides).

In the back, glue and slide into place two part Es.

Step 5: Add engine props

On both sides of the engine, slide and glue into place a part F (engine side) followed by a part G (propeller) and H (propeller cap) in the front.

Do not glue the propeller in place, and leave enough room between the cap and the engine edge so that it can freely rotate. Duplicate steps 4 and 5 to make another engine.

Step 6: Add the wing

Slide and glue into place part K (wing) and then glue into place all three part Js (top caps) and Is (side caps) to complete the engine assemblies.

Step 7: Perform some finishing touches

Glue and slide into place the finished cockpit and part L (tail) in the back. Allow everything to dry overnight.

There is no need to apply any preservatives or paint; feel free to sand any sharp edges down. ENJOY!

## Parasaurolophus

Completed 17 piece model dimensions @ 6 mm L: 36 cm, W: 4.5 cm, H: 17 cm
(1/4 inch L: 14-11/64, W: 1-49/64, H: 6-11/16)
Laser version found in WoodMarvels.com Volume 2: 50 Laser Cutting Plans (page 27)
Dowel version found in WoodMarvels.com Volume 1: Top Sellers (page 33)

Product Renders

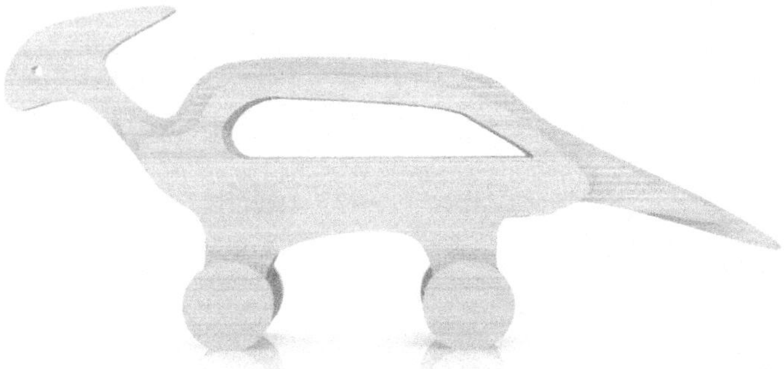

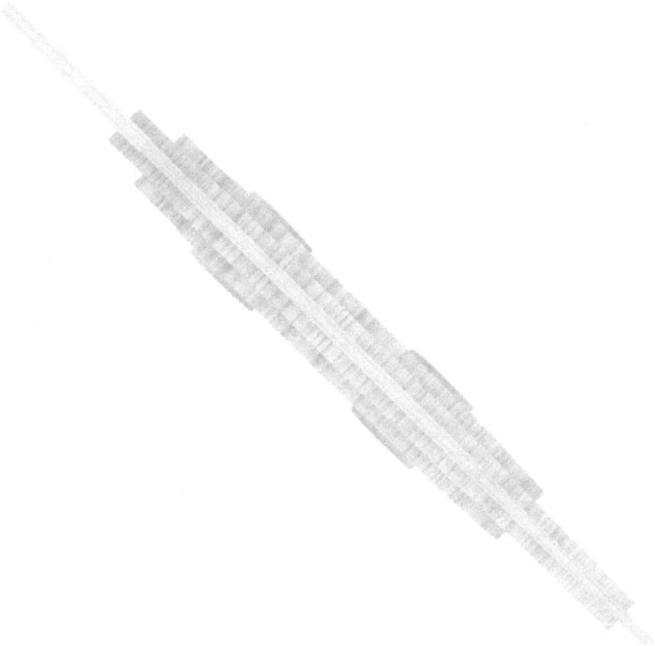

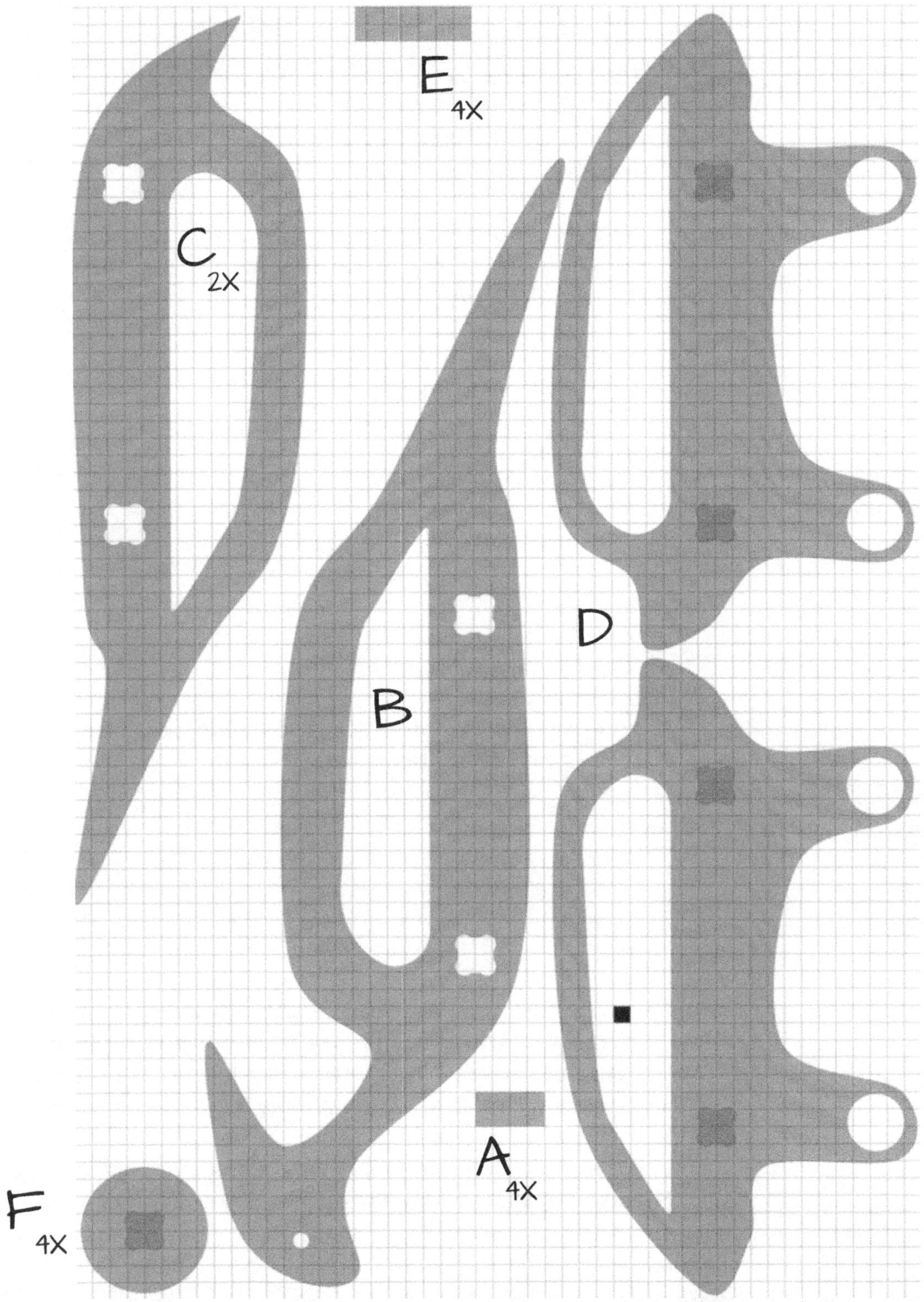
E
4X
C
2X
D
B
A
4X
F
4X

# Step-by-Step Assembly Instructions

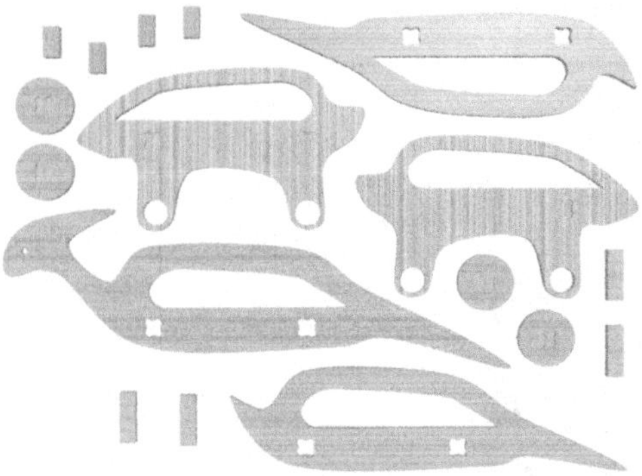

Step 1: Label and check inventory

Once you have all the pieces in front of you, the process of putting them together is no different from assembling pieces of Lego.

Label, using a pencil, the various parts and double-check your inventory. Take your time and enjoy the building process! Before jumping to the glue, be sure to do a dry run.

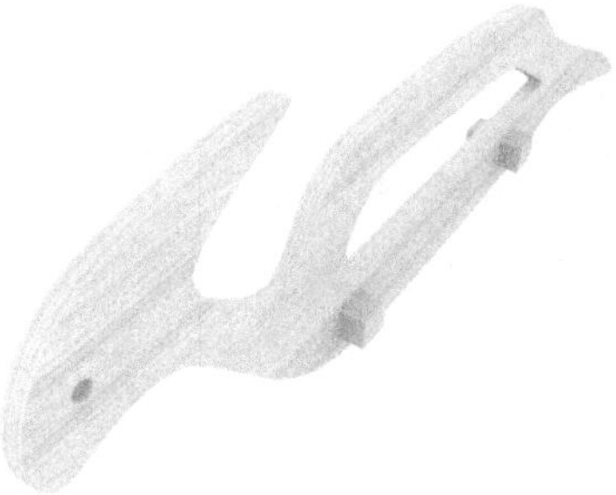

Step 2: Start the body

Slide and glue into place part A (guides) into part B (body with head).

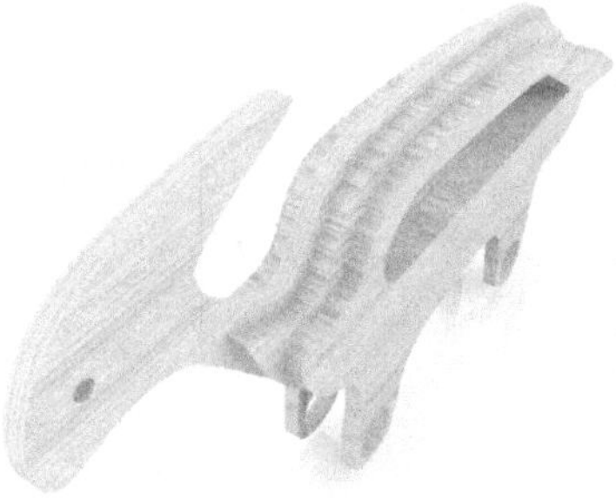

Step 3: Add sides

Using part As as internal guides, slide and glue into place parts C and D on both sides.

Step 4: Perform some finishing touches

In the holes found in the part Ds, slide into place doubled up part Es (axles) with part Fs at both ends (wheels). Allow everything to dry overnight.

There is no need to apply any preservatives or paint; feel free to sand any sharp edges down. ENJOY!

# Pen and Pencil Holder B

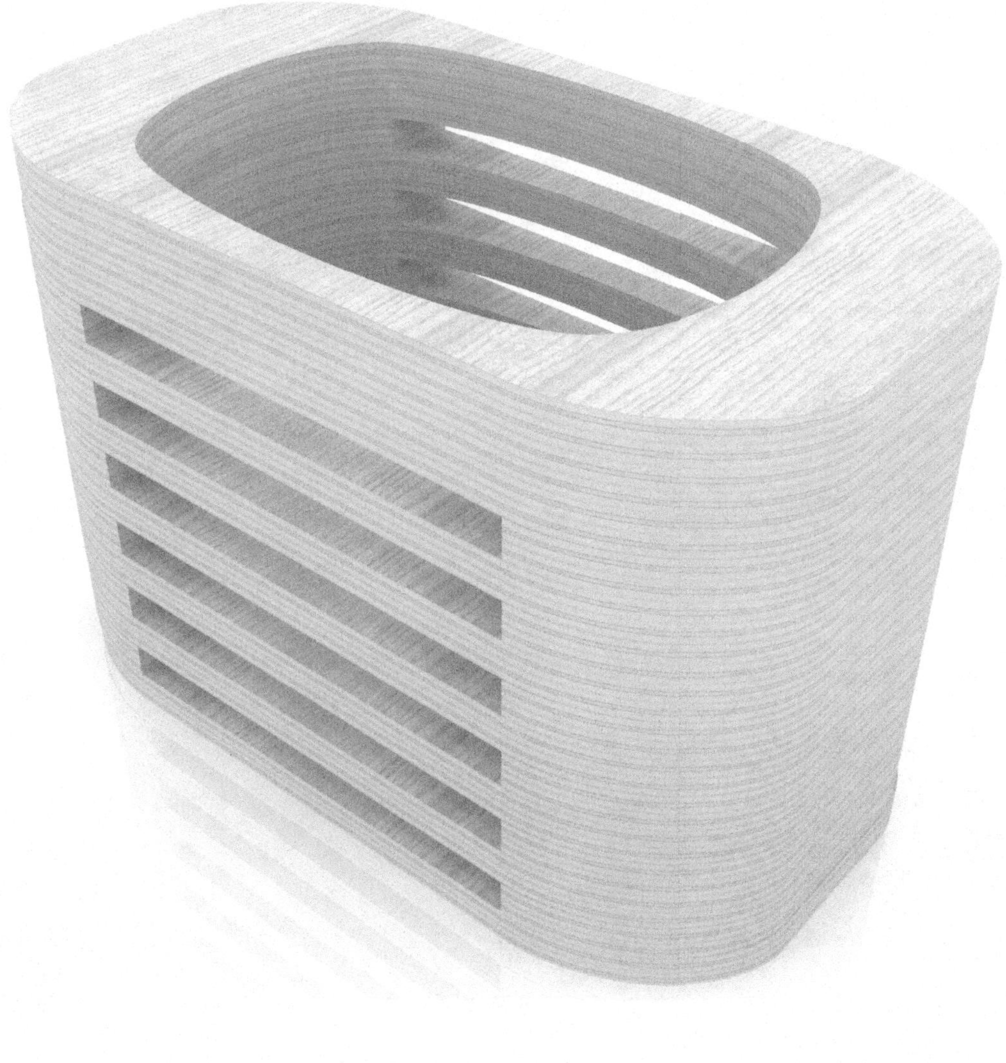

Completed 19 piece model dimensions @ 6 mm L: 13 cm, W: 7 cm, H: 9 cm
(1/4 inch L: 5-1/8, W: 2-3/4, H: 3-35/64)
Dowel version found in WoodMarvels.com Volume I: Top Sellers (page 36)

Product Renders

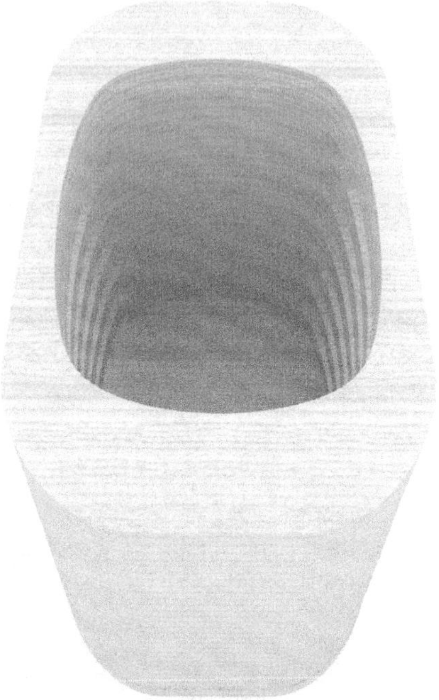

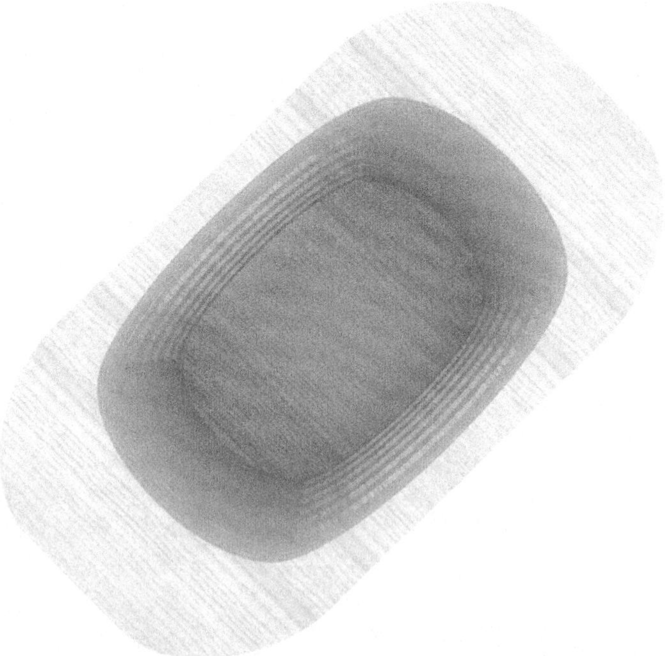

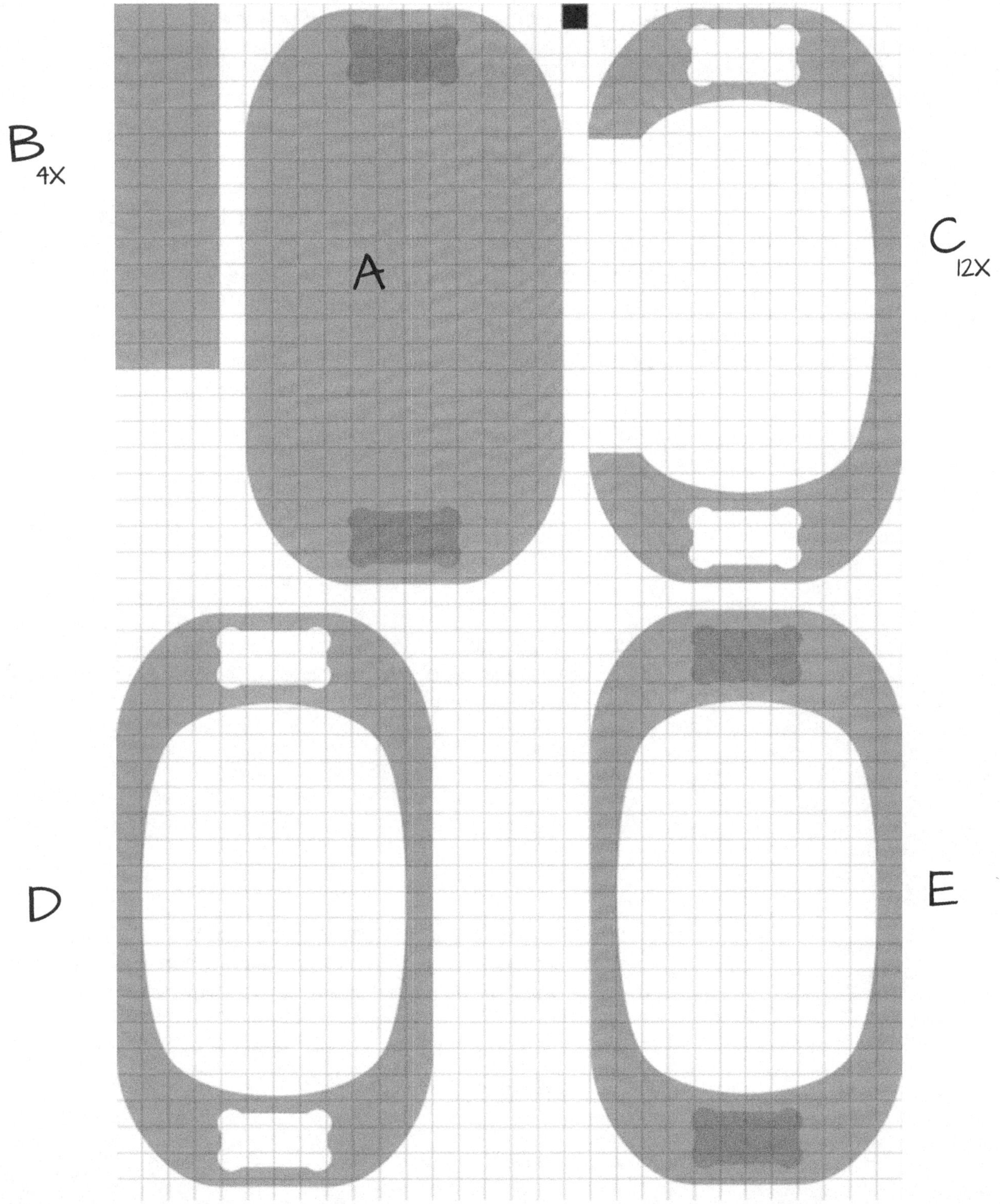
B
4X
A
C
12X
D
E

# Step-by-Step Assembly Instructions

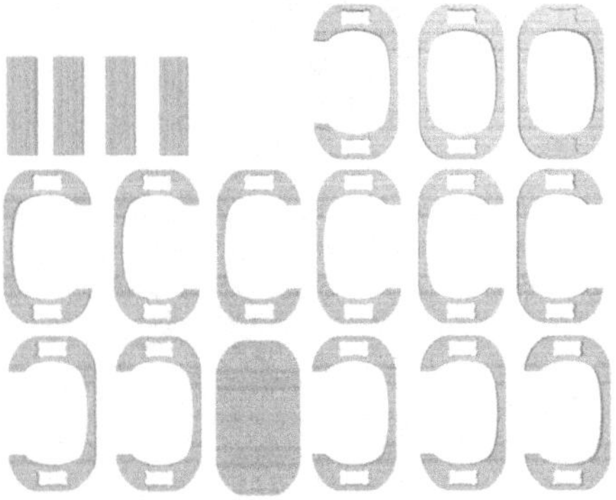

Step 1: Label and check inventory

Once you have all the pieces in front of you, the process of putting them together is no different from assembling pieces of Lego.

Label, using a pencil, the various parts and double-check your inventory. Take your time and enjoy the building process! Before jumping to the glue, be sure to do a dry run.

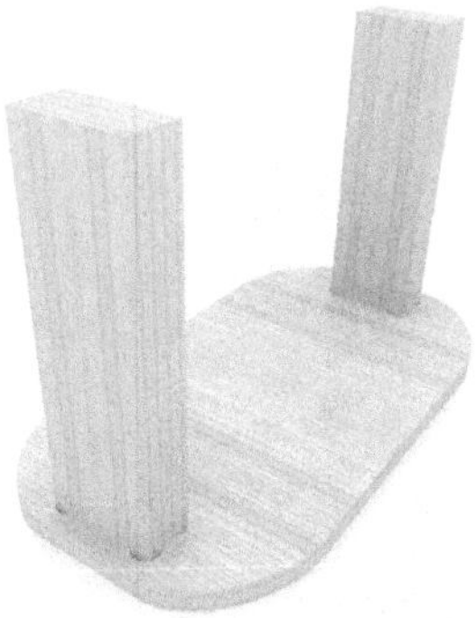

Step 2: Start the base

Slide and glue into place part A (base) followed by all four part Bs (guides).

Step 3: Add sides

Slide down and glue into place all 12 part Cs (half sides) in alternating fashion.

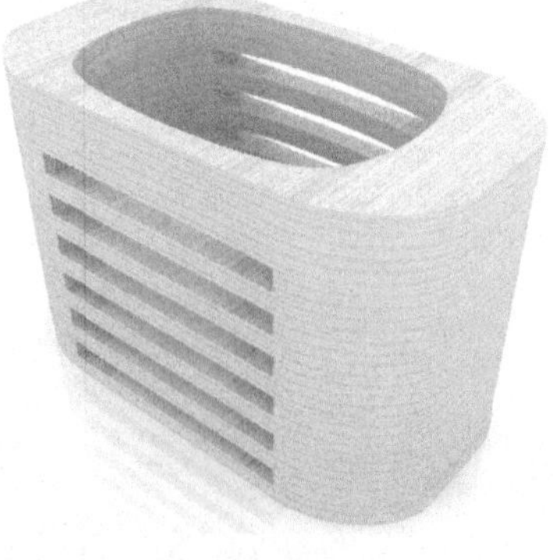

Step 4: Perform some finishing touches

Slide into place, with glue, part D and then E. Use clamps to keep everything tight as it dries overnight.

There is no need to apply any preservatives or paint; feel free to sand any sharp edges down. ENJOY!

# Platform Crane B

Completed 41 piece model dimensions @ 6 mm L: 47 cm, W: 32 cm, H: 25 cm
(1/4 inch L: 18-1/2, W: 12-19/32, H: 9-27/32)
Laser version WoodMarvels.com Volume 3: Evolution of Wooden Designs (page 68)
Dowel version found in WoodMarvels.com Volume 1: Top Sellers (page 92)

## Product Renders

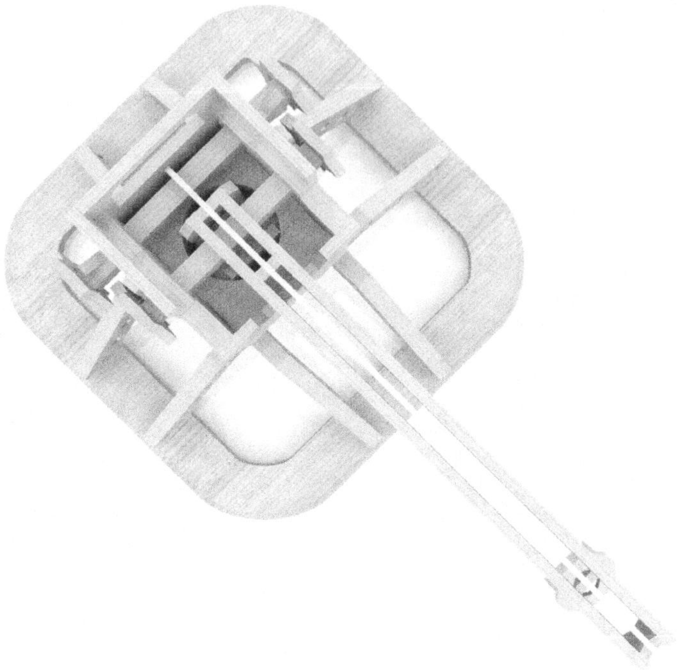

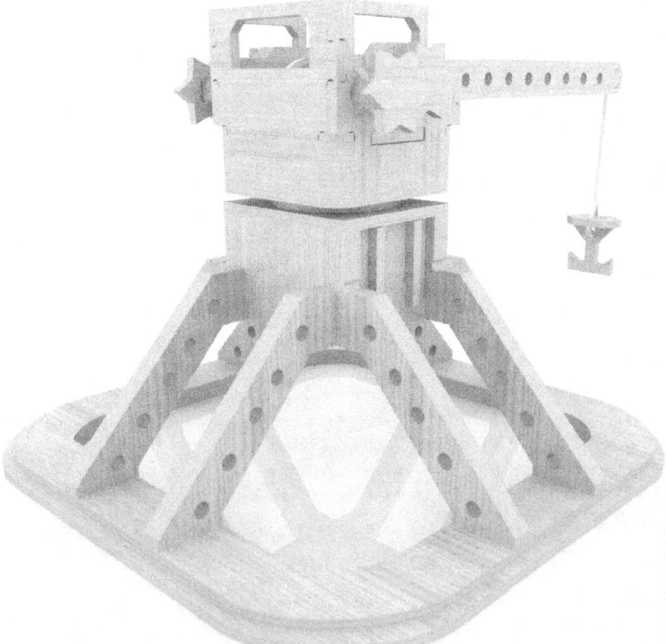

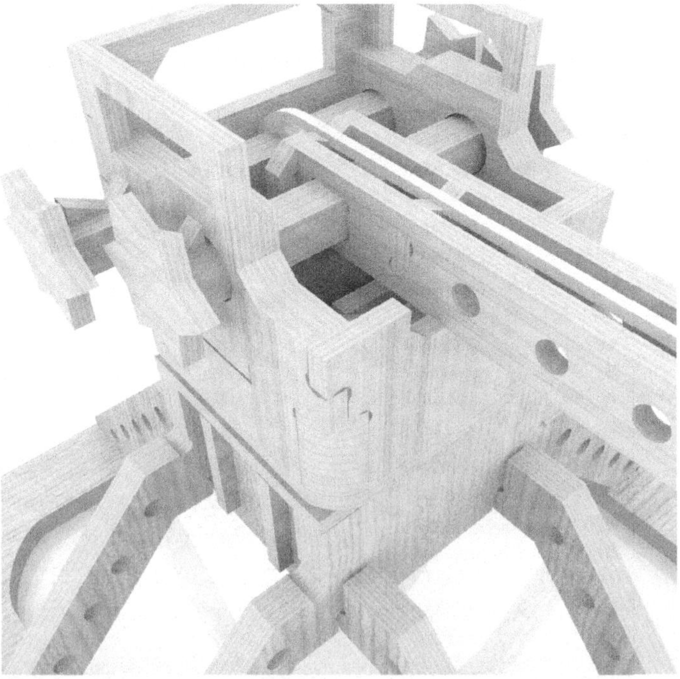

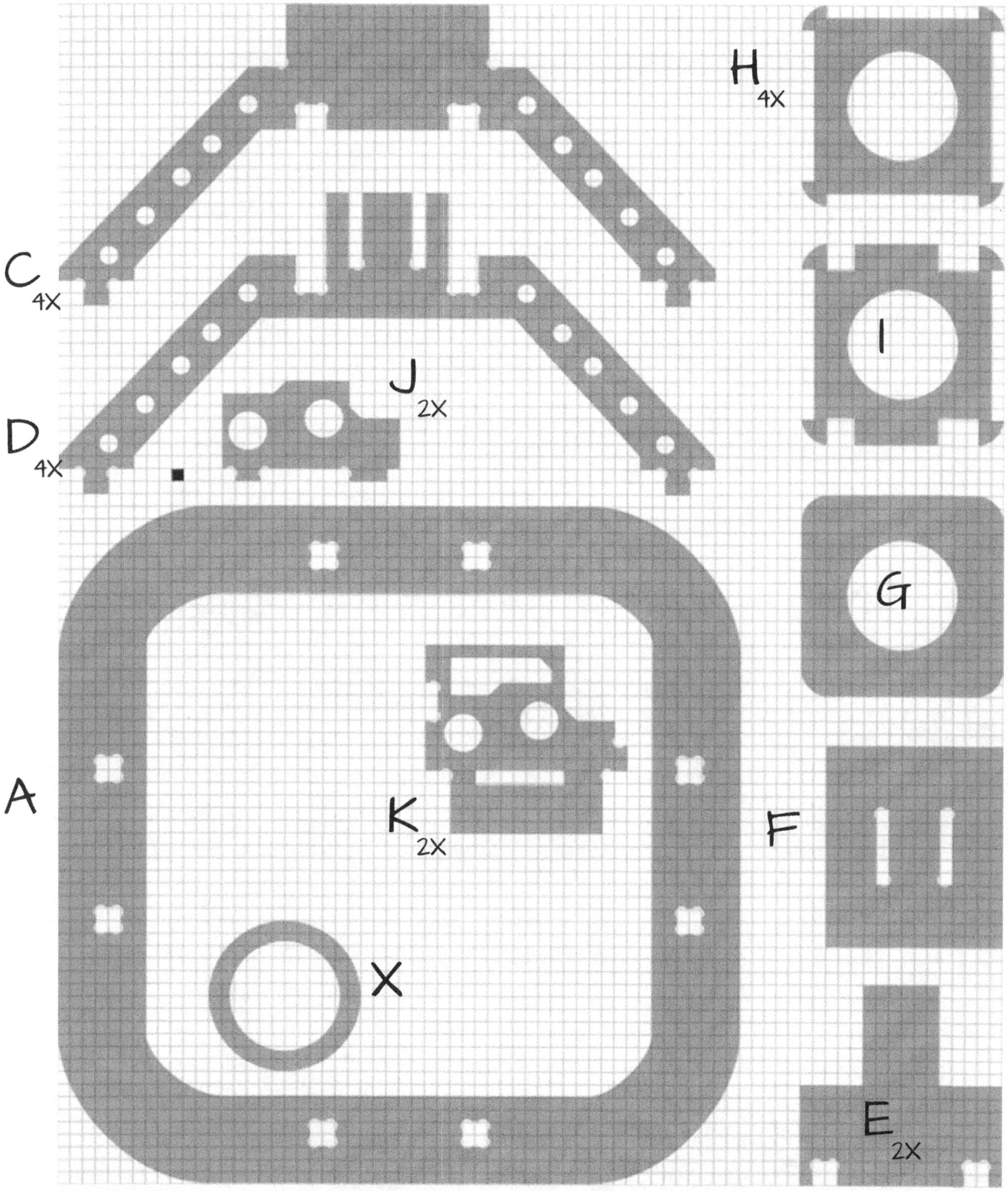
H
4X
C
4X
I
J
2X
D
4X
G
A
K
2X
F
X
E
2X

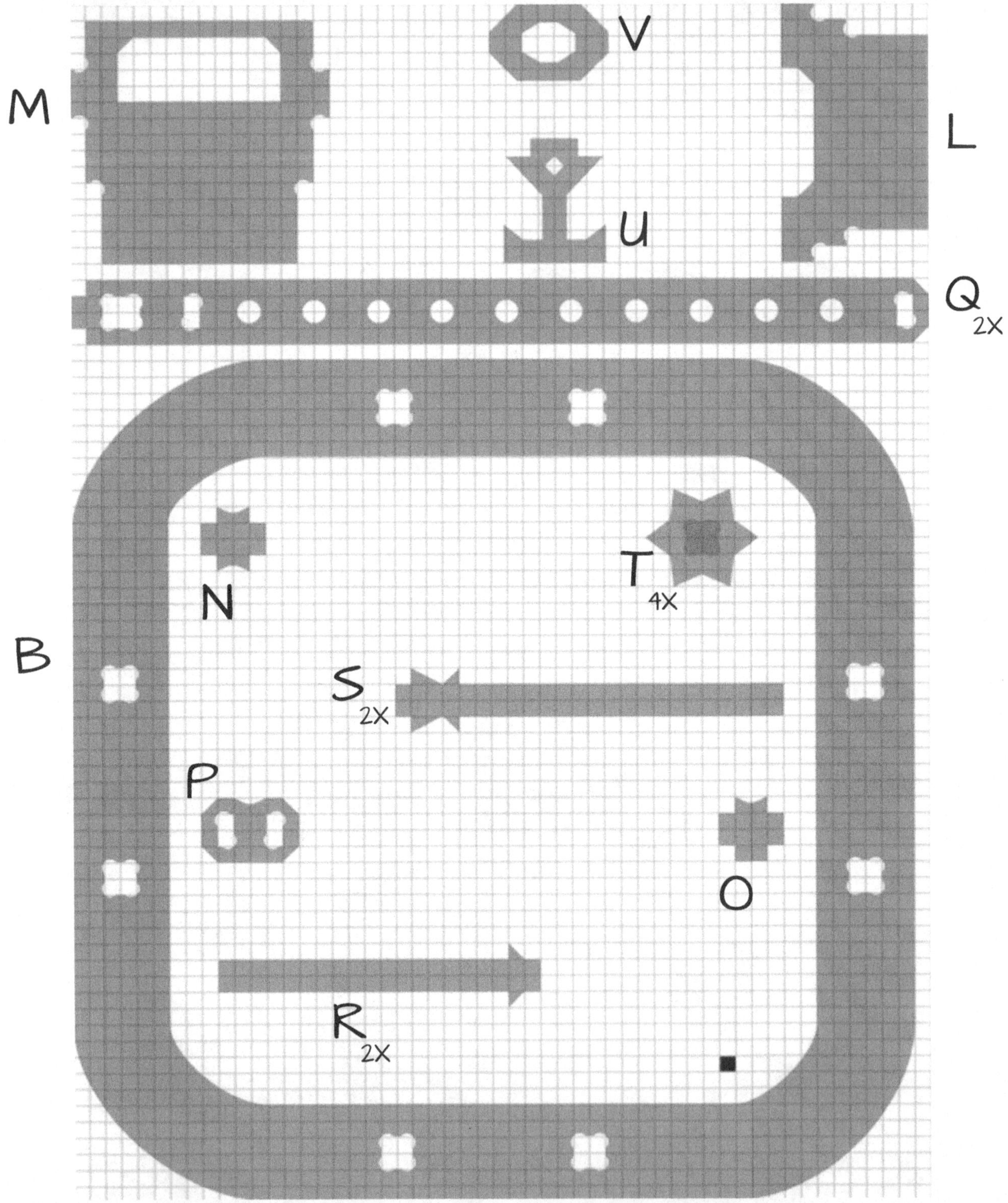
M
V
L
U
Q
2X
B
N
T
4X
S
2X
P
O
R
2X

# Step-by-Step Assembly Instructions

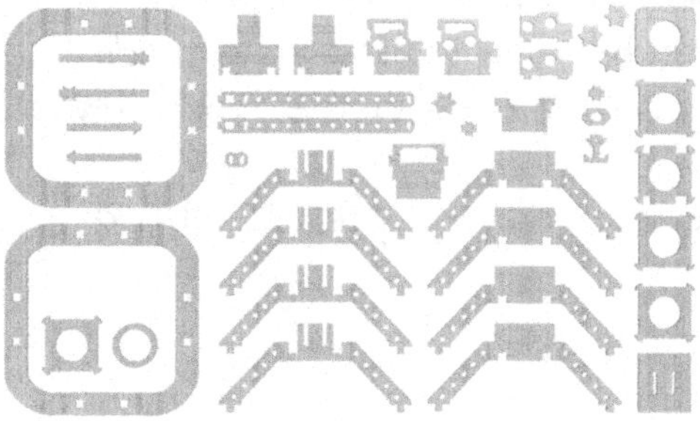

Step 1: Label and check inventory

Once you have all the pieces in front of you, the process of putting them together is no different from assembling pieces of Lego.

Label, using a pencil, the various parts and double-check your inventory. Take your time and enjoy the building process! Before jumping to the glue, be sure to do a dry run.

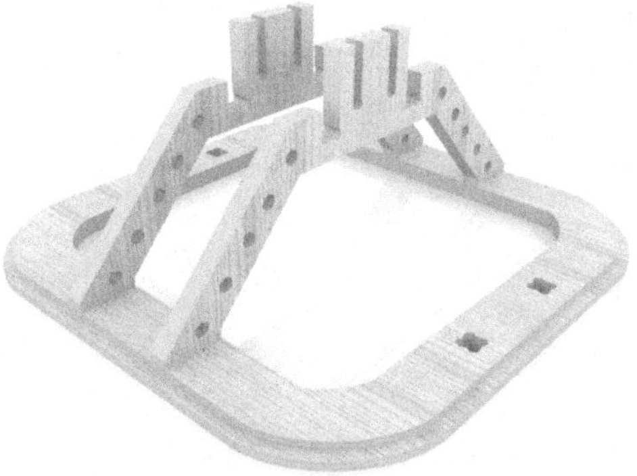

Step 2: Start the base

Slide and glue into place parts A and B (bottom platforms) followed by part C (vertical support leg).

Step 3: Complete the base

Slide and glue together part Ds (alternate vertical support leg) followed by both part Es (vertical tabs) and F (platform with two holes).

Allow this to dry overnight.

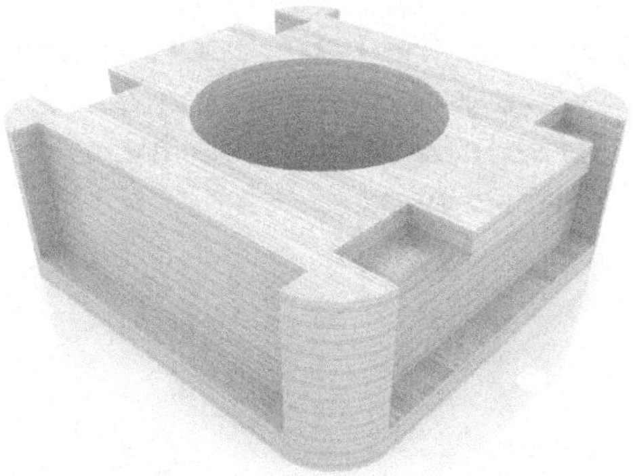

Step 4: Start the control cab

Starting with part G at the bottom, glue into place, vertically, all four part Hs. Cap with a part I.

Step 5: Add walls to the cab

Starting from the sides, slide and glue a small part J followed by a part K. These two parts help to provide additional support to the incoming rods in the next step.

On the front, slide and glue into place a part L with a part M on the back.

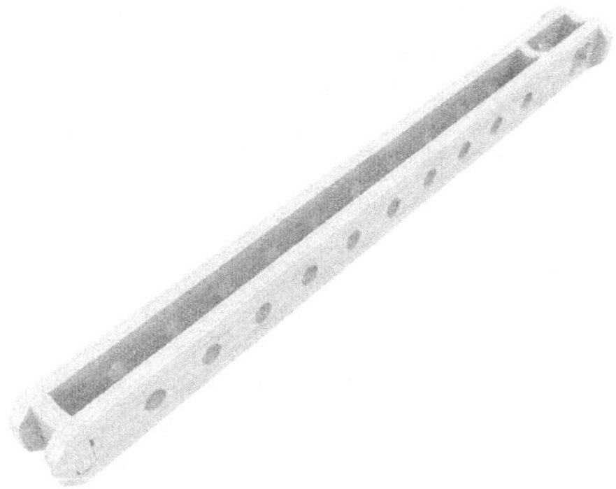

Step 6: Build the crane arm

Slide and glue together parts N, O, and P in between part Qs (arm).

Step 7: Complete the cab

Slide (glue is optional) the arm you built in the previous step into the cab. To lock the arm into place, slide into place two part Rs (small rods) through it; then using glue, slide into place both part Ts (caps).

Further back, repeat the process using both part Ss. This is where you will glue the rope.

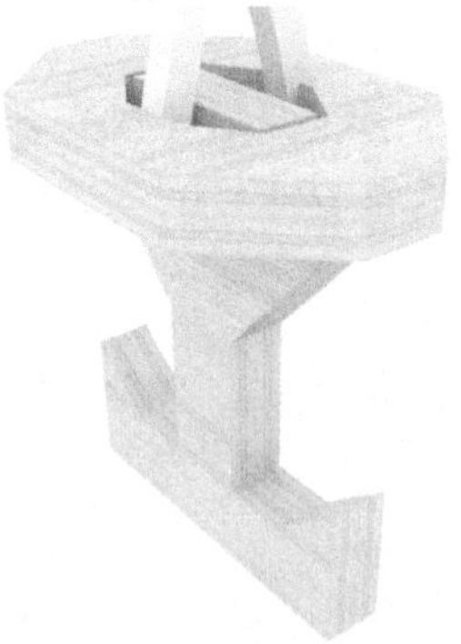

Step 8: Rope the end

Slide and glue into place, along with the rope, parts U and V. You will need to allow this to dry and, depending on your setup, add sawdust to create a solid fill so there isn't any movement between these two pieces.

Step 9: Add the string to the cab control

Wrap the string, with glue, around and in between the part Ss so the string does not slip.

Allow this to glue overnight; again, feel free to add sawdust to create additional bonding and friction along the rope so that it doesn't slide in place when turning the rod.

Step 10: Put it all together

Slide in between the base and the cab a part X. Do not glue it in place, as it provides a method for the cab to freely rotate around the base.

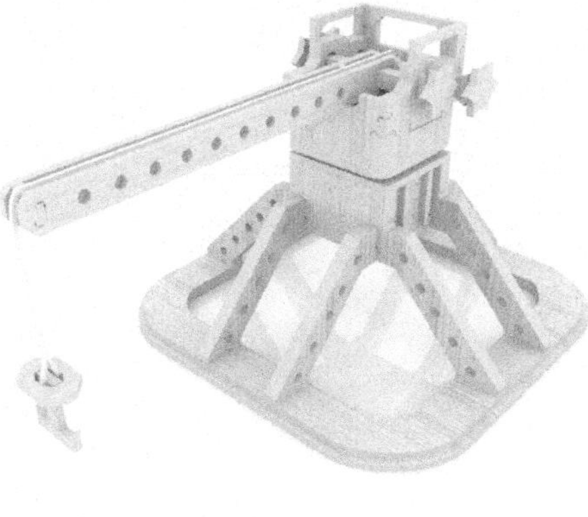

Step 11: Perform some finishing touches

Test out the various mechanisms. The top is not meant to be glued in place; it just hold the knobs firmly. It will be able to lift small blocks and objects without tipping over.

There is no need to apply any preservatives or paint; feel free to sand any sharp edges down. ENJOY!

# Pterodactyl

Completed 14 piece model dimensions @ 6 mm L: 40 cm, W: 25 cm, H: 11 cm
(1/4 inch L: 15-3/4, W: 9-27/32, H: 4-21/64)
Laser version found in WoodMarvels.com Volume 2: 50 Laser Cutting Plans (page 96)
Dowel version found in WoodMarvels.com Volume 1: Top Sellers (page 82)

## Product Renders

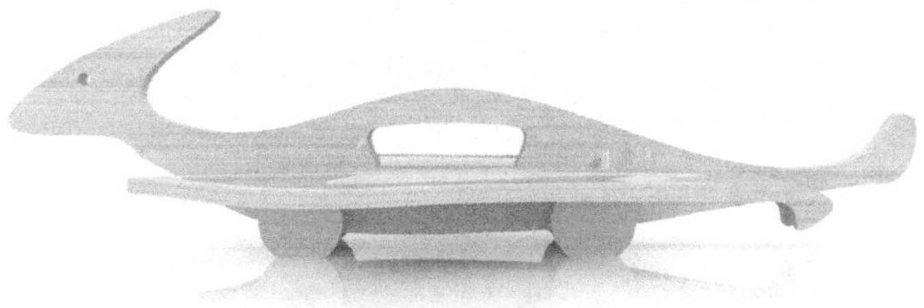

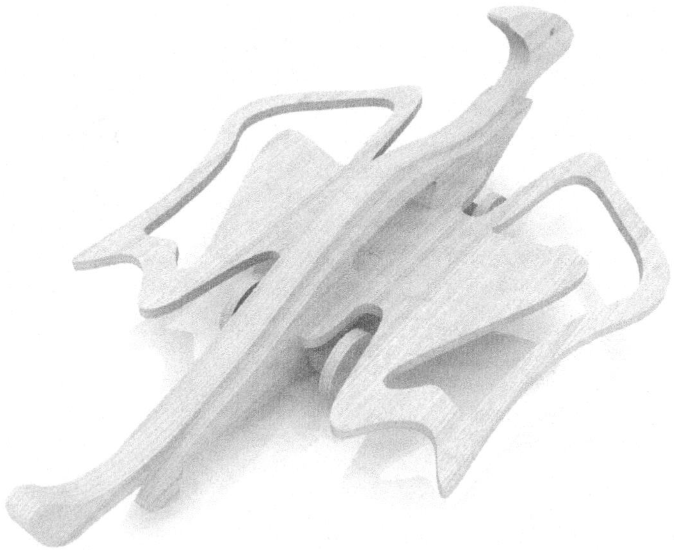

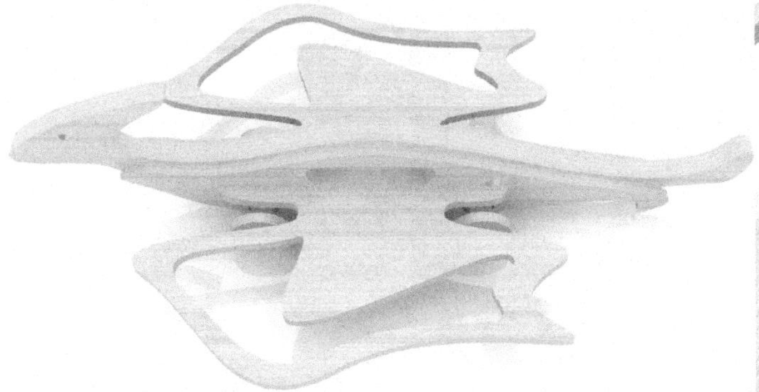

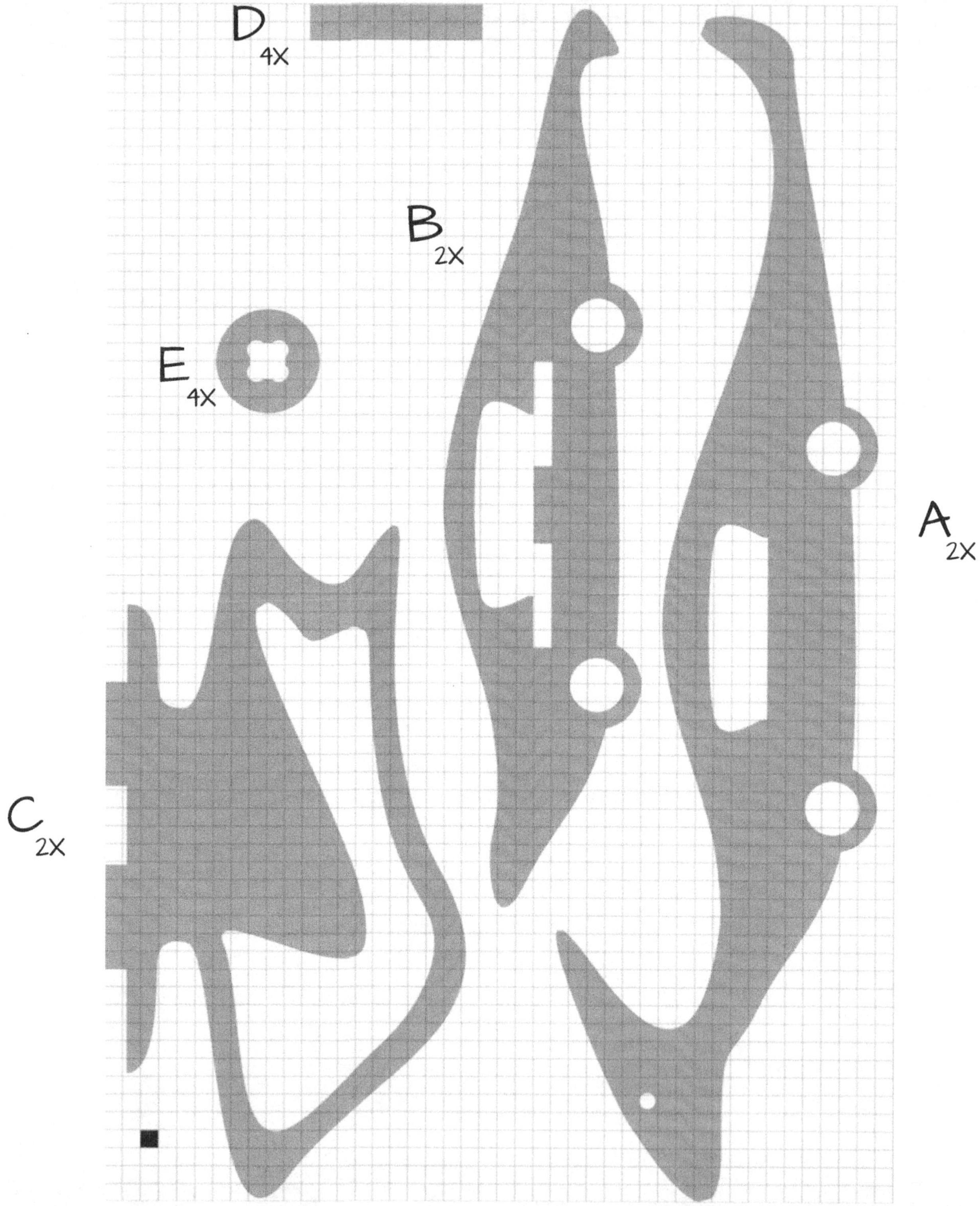
D 4X
B 2X
E 4X
A 2X
C 2X

# Step-by-Step Assembly Instructions

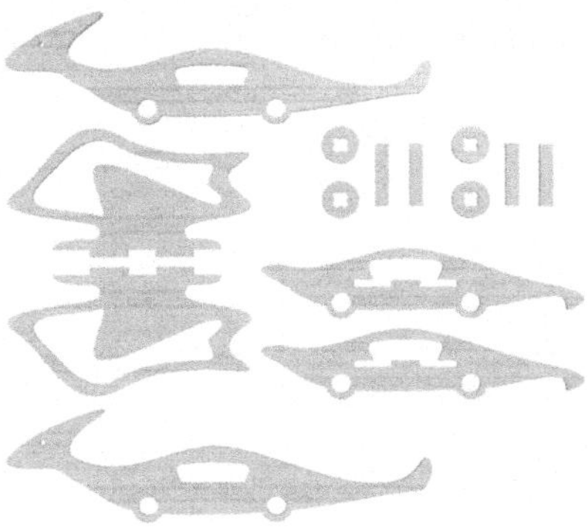

Step 1: Label and check inventory

Once you have all the pieces in front of you, the process of putting them together is no different from assembling pieces of Lego.

Label, using a pencil, the various parts and double-check your inventory. Take your time and enjoy the building process! Before jumping to the glue, be sure to do a dry run.

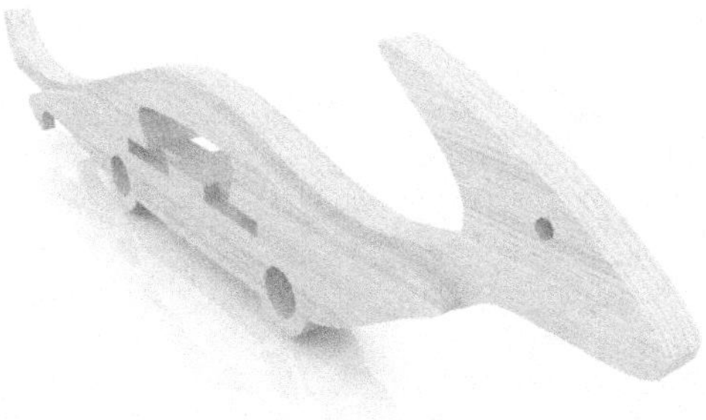

Step 2: Start the body

Slide and glue to one another both part As (body with head,) with a part B (body without head) on both sides.

Step 3: Add wings

Glue into place a part C on both sides of the body.

Step 4: Perform some finishing touches

Glue and slide into place doubled up part Ds (axles). On both sides, slide into place part Es (wheel).

There is no need to apply any preservatives or paint; feel free to sand any sharp edges down. ENJOY!

# Rhinocerous

Completed 25 piece model dimensions @ 6 mm L: 20 cm, W: 6.5 cm, H: 13 cm
(1/4 inch L: 7-7/8, W: 2-9/16, H: 5-1/8)
Laser version found in WoodMarvels.com Volume 2: 50 Laser Cutting Plans (page 36)

## Product Renders

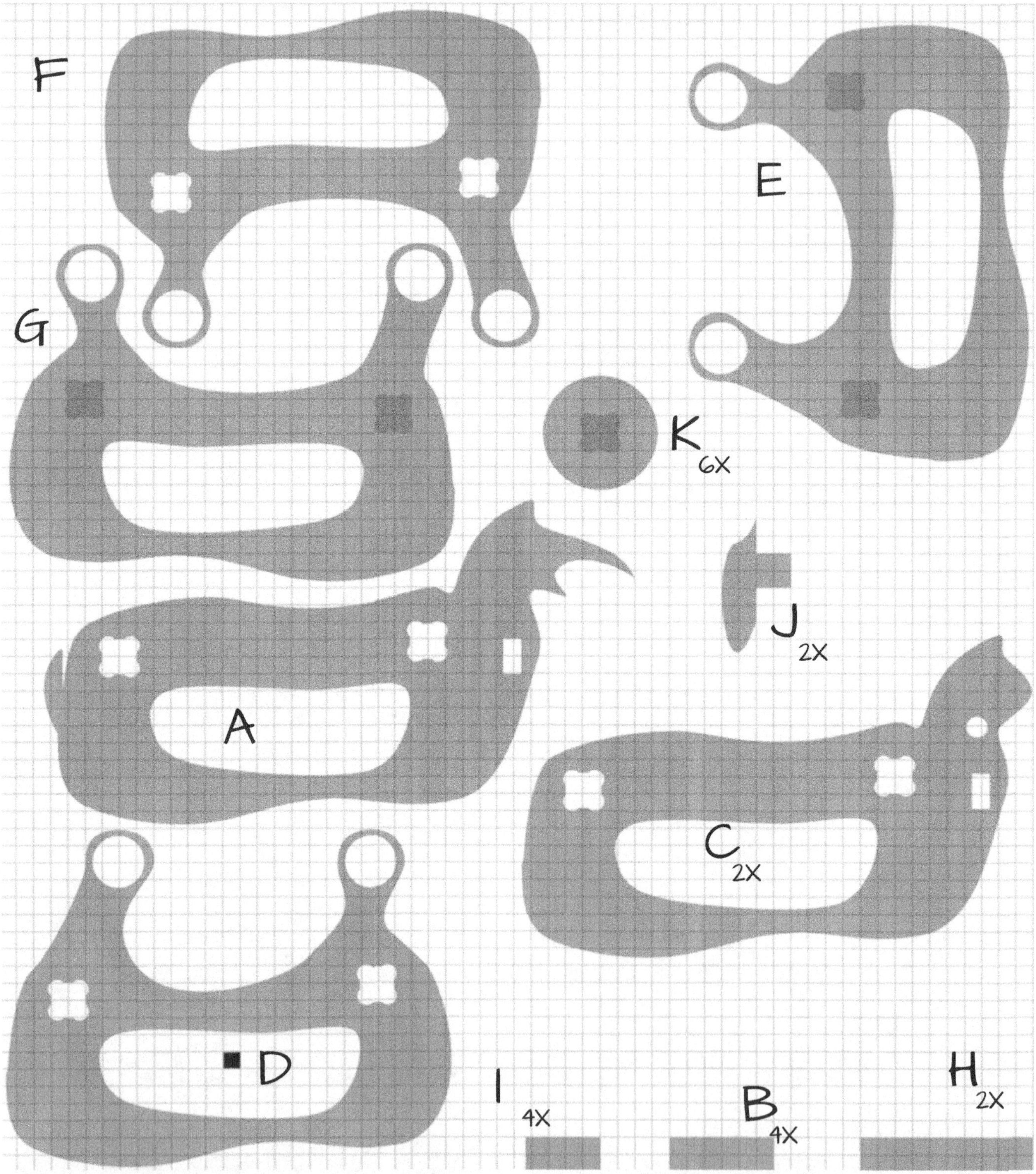
F
E
G
K 6X
J 2X
A
C 2X
D
I 4X
B 4X
H 2X

## Step-by-Step Assembly Instructions

**Step 1: Label and check inventory**

Once you have all the pieces in front of you, the process of putting them together is no different from assembling pieces of Lego.

Label, using a pencil, the various parts and double-check your inventory. Take your time and enjoy the building process! Before jumping to the glue, be sure to do a dry run.

**Step 2: Start the body**

Slide and glue into place part A (body with horn) with all four part Bs (guides) within the body. Afterwards, slide and glue into place, on both sides, a part C (body with head).

**Step 3: Complete the body**

Slide and glue together parts D and E on one side and F and G onto the other.

On the bottom, in the front, slide (without glue) a doubled up part H (front axle) and two sets of doubled up part Is (back axle).

**Step 4: Perform some finishing touches**

On the head, slide and glue into place a part J (ears) on both sides; and on each axle below, a part K (wheel). Feel free to use clamps to keep everything together while it dries.

There is no need to apply any preservatives or paint,;feel free to sand any sharp edges down. ENJOY!

# Stegosaur

Completed 17 piece model dimensions @ 6 mm L: 41 cm, W: 5 cm, H: 14.5 cm
(1/4 inch L: 16-9/64, W: 1-31/32, H: 5-45/64)
Laser version found in WoodMarvels.com Volume 2: 50 Laser Cutting Plans (page 39)
Dowel version found in WoodMarvels.com Volume 1: Top Sellers (page 51)

Product Renders

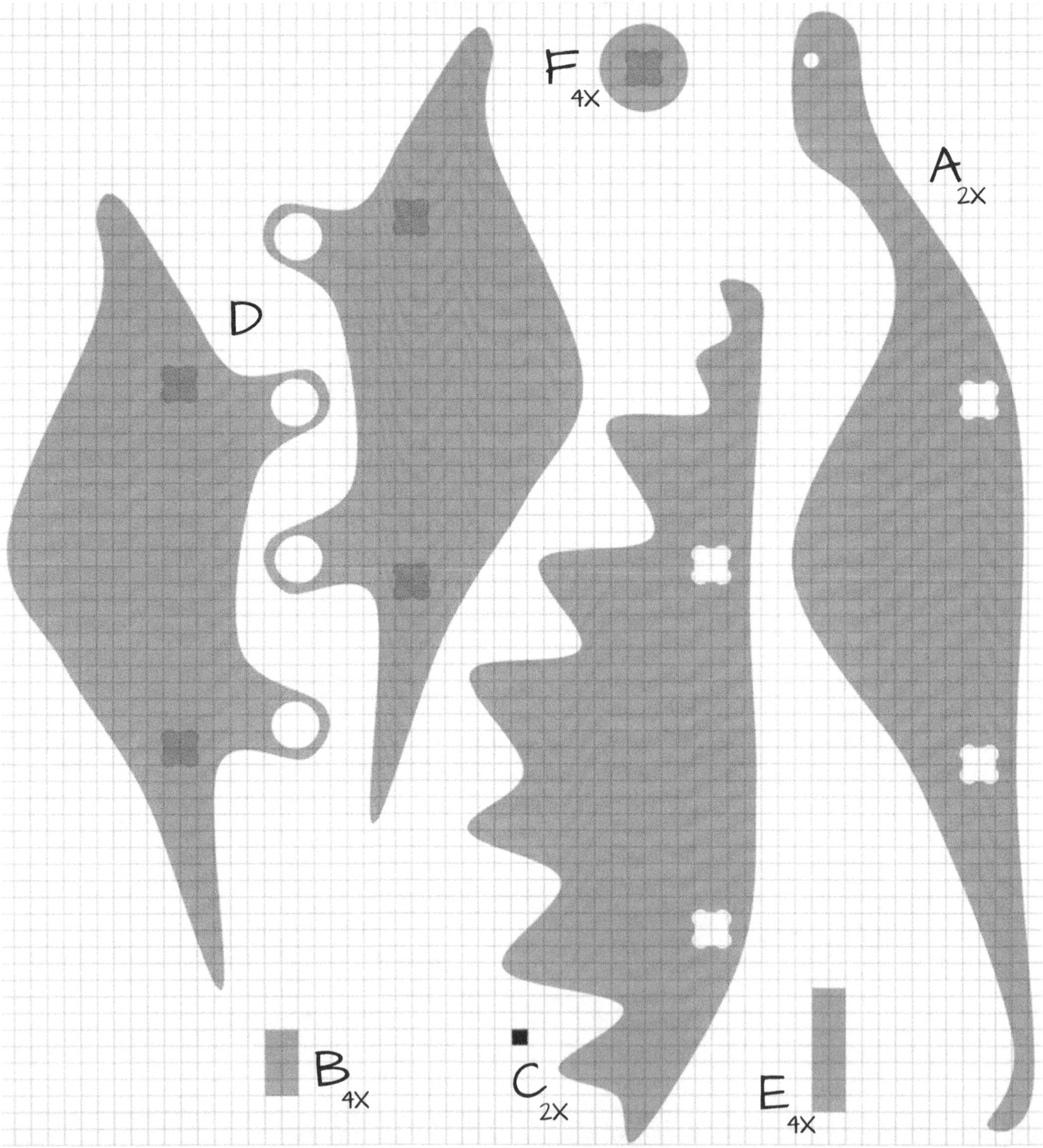
F
4X
A
2X
D
B
4X
C
2X
E
4X

# Step-by-Step Assembly Instructions

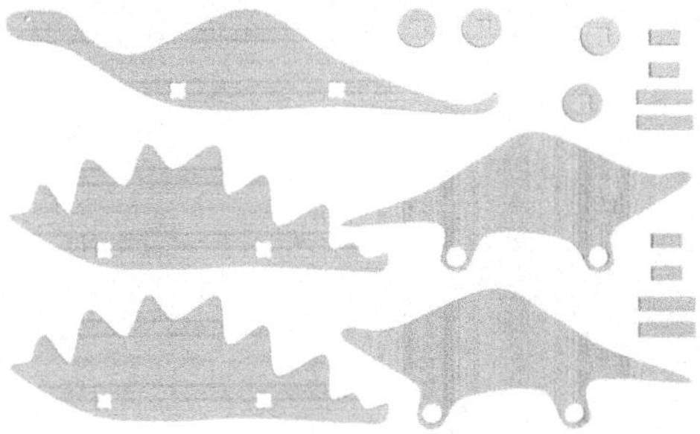

Step 1: Label and check inventory

Once you have all the pieces in front of you, the process of putting them together is no different from assembling pieces of Lego.

Label, using a pencil, the various parts and double-check your inventory. Take your time and enjoy the building process! Before jumping to the glue, be sure to do a dry run.

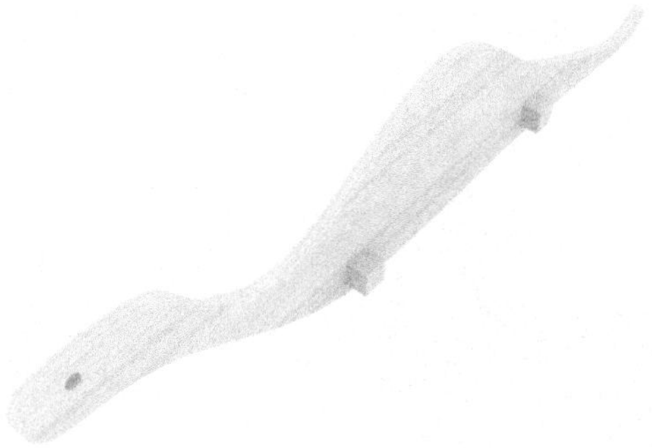

Step 2: Start the body

Slide and glue into place part A (body with head) with all four part Bs (guides).

Step 3: Complete the body

Slide and glue together a part C (body with spines) and D (body with legs). Use clamps to hold the model together tightly until its body is dry.

Step 4: Perform some finishing touches

In the front and back, slide into place part Es (axle) in between part Fs (wheel).

There is no need to apply any preservatives or paint; feel free to sand any sharp edges down. ENJOY!

# Straddle Carrier B

Completed 125 piece model dimensions @ 6 mm L: 32 cm, W: 25 cm, H: 28 cm
(1/4 inch L: 12-19/32, W: 9-27/32, H: 11-1/32)
Laser version found in WoodMarvels.com Volume 2: 50 Laser Cutting Plans (page 174)
Dowel version found in WoodMarvels.com Volume 1: Top Sellers (page 97)

## Product Renders

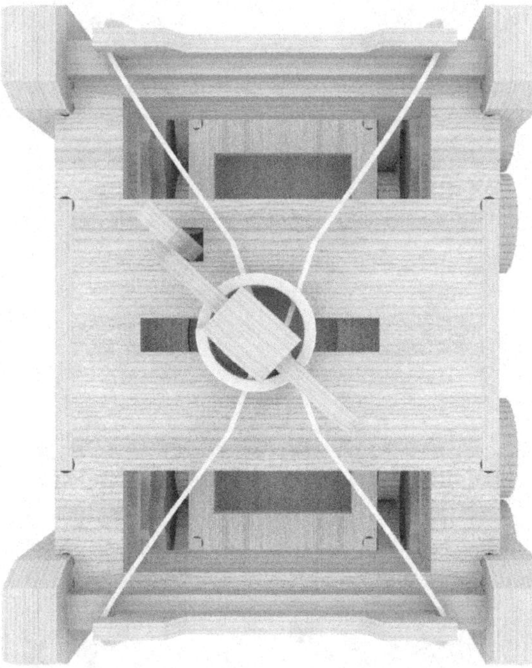

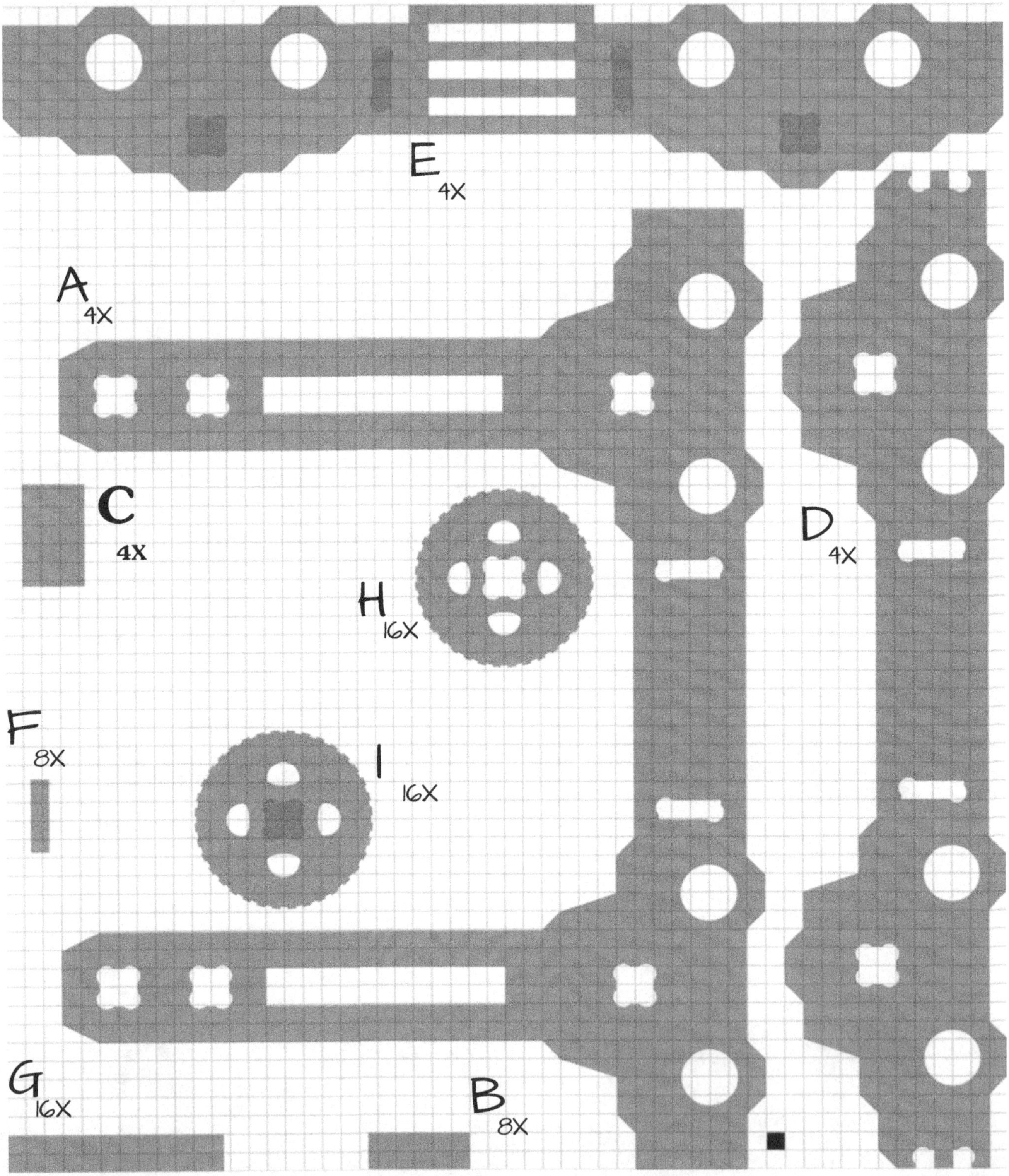

E 4X
A 4X
C 4X
D 4X
H 16X
F 8X
I 16X
G 16X
B 8X

# Step-by-Step Assembly Instructions

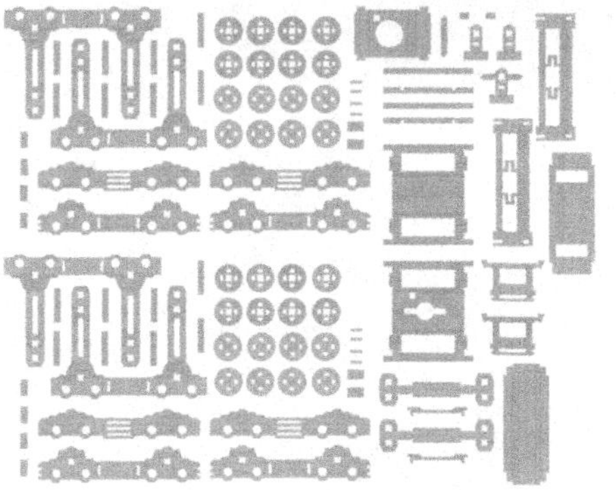

Step 1: Label and check inventory

Once you have all the pieces in front of you, the process of putting them together is no different from assembling pieces of Lego.

Label, using a pencil, the various parts and double-check your inventory. Take your time and enjoy the building process! Before jumping to the glue, be sure to do a dry run.

Step 2: Start the body

Slide and glue together both part As (body with towers,) two sets of doubled up part Bs (small guides) horizontally, and two sets of doubled up part Cs (wide guides) vertically.

Step 3: Complete the body

Slide and glue together on both sides, a part D (short body with holes) and mirrored part Es (short body with pockets). At the ends of each body, glue into place part Fs (ladders).

Step 4: Finish the body

Slide into place doubled up part Gs (axles,) followed by a part H (holed wheel) and I (capped wheel).

Repeat these four steps to complete the other tower.

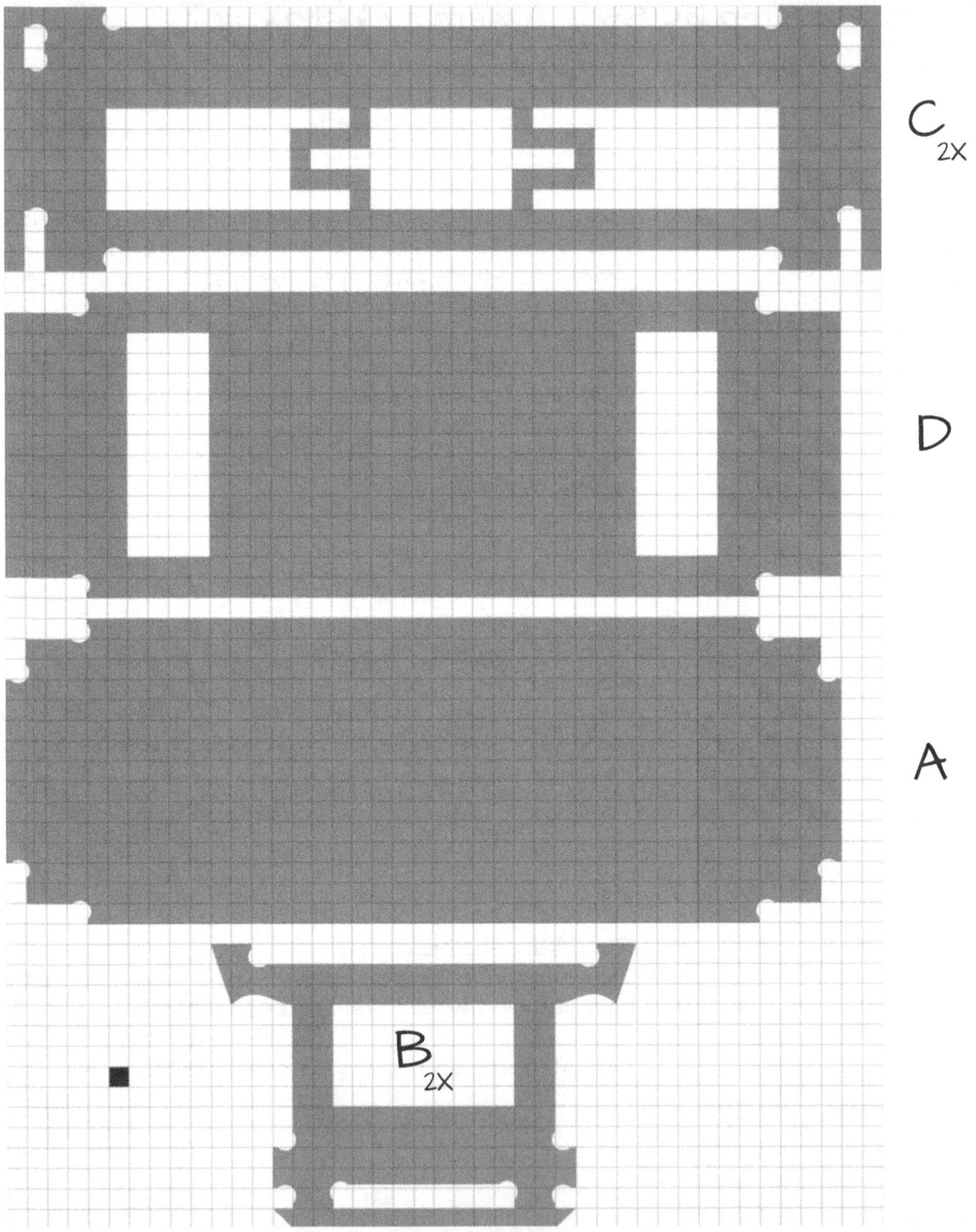
C 2X
D
A
B 2X

Step 5: Start the container

Glue and slide into place a part A (floor) between two part Bs (front/back).

Step 6: Add sides

Slide and glue on both sides a part C (windowed sides,) and allow the assembly to dry overnight.

Step 3: Complete the container

You can now slide into place part D (lid,) but don't add any glue as you want to be able to easily add and remove it.

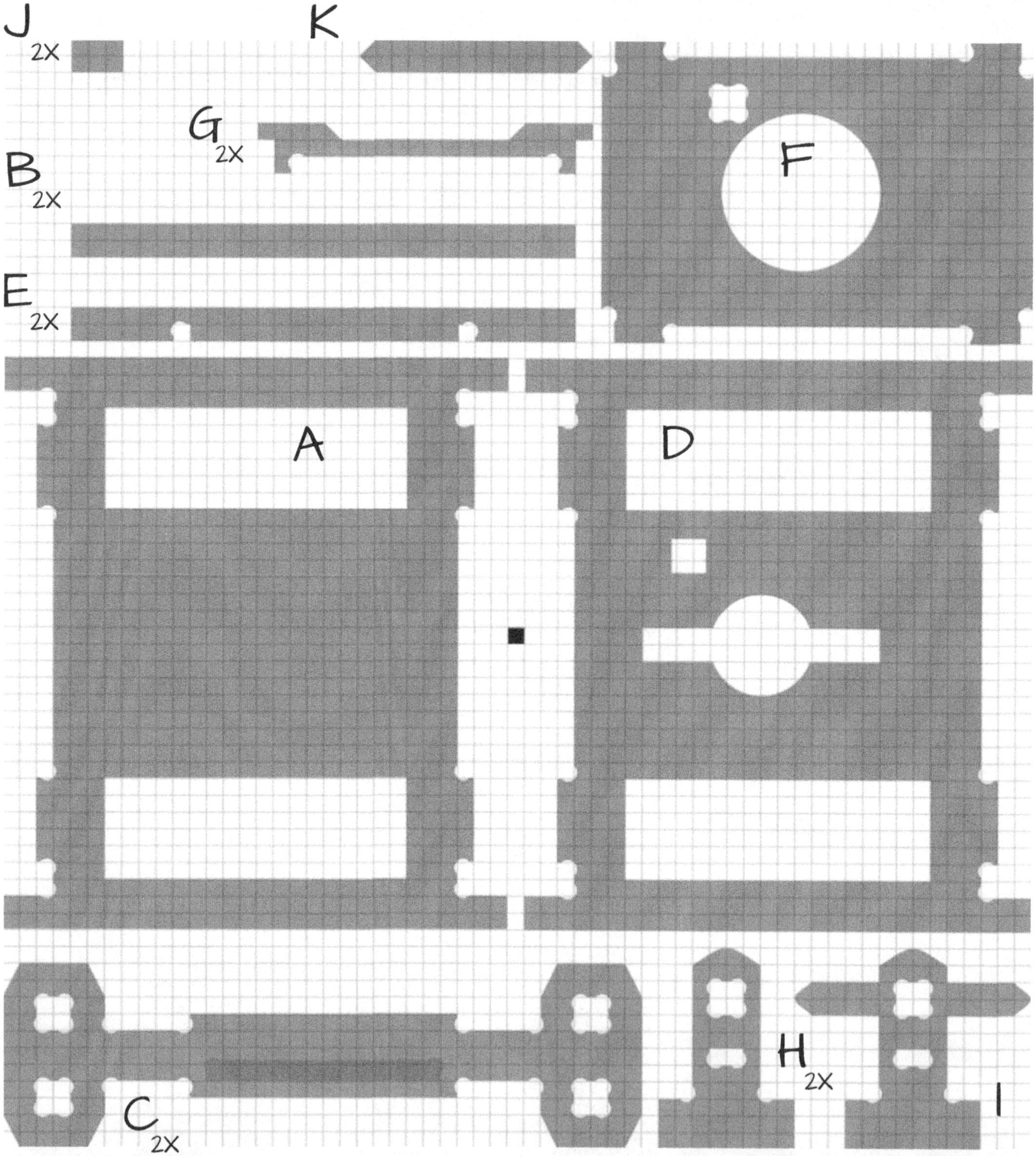
J
2X
K
G
2X
B
2X
E
2X
F
A
D
H
2X
C
2X
I

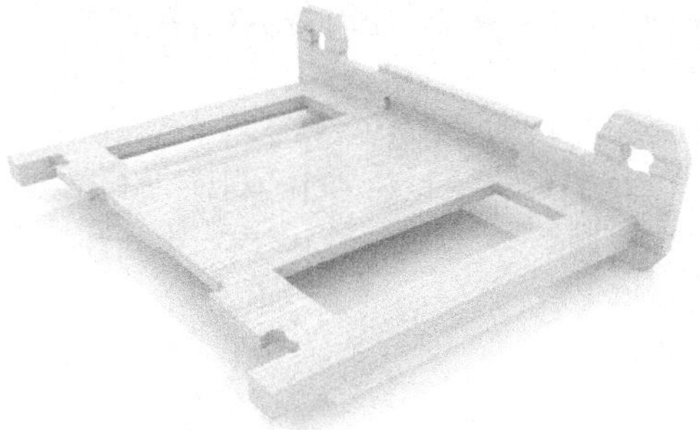

Step 8: Start the base
Slide, with glue, a part A (foundation) with two part Bs (horizontal cross-members) under it into a part C (vertical side).

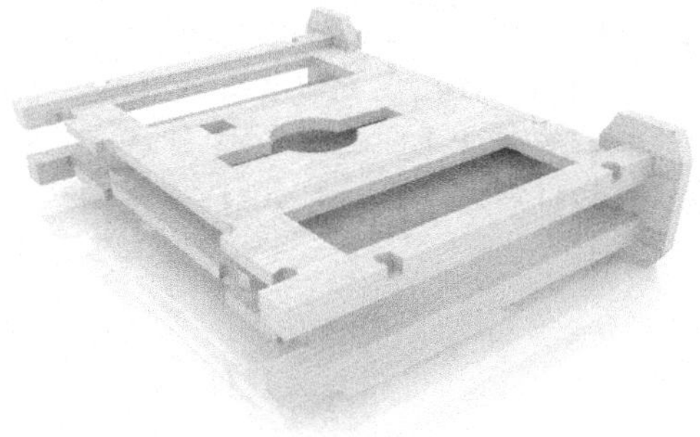

Step 9: Add the top
Within the top hole of part C, glue and slide into place a part D (ceiling) followed by both part Es (guide holders). In between, into the hole found in part C, glue and slide into place a part F.

Step 10: Complete the top
On the opposing side, glue and slide into place the other part C followed by a part G (long Ts) on both ends, to help guide the rope below.

Step 11: Add the winder
Into the hole in the top, slide into place a part H (tab without arms) and then, applying glue on both sides, a part I (tab with arms). On the top, glue and slide into place a doubled up part J.

Be sure to rotate the winder while the glue is still setting. At this time, insert (without glue) part K (so you don't lose it).

Step 12: Add the top rope

Within the winder, you will see a hole in the bottom section below the doubled up part Js. What you want to do is slide two pieces of rope through this and cross them so that one end goes on the left front while the other goes on the right back. Beforehand, wrap them around the winder a few times.

Step 13: Finish the container rope

At the other end of the ropes, make a knot and apply some glue so that it stays tied together. You can either have individual loop segements for each guide on the container or one large rope holding both (as you see in my live videos).

Step 14: Perform some finishing touches

There is no need to apply any preservatives or paint; feel free to sand any sharp corners. Enjoy!

# Submarine B

Completed 20 piece model dimensions @ 6 mm L: 36 cm, W: 5 cm, H: 14 cm
(1/4 inch L: 14-11/64, W: 1-31/32, H: 533/64)
Laser version found in WoodMarvels.com Volume 2: 50 Laser Cutting Plans (page 41)
and variant WoodMarvels.com Volume 3: Evolution of Wooden Designs (page 32)
Dowel version found in WoodMarvels.com Volume 1: Top Sellers (page 51)

Product Renders

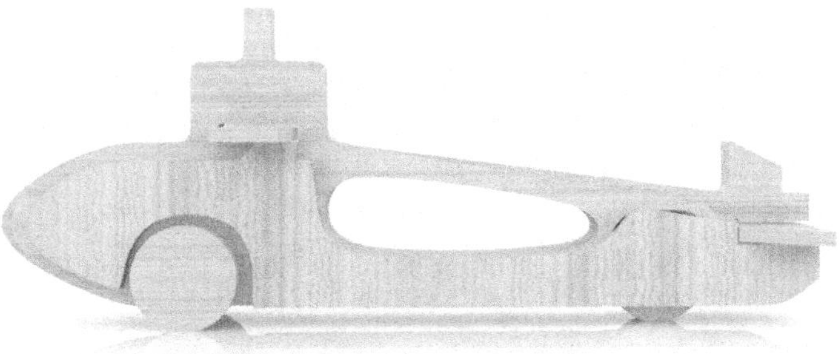

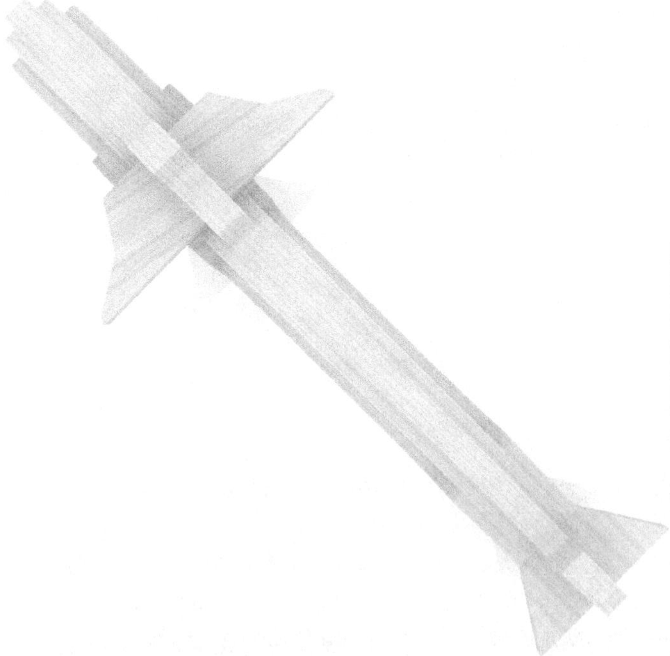

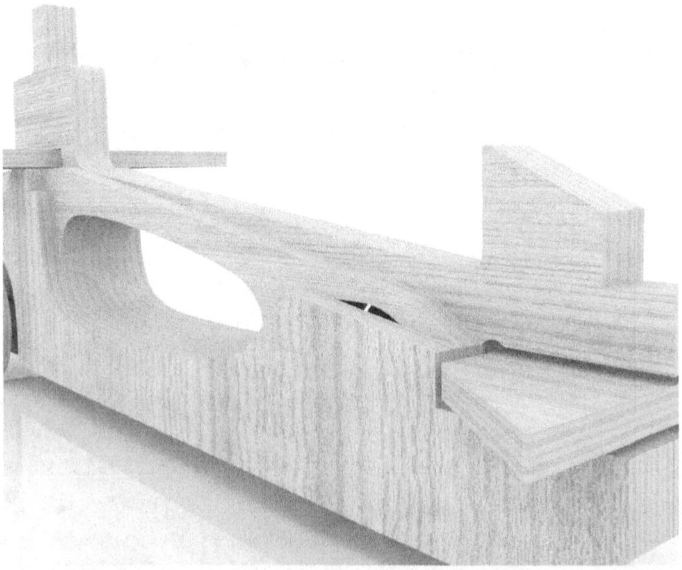

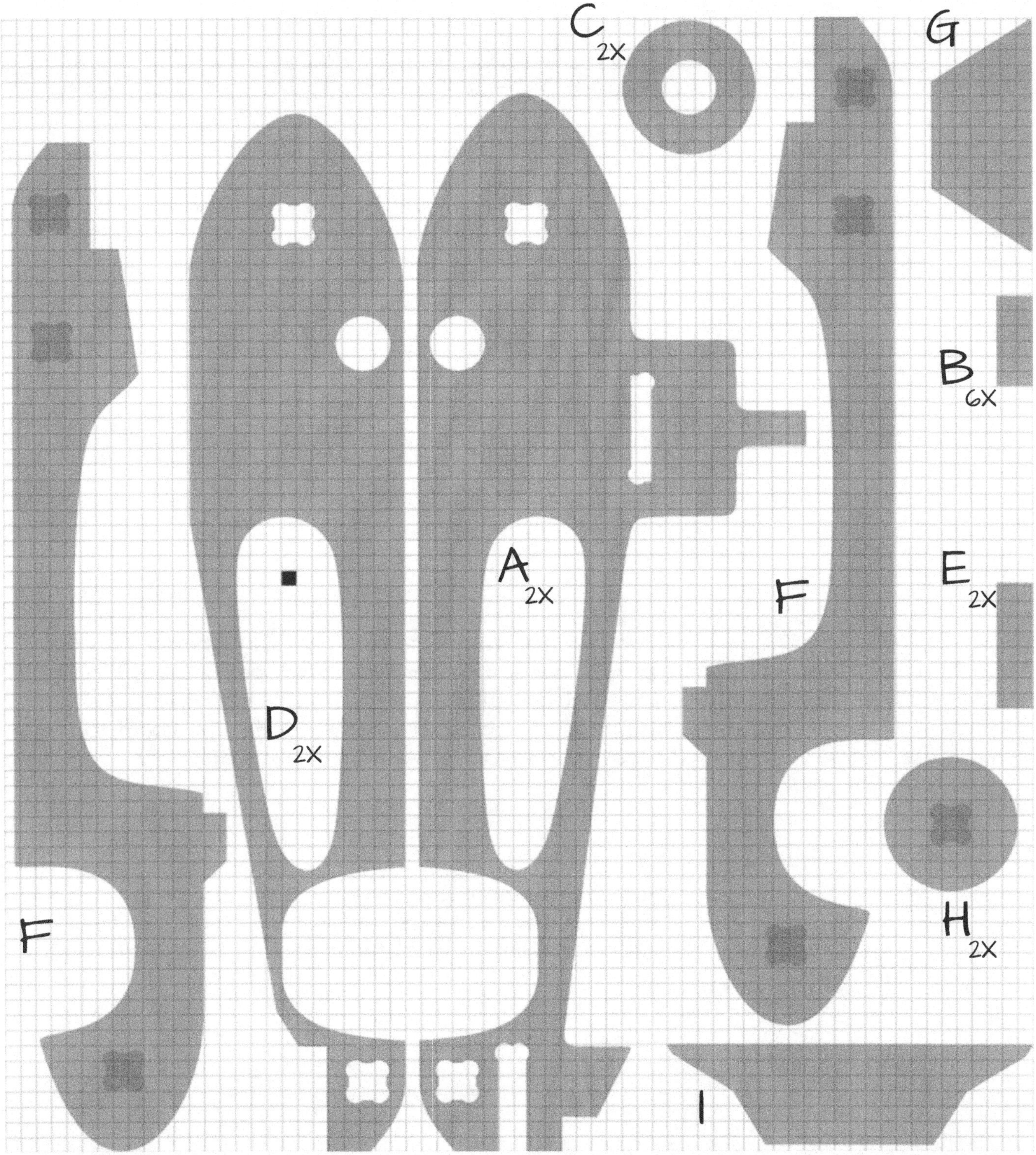
C
2X
G
B
6X
A
2X
F
E
2X
D
2X
H
2X
F
I

# Step-by-Step Assembly Instructions

Step 1: Label and check inventory

Once you have all the pieces in front of you, the process of putting them together is no different from assembling pieces of Lego.

Label, using a pencil, the various parts and double-check your inventory. Take your time and enjoy the building process! Before jumping to the glue, be sure to do a dry run.

Step 2: Start the body

Slide and glue into place part A (large body panels) with all four part Bs (guides).

Step 3: Add back wheels

Slide and glue into place two part Cs (wheels) with two part Bs going through them. Take special care, though; you want one of the part Cs to be about half the thickness of the other so they both have room to spin once the submarine's sides are glued into place.

Step 4: Glue more panels

Slide and glue into place both part Ds (sides) and slide into place (without glue) two part Es (axles) in the front.

Step 5: Add tail and sides

Slide and glue into place part F (side half panel) and G (back tail).

Step 6: Perform some finishing touches

On the front, slide and glue into place both part Hs (wheels) and part I (fins).

There is no need to apply any preservatives or paint; feel free to sand any sharp edges down. ENJOY!

## Steve's Osage Orange Box with Live Edge

This is a series of boxes made from a storm damaged tree near my home in Dallas County, Texas. The tree was cut into manageable size pieces by the home owner (or someone he hired) and set out to be hauled off. Bodark, as Texans call Osage Orange, is a very dense, heavy wood, especially when wet. Consequently, "manageable" size is quite small, but more than adequate for the use I intended. I cut the chunks of wood into 1" and 2" lumber with my 18" Jet band saw, equipped with a sliding carriage of my own design.

To control checking of the lumber while drying, I coat the ends of my boards with a heavy coat of paint. Putting the stickered and banded stacks of lumber in the attic is the last attention they get from me for a long time ... at least 1 year per inch of thickness. I don't own a moisture meter but think I will soon. Since I have multiple years between acquiring and using the lumber, I sometimes lose track of what is available to use. Lumber has to be well-marked. I use a permanent marker to record the cut month, year and species.

To assure a continuous supply of wood, I'm always looking and thinking "is that usable" when I see trees put on the curb for removal. Even though my attic "kiln" looks quite full, constant new wood is the only way to assure a future supply. Wood from branches, even very large branches, is unusable due to the stresses built into the wood while growing. This "reaction" wood is easily identified by the off-center growth rings. This series has five boxes with several things in common.

- All from the same tree
- All have one edge with the tree bark still on, or a "live edge"
- All have one end that is "as found". That end still has the chain saw marks from the tree harvesting and the paint I used to seal against checking.

The series was originally seven boxes but two were lost during manufacture. Of those two, one had hidden structural issues that made it unusable and the other "got away" while routing.

The first step when making this style box is to analyze the candidate wood pieces... Make mental notes of splits, checks, knots, and other defects; you'll want to work around these. You want the finished box to have pleasing proportions, interesting grain, coloration and perhaps even a function! Function? This type of box is art in and of itself ... But, in my mind nothing is more pleasing than functional art! If made to order, the interior can be made to hold a keepsake, tool, instrument, or even a pad of sticky notes.

Step two is to reduce the rough stock to approximately finished size. Air dried lumber, especially small chunks, doesn't have any flat reference surfaces and it's too dangerous to run these small pieces over a jointer.

A CNC router can be used to get a flat surface on irregular or awkward shaped pieces, but it was easier to hand hold these parts up to my 6" belt sander. It may not be important, and may even be desirable for the top and bottom of this "artsy" type box not to be parallel. If you want them parallel, the CNC will make quick work of it. If the perimeter is to be straight sided, you can cut the finished exterior on the table saw or use the CNC router. For irregularly shaped perimeters the CNC can't be beat!

After you have the wood cut to the finished exterior dimensions, the next step is to cut the top or lid off. Before you do that, stabilize the live edge if it needs it. I coat it with lacquer until it stops soaking in and use super glue or epoxy anywhere it seems necessary. You want the bark as tight as possible before you cut the top off, otherwise you may lose chunks of it at this step.

For this project, I band sawed 1/2" off the top to make my lids. Belt sanding removes the saw marks, but I wait until after I have hollowed out the interior of the box.

That way if any cupping, bowing, warping or other movement happens due to stresses being released, I can still sand a flat mating surface on the box and lid!

Now it's time to clear out the interior of the boxes with the CNC router.

This is where you can show your creativity! Interiors can follow the exterior... or not. Partitions, hearts, stars, sea shells ... your imagination is your limitation!

As with any tool, there are many lessons, tips and tricks to learn on the way to becoming a skilled CNC router operator.

There are right and wrong times to use up spiral bits, down spiral bits, climbing cuts, conventional cuts, "O" flutes, compression flutes, ball end and FEM end bits. Then you must calculate proper chip load, RPM and move speed of the X,Y and Z axes.

When you get past the learning curve, you'll wonder how you ever worked in a shop without one! Tool paths of modern CNC routers are created with software packages that have user-friendly interfaces and render images of what will be created from the specifications you give it.

This rendering is amazingly accurate. If you don't see what you want, you'd better try the design process again! For this series of boxes, I choose to basically follow the exterior and to mirror the base cutout in the lid. This is a personal preference and requires separate tool paths and locating of top and bottom, as the boxes are not symmetrical.

# Tool Path strategies for a live edge box...

A live edge box is a great CNC beginners project because the exterior of the box is cut with traditional tools... This simplifies the hold-down problem and reduces tool pathing to interior material removal. Even though we have eliminated some problems there are still some things to keep in mind when planning cutting strategies to increase our chances of success!

The biggest issue in material removal is chipping/splintering. Often the cut material wants to follow the grain and "run" ahead of the bit, possibly ruining the project by going where we had not intended to remove material... There are some things we can do to help prevent this. **First**... Do roughing cuts as a climb cut... Climb cuts (see illustration) have less propensity to chip out chunks of material, however they are not the smoothest cut in our arsenal. I like to leave .015" to .020" allowance on my roughing cuts and come back with a second pass in conventional cut mode to finish the job... this is called an onion skin cutting strategy. **Second**... When pocketing, you can choose to do a profile pass of the pocket first or last... By choosing "first", any splintering of the ensuing pocket is stopped by the perimeter profile cut. **Also**... Choose a Router bit that is going to work with you instead of against you! I would choose to use a "down cut" bit in this situation as its downward forces will work with you to help hold the material in place... A spiral flute bit will have less "hammering" or vibration as its cutting is spread out over the diameter of the bit and a "chip breaker" geometry will help prevent chips that want to run with the grain. **Last**... But actually before you even start, be smart with your design! Box walls in the end grain portion of your design will be the weakest. If a 1/4" end grain wall survives manufacturing... It won't survive that first inevitable drop.

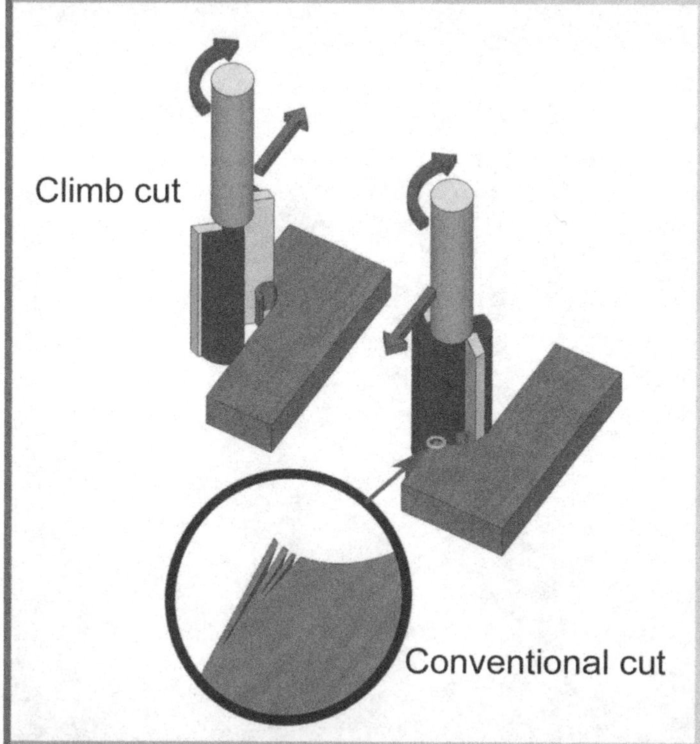


This is the Down cutting, spiral flute, bottom cutting, chip breaking bit that I choose for this job

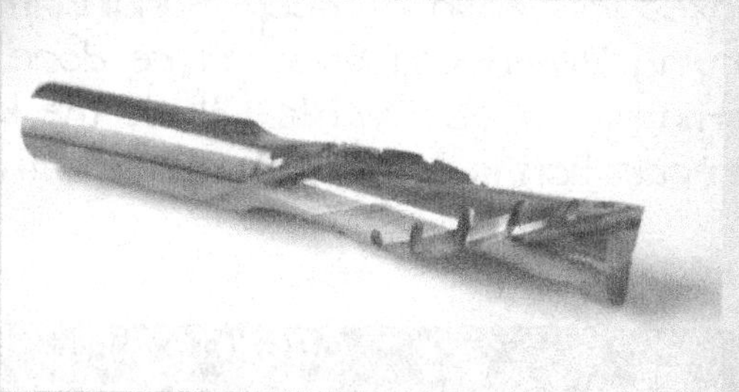

Some CNC Artisans consider material hold-down to be the most challenging part of working with a CNC router.

High volume shops working mainly with sheet-goods (cabinet shops) tend to use vacuum tables to hold materials in place. Other options include double stick tape, screws in places that will be removed in future operations, clamps in places not being cut, plastic nails, and cam clamps to name a few. I chose cam clamps for this project as I felt it was my best choice to counteract the high cutting forces developed by the 1/2" rougher/finisher bit I was using. The one that "got away" was clamped with the cams clamping directly on the bark face. This slightly spongy surface proved inadequate to hold the blank against the forces of cutting ... and the rest is history.

Machine and hand sanding, the last step before hardware and finish, is the hardest part of this project, physically. Good choices with both router bits and saw blades can minimize this chore. I have a palm sander I've modified to help with sanding the interior of boxes less than 1.5" deep. Sanding the bottom of irregular shaped pockets can be very trying. When you think you're done sanding, look at it again in different light and after letting it sit for awhile. I think the best way to see if a box needs more surface preparation is to photograph it ... the camera doesn't lie!

"Stop" hinges hold the lid open just past 90°. It requires a gain cut for the hinge ... Just another chance to mess up!

A "quick and dirty" solution to keeping the lid from flopping 180° open. Using round head screws with a small hinge causes interference and acts as a stop hinge.

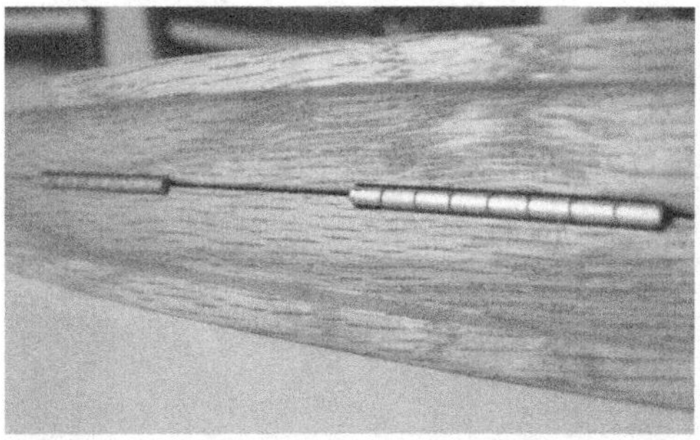

An elegant solution to the stopped hinge dilemma. Here we use a design feature of the box -- lips, to contact at the desired amount of "open".

This example also shows Kerf hinges being used. Gain, kerf and shim are all cut on the CNC router at time of production, making assembly a matter of tapping everything in place.

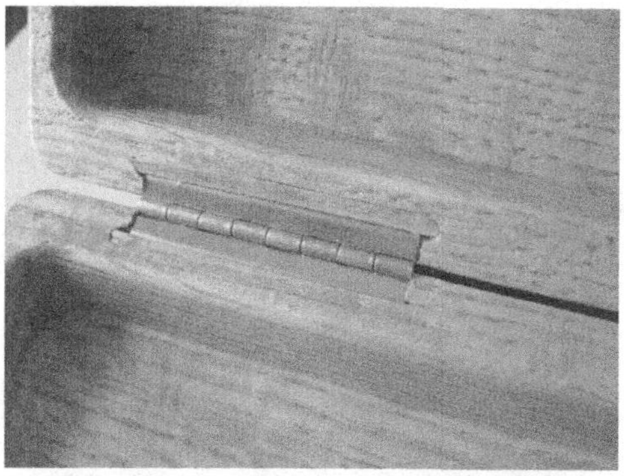

**Hardware and finish**

My favorite source for small box hardware is craft-inc.myshopify.com. They have many hinges available, reasonably ... The L3237 Mini is a favorite of mine.

Generally people don't like the lid of their boxes to flop completely 180° open. You can get "stop" hinges that only open 90°, use expensive "quadrant" hinges, use cable stops or like me ... use round head screws where flat were expected. With these mini hinges the round heads hit each other and stop the lid at a nice angle! Use plated steel screws. Brass screws are soft and will shear off at inopportune moments. In any case, pre-drill screw holes. There's nothing like having that last screw shear off, preventing delivery of a perfect project to a impatient, important customer.

I chose to not put latches on this series of boxes... The live edges made traditional latches/catches a poor match. I also did not embed rare earth magnets as I so often like to do because I didn't want the eventual owners tugging on the bark to overcome the mated magnets' attraction to each other.

I tend to use rattle can finishes as I dislike cleaning my gun for small projects. Typically after belt, flap, and hand sanding down to 220 grit paper I apply three to four coats of gloss lacquer sanded with 300 and 400 grit paper between coats. I'm partial to a satin finish and tend to use a Deft clear satin finish as my last spray coat. After several days of curing, I apply a coat of beeswax dissolved in turpentine, let it flash off, and buff it with a power buffing wheel. This procedure gives me a finish with a warm glow and zero tendency for the lid to stick to the box after sitting for extended times.

I started incorporating CNC equipment in my shop in 2006. CNC equipment frees me from non-creative tasks and changed how I think, design and work.

I'm retired, but the truth be known, once you have been self-employed the fever never quits. Right now I'm working with two start-ups and the "retired" status could be in jeopardy! For other examples of my work see: TreasuresFromTrees.com.

# Interview with Ralph Bagnall
# Owner of ConsultingWoodWorker.com

Ralph is a well known woodworking expert with over 20 years of experience across the industry, from hand tools to CNC machines. He travels extensively, helping manufacturers and shop owners get the most out of their operations using lean manufacturing concepts, and consults well known brands testing and improving their products before they go mass market.

Find out more at ConsultingWoodWorker.com.

**Can you give me a brief background of your experience in the woodworking field and how you eventually got into CNC machines?**

I started woodworking in the basement of a rented home. My wife and I wanted nicer things than we could afford as a young married couple. My first "real" woodworking project was a swinging crib for my daughter. I kept the home woodshop over the years but also started working in professional shops. In 1990, I was working in a company that decided to acquire a CNC, and knowing that this was the future of the industry, I asked to be one of the people trained to program and run the machine. In about a year I was pretty much the only person working with the CNC.

**You actually had to learn G-code to run your CNC machines -- how long would it take you to program a project, and can you give me an idea of the skills required compared to the seeamless software emulators in use today?**

Writing a successful program to create a typical base cabinet side could take several hours. Literally, the designers would draw the side on their computer, print it out on paper and hand me the paper to enter the information on my computer! The programming had certain features to assist, but they were primitive. So I had to enter in each line of code specifying the X,Y coordinate that was the end of the next machining step. To program a radius, I had to enter the end point of the arc as well as the center point.

While this took time, and today's software is indescribably faster and more powerful, there were some advantages to the old system. For example, I could write subroutines in the programs.

This would allow me to create one program that could then be used by simply typing in parameters at the start. For example, I wrote a program to cut half laps in the mullions for a divided light store front window. We were doing several mall stores for this customer, all looking the same, but each needing to fit a specific opening size. Once the program was written (about 15 hours) all I had to do was enter the number of panes in the window and the program adjusted the half lap spacing to suit.

Operators today do not have to learn G-Code at all in order to make money with a CNC. This is a great thing in general, but knowing G-Code makes it a LOT easier to troubleshoot problems, or 'tweak' programs to minimize run times.

**Eventually you got into direct sales of specialized woodworking CNC machines, how much of a transition was it from operator to sales? What were some of the challenges you had to overcome?**

Obviously, selling requires VERY different skills, so learning those selling skills was a challenge. The advantage I had was that coming off of the production floor, I knew the challenges that the customers were facing. Many CNC sales people at the time had never actually made anything from wood, with or without a CNC. The biggest challenge for me was figuring out that each client had a particular 'button' that would get them to close the deal.

This 'button' is different for each client, based on their needs, personality, whatever. Learning this was challenge #1, then figuring out the 'button' for a particular client was an ongoing challenge.

As an example, the company I worked for had been trying to sell this one client for years before I started working there. I made several visits with my supervisors to this client. There was no question that a CNC would save them money. They understood this and we all agreed on the cost justifications that were put together, but they just would not sign the contract.

Because of the old equipment they were using, the set up time required that hundreds of parts be made with each set up, so at any given time, they had a massive inventory of parts waiting to be assembled.

As we were leaving their plant at the end of yet another visit with no sale, I asked the CEO how much money they had tied up in all those stocked parts. The answer was in the millions. I reminded him that their own numbers showed that the CNC would reduce that inventory by about 75%, but even if it were only 10%, that would pay for the machine. We left, but got a signed contract back from them the next day.

**With all the experience you have gathered through the years, you began consulting companies on woodworking products, increasing shop efficiency and employee training. Can you share a few notable stories of some of the work you've done in this field and the results you've been able to achieve?**

My favorite story to tell is about my first client in Jamaica. On the first visit, they showed me some furniture designs for a new hotel being built on the island. The hotel wanted modern 'slab' style furniture made from thick slabs of solid wood. My client was set up for panel processing, not solid wood. I helped them set up their CNC to miter fold plywood parts. When assembled, they looked just as the hotel wanted. They only trick was to convince the client that this would work as well as solid wood.

To seal the deal, my client erected stud walls and actually built the hotel room to exact dimensions in their rather large lobby, built samples of all the furniture and put them in the "room" with a fake TV, pictures on the wall and even curtains for the windows. When the hotel clients arrived, we met with them in a full scale mock up of their hotel room!

Not only did they get to see and use the furniture, but they actually made some changes to the furniture placement based on being in the room! My client got the job, and the best part is that two years later on another visit to work with my client, I actually stayed in one of the hotel rooms that I helped them build!

**From my experience, there seem to be two solid camps in the woodworking field: those who stick to traditional non-CNC working methods and those that actively embrace new technology coming into their shop. Can you help explain this disconnect and possible solutions to get more CNCs in traditional woodworkers' hands?**

Personally, I find this 'disconnect' amusing. Those who claim that "the old masters didn't use CNC's" fail in their historic knowledge. For example, the circular saw blade was invented by a Shaker woman and adopted very readily and quickly by Shaker craftsmen. Frye's Measure Mill in Wilton, NH is the oldest water powered mill still operating in the US, and what those folks back then did with no electricity would blow your mind! Variable speed drill presses, gang-rip saws, a veneer cutting lathe, you name it, they automated everything that they could imagine. Men like Thomas Chippendale made everything by hand because they had no choice. But make no mistake, they were business men first and foremost, and I believe that they would have adopted new technology were it available to them.

There is always a place for hand working skills, even in a CNC equipped shop. When I worked at a stair company, I might spend my morning drawing 3D computer models of parts and cutting them on the CNC, then spend my afternoon squaring up a spiral staircase handrail using a Stanley #113 compass plane. Cutting a square profile on a corkscrew shaped laminated block of wood requires a level of control not possible with power tools, but that Stanley plane was perfect for removing just the right amount of material. To this day, in my own shop, I do most of my work with power tools, and use hand tools to 'sweeten' the fit and finish.

**You do a lot of public speaking and outreach within the woodworking field and even managed to get Rockler to sponsor your endeavors. How did this all come about? Was this a conscious decision or did it just evolve over time?**

The decision to start was conscious, the end result I certainly did not plan! I had always enjoyed reading various woodworking magazines, but as my skills grew, I began to find that fewer and fewer of the articles were intriguing or challenging. So I decided that I wanted to start writing my own articles. I created a sample article that I could send out, and then I contacted every editor of every magazine that I could get an address on. This was 1999, so not every magazine even had an online presence yet. I sent out dozens of letters. Most I never heard back from, a couple sent me very nice letters explaining how impossible it was to break into the business, but Woodworker's Journal asked to see my sample. They published it in 2000, and I have been writing freelance articles ever since.

Being published in a national magazine helped to open doors. I now had 'passed muster' with a group that was well known in the industry, so when I talked with others in the field, I was taken more seriously. I have been published now in WWJ, Woodcraft, Wood and even Woodshop News. I began running my website to share some of the tips and tricks I have learned over the years. My relationship with Rockler began in 2009 when they introduced their Bench Cookies at the AWFS trade show in Las Vegas. They were looking to spread the word, so I asked them to send me some samples to review, which they did.

At the time, there was a lot of talk online about the cookies, and one of the questions that kept coming up was; "How good will they hold when they get covered in dust?" So when I got my samples, I created a video review and tested that question. (They work just fine) Rockler liked the video well enough to ask if they could post it on the product page for the Bench Cookies. I said yes. They liked my work well enough that I began producing product videos and blog posts for them on a regular basis. Many of the product pages on their website have videos I produced for them. Having them sponsor my site was a fairly natural progression from there.

**You recently began producing videos sharing your woodworking knowledge and have published two books thus far, what aspect social media teaching do you enjoy most?**

So far, I like all of it! Woodworking as a hobby is great, but often you work alone, and finding other woodworkers to chat with and share information with can be challenging. The internet and social media have made this much easier. I still don't often get to meet other woodworkers face to face, but I have many friends online that I converse with regularly through Twitter, Facebook and online forums. Honestly though, I love it when a woodworker shares one of my articles or projects that they have built. I really enjoy seeing how they interpret my designs.

**With all the knowledge you've been able to accummulate thus far, what are some of the biggest challenges people new to the CNC field need to overcome? What are the common issues you see time and again?**

The biggest challenge has not changed much over the years I have been working with CNC machines, and that is that the basics never change. One question often asked by company owners about to buy a machine is: "Do I need to hire a computer person to program the machine?" The answer is "No." The way that the bit in a CNC cuts wood is no different than the way it cuts in a hand router. I have always found that it is easier to teach a woodworker how to run the computer than it is to teach a computer person how to work in wood. The fact that the software gets easier to use every year only reinforces this.

**Are there some features you see on other woodworking machinery that you wished would make their way onto CNC machines such as table routers?**

CNCs, even entry level ones, are getting better almost every day. Rotational devices are even now available for standard 3 axis machines that allow for milling circular parts like spirals and screw threads. The only thing currently lacking in my estimation is the ability to actually program in all three axes at once. Typically, what are called 3 axis machines are actually programmed in "2 1/2 axes." You can cut in X,Y or X,Z, or even Y,Z but making the machine cut in X,Y and Z at the same time is not readily possible in most of the current software sold with typical CNCs.

If you want to cut a shoulder into a disk so that the shoulder starts at 1/8" deep, drops to 3/8" deep at the half way point and returns to 1/8" deep, that is difficult within the program itself. The machines and programs are certainly capable of it since a spiral ramp cut is offered as an entry option, but that is set up as a subroutine, not an active programming step.

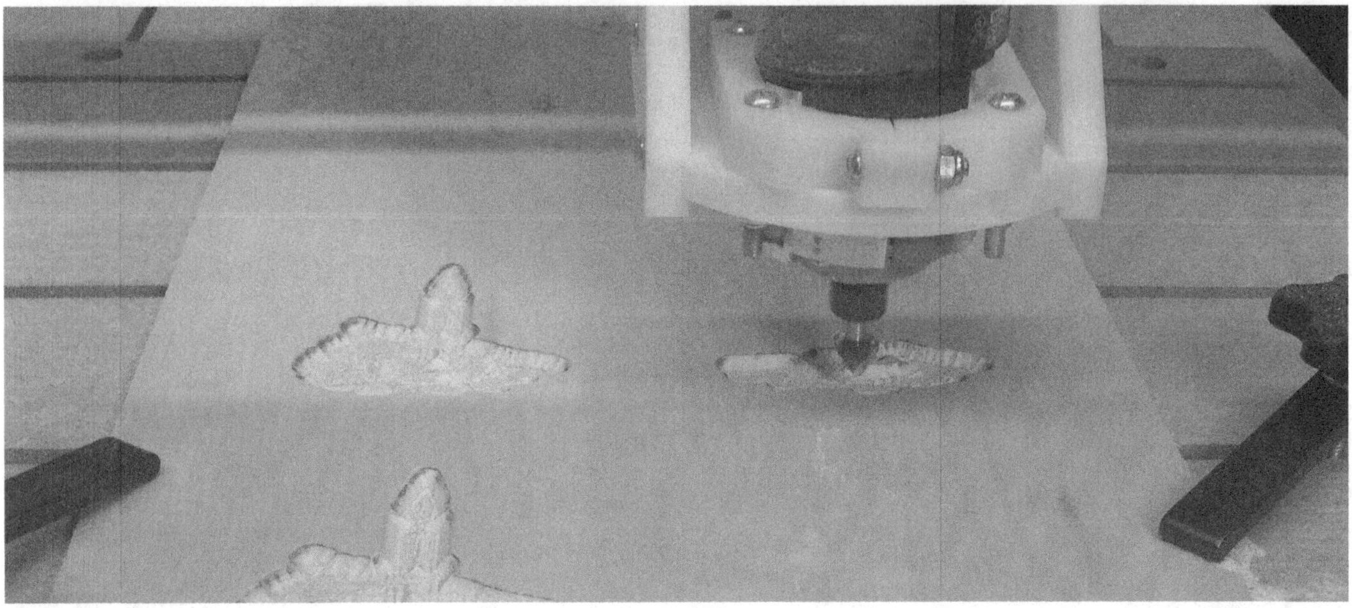

**How do you see the woodworking field -- specifically related to CNC adoption -- evolving over the next few decades? Will most subtractive technologies be replaced with additive ones (3D printers)?**

Interesting question! I just acquired a 3D printer myself, and I have been writing and talking about the future of manufacturing for some time now. While 3D printing (additive manufacturing: parts are 'built up' from raw materials with very little waste) may well replace milling (subtractive manufacturing: parts are cut out of larger raw material blocks) in time, there is plenty of crossover between the two. The learning curve for the software is shrinking all the time, but the mental skills to be able to envision how a part should be made are transferrable across manufacturing platforms. It is a small step from being able to efficiently program a traditional CNC to programming a plasma cutter, laser, or even a 3D printer. The real skill lies in recognizing the capabilities of the system at hand, and maximizing them. For example, a 3D printer can often create parts already assembled like a gear on a shaft.

I really see a "Cottage Industrial Revolution" beginning. For about the price of a new mid-sized car, one can create a manufacturing facility in their garage capable of making at least half of the common items in the home. These can be easily custom made to order, do not need to be transported halfway around the world, and require no packaging that needs to be disposed of. No excess inventory is produced that is scrapped when it does not sell, and no hassles with trying to figure out how many to make and store and where to keep them. It is environmentally sound, and people will be able to know and talk with the folks who make the items that they use daily.

## CNC'ed Guns: Unwarranted Fear and Paranoia

If you watch American news for any extended period of time, two things will become abundantly clear. The first, the incredible amount of gun crimes and the second, the "weekly fear" that is pushed by media to scare the populace. It's quite incredible to watch this and compare it to news coming from other countries in the world. Not a week goes by where somebody didn't get shot in just about any major city; contrast this to Japanese news and you'd be hard pressed to hear about anybody getting shot anywhere across the entire country over the same period of time!

What came to the forefront this year was the fear, instilled by legislators and media pundits, about the possibilities of some psycho getting their hands on digital files and 3D printing their own gun. The most common situation that is expressed is somebody circumventing metal detectors at an airport or important function with their plastic gun and hurting people with it. Let's de-construct this argument a little with logic.

1) REAL guns are cheap and not difficult to find in America
You can buy them at gun shows, second-hand, or online; or, if you are so inclined, steal one if given the opportunity. Police say that the cost for a black market used gun is as little as a hundred dollars if not less. The economics don't make sense to spend a few thousand dollars to buy your own CNC equipment, spend time to learn how to use it, and THEN make a gun. The ROI on your equipment would be horrendous vs just buying one!

2) Guns require metal bullets
All bullets are made of metal with gun powder used as the propellant. This WILL be found with a metal detector and at airports. They have explosive testing machines as well. Even if you did get through with a plastic gun, the bullet most likely wouldn't.

3) You don't need a CNC to make a lethal device
Guns and other lethal devices can be easily made with pipes and other devices found at most hardware stores. Plans are easily found across the internet for such contraptions. If you wanted to be very creative, you could seriously hurt somebody using nothing other than a blunt pencil eraser. Easy access to a CNC doesn't produce new malicious ideas.

4) CNC guns aren't accurate or powerful
If your goal is to circumvent a metal detector, your entire gun will have to be made out of plastic, wood or some other benign consumer-friendly material that simply doesn't have the strength required to shoot anything with any type of accuracy over a meaningful distance. There is incredible pressure created in the barrel to shoot the bullet, something that can't repeatedly be generated adequately using material weaker than metal for the barrel.

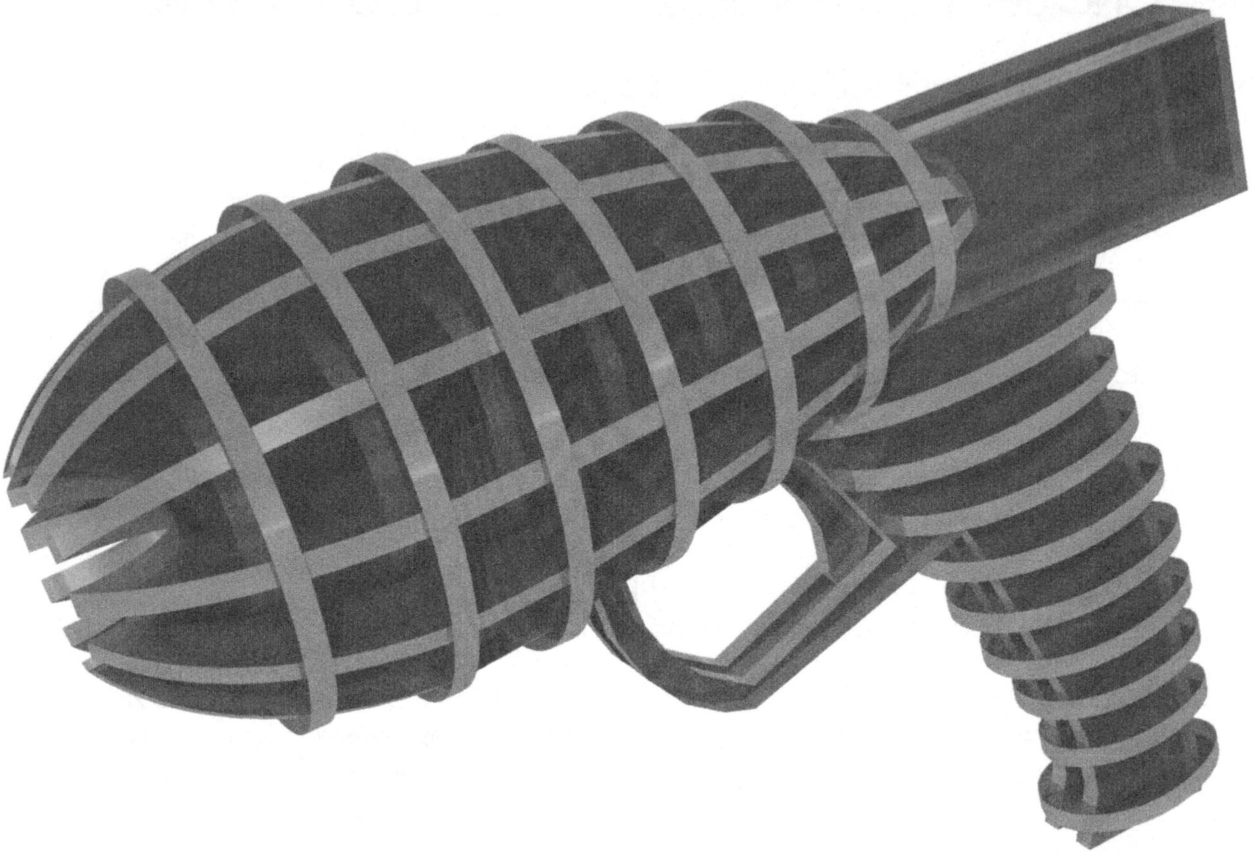

Although I could go on, I'd suggest that there are far more dangerous habits and creative mechanisms in this world to hurt other humans than the sudden availability and proliferation of consumer friendly CNC machines around the world. Fear, especially misplaced fear, shouldn't be allowed to propagate and hold a new technology hostage.

When I was a kid, I spent lots of time in the woods making bows that were actually pretty strong and could shoot arrows up to 100 meters away using nothing but a shoelace. Growing up, I saw lots of craft shows selling wooden guns and other elastic powered mechanisms which I found quite fascinating.

I fully plan on developing a whole series of CNC guns for CNCKing.com, but they will be elastic powered and, instead of bullets, use small wooden projectiles or elastic bands -- they are great fun, they are educational, and I regularly get requests to make such devices. Laser cut, table routed, even 3D printed and plasma cut guns are one of the areas I'd really like to explore further. I won't allow people who dispel uneducated paranoia about CNC equipment to deter me in my endeavors.

# Brian Lawrence's Design Process

Brian standing next to Hideaki Yoke, or Yoke San at Botcon 2008 in Cincinnati Ohio. He's one of the original creators of the Micro Naughts and Transformers.

Inspired by Star Wars from a young age, Brian Lawrence knew he'd be designing cool stuff later in life.

After attending Columbus College of Art and Design straight out of high school, he was hired by Kenner / Hasbro to work on a Batman 3 Batwing toy. He later designed additional items for such notable world brands as Star Wars, Jurassic Park, Nascar Racers, and Transformers!

As for tools, Brian prefers to start with pen and paper, then move on to Illustrator, followed by Photoshop and Sketchbook for added depth and visualization.

He prefers to use acrylic and plywood for testing -- using a table router for larger models and a laser for more delicate cuts.

His advice to young designers is to follow your passion, and I'm very happy that he accepted my invitation to showcase his design process in this book.

He approached me after visiting CNCKing.com, and his talent has already inspired many people around the world.

You can find more information about him and his company Red Lion Design Studio at RedLionDesignStudio.com.

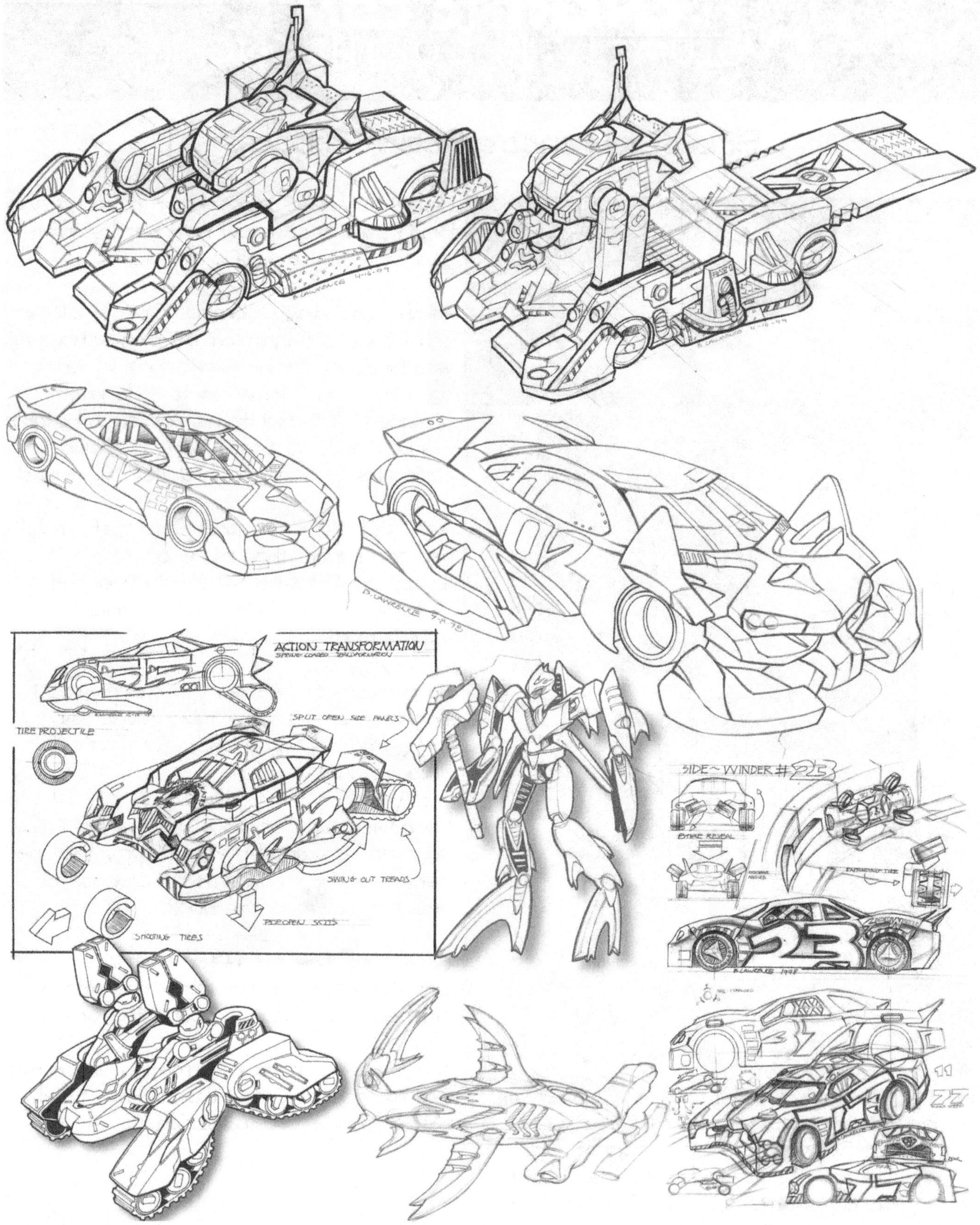
ACTION TRANSFORMATION
SPRING LOADED TRANSFORMATION
TIRE PROJECTILE
SPLIT OPEN SIDE PANELS
SWING OUT TREADS
SHOOTING TIRES
SIDE~WINDER #23
EXPANDING TIRE

## Delphinus 7
### 123D Sculpt 3D file

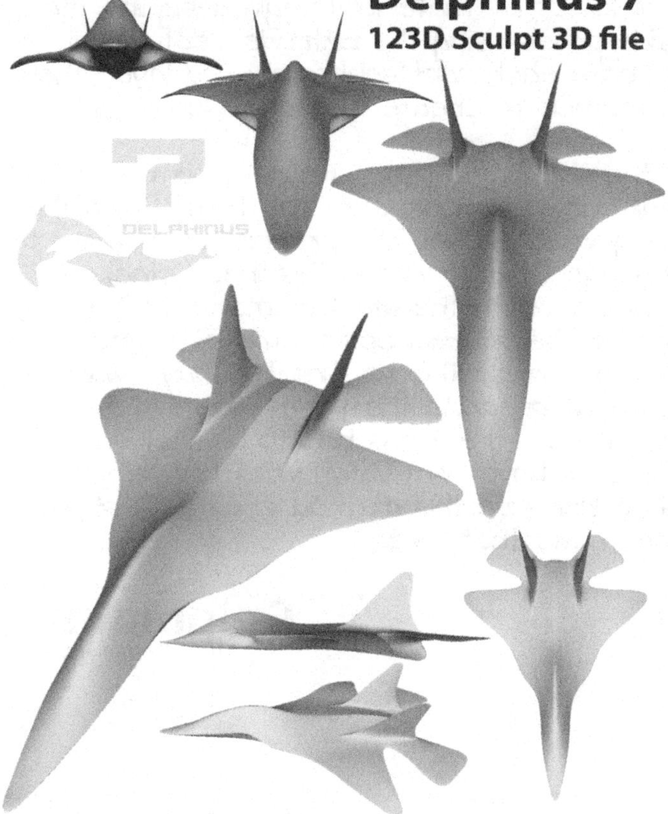


This is a file I created using 123D Sculpt on an iPad. This program is like sketching in 3D and yields very cool results. By pinching and pulling and adding and subtracting material, I was able to come up with a basic form for the jet. Rotating the mold in 3D and being allowed to sculpt sketch is a new experience for me, especially when I can use the mirror feature and get symmetry. Refining the form and smoothing out the rough spots with one of the specified tools was very nice. There is a limit to what can be done using this program, Creating airfoils is very difficult. I then wondered what it would look like in my hand.

Using 123D make, I explored different options to create the model using their methods. I wanted to see if the layered sections could be converted into files to cut birch plywood for a more sturdy model. The program limits the direction of the layering to one direction, side to side or front to back.

Because of the angles of the tail fins, the side by side profile cardboard models look strange, while the front to back looks more complete. I preferred the side to side. So thus begins the modification of the design to allow flat wings suggesting the form.

**Delphinus 7**
Isometric Layer Studies

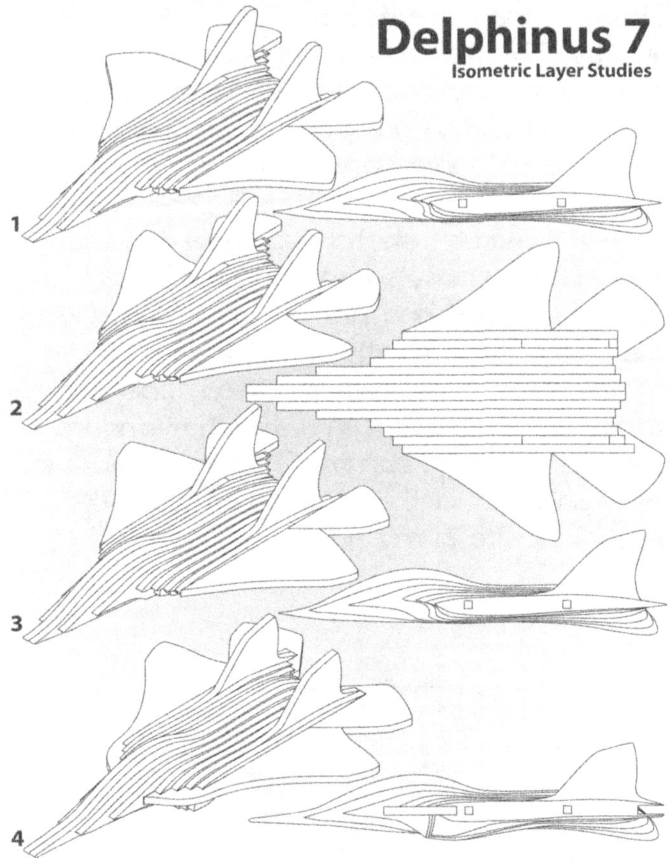


Utilizing Adobe Illustrator and having the thickness of material, I begin designing the side panels using the Extrude feature to create each part separately and layer them together to create a solid model.

I then began to play with different wing and tail configurations, changing them from the original 3D model. The proportions needed to change when it was in this linear stacked format. With the stacked method planes had to be 90 degrees from each other; I couldn't angle the tail fins the way I wanted without breaking the form where they exist. Also, the wings had to be flat and couldn't have bent down wing tips. Which is fine because this method creates a stylized aesthetic that I like.

**Delphinus 7**
Refining Chaos

This image shows the refining process in a wide lens view. Adding landing gear and possibly an opening canopy were all explored on this page. This is typically how I work in Illustrator to get to the final result. It creates a history of how I made a part, which becomes very useful when I need to replicate a similar part.

Chaos it seems, but each part was individually designed, then modified to get to the final result ... then tweaked and changed until a page full of modifications appear. I always keep these files to remind me of the creative process. About two days' worth of work in between other projects.

**Delphinus 7**
Assembly

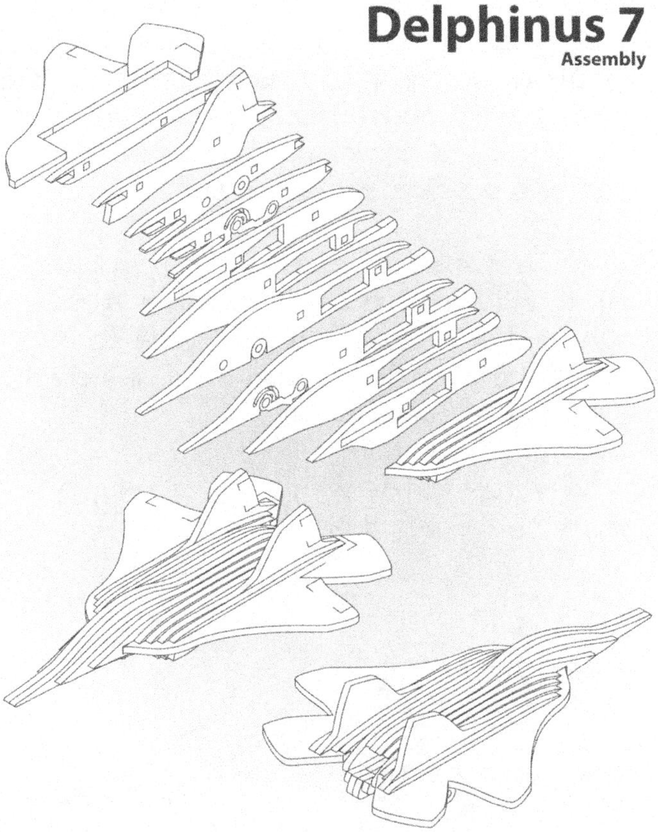

This drawing shows the layers in isometric so that I can test fit before making a single cut. The assembly also shows the early design of the landing gear that didn't work and click into place as I wanted. Using this layered method I was able to see where the airflow would be from the intakes to the thruster.

This allows me to test the alignment of each part testing the fluid design once assembled. Using the Isometric feature in Illustrator offers a low tech alternative to rendering it in 3D.

**Delphinus 7**
Landing Gear
& Decoration Studies

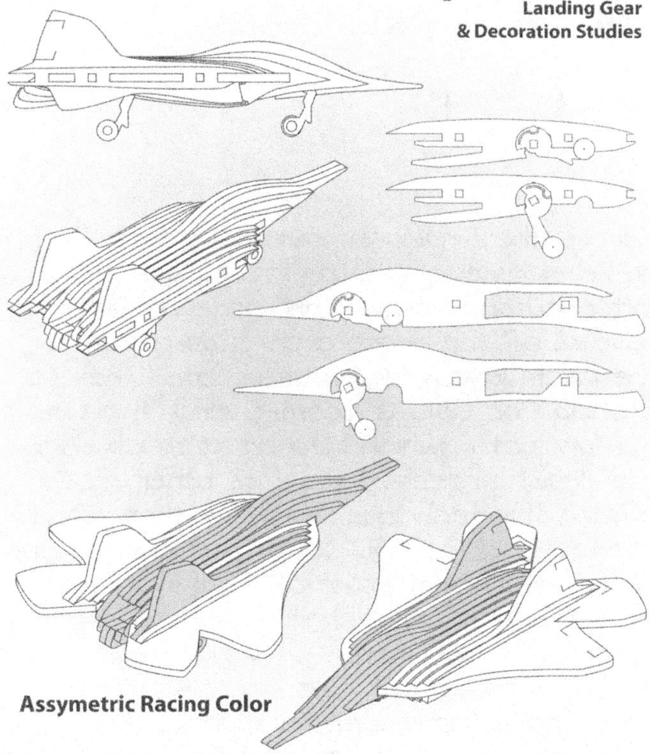

Assymetric Racing Color

Here is the refined working detented landing gear with Isometric rear view showing extended landing gear. The new landing gear design will snap closed and extended by use of a living hinge spring. I wanted to make a jet with retractable landing gear as I loved toys like that when I was a kid. Also shown is a decoration study at the bottom in which stain or paint could be applied in the final model.

This laser cut model has become more artistic than a dead-on accurate representation of the original. This is one of the joys of exploring different alternatives and coming up with a new style.

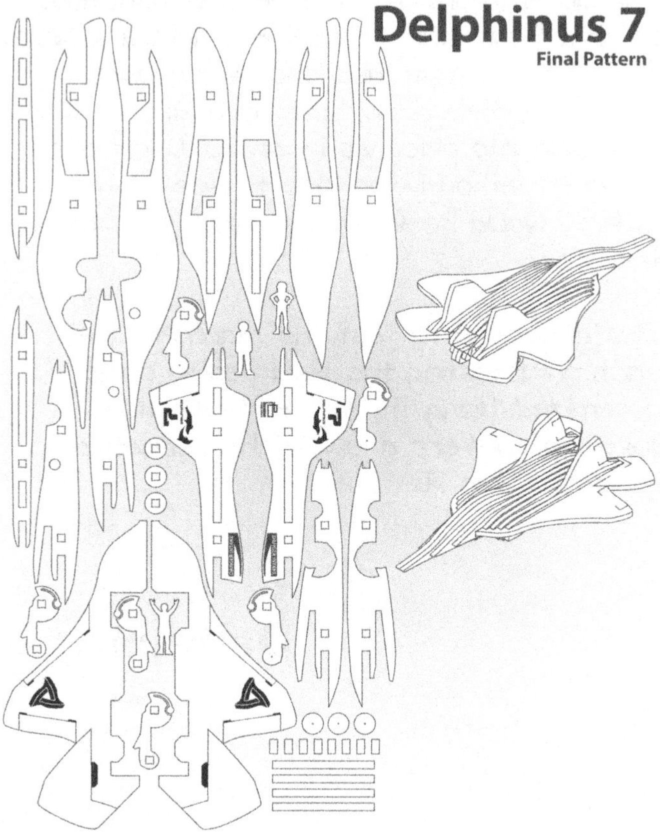


Here are all of the pieces to assemble the model, including pilot figures. I included two sets of landing gear, one with possible rolling wheels using a counter-sunk screw and the other a pegged non-rolling wheel. Also, the file shows possible etching in black.

The etching is cool because it can still be painted and show up because it has a good depth to it. It can allow a paint wipe to be laid into the etch that would be a different color than the base color.

So many possibilities with laser cutting in giving you a finished product!

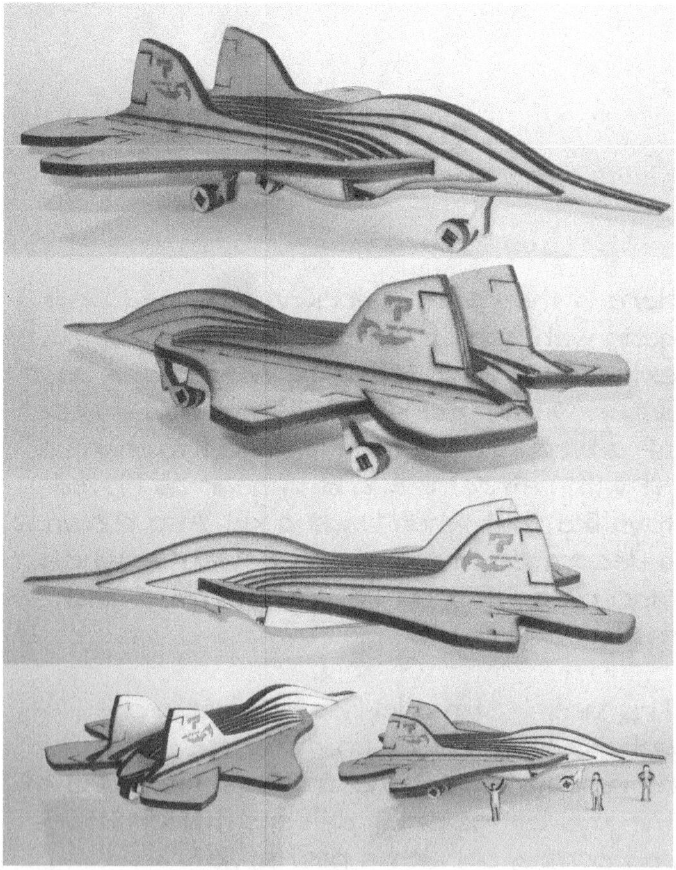

Here is the final laser cut model with different views showing landing gear extended and closed. This model doesn't have all of the photo etching as shown in the parts layout, as I have updated the etching for CNCKing.com. Overall, this was a fun project in which I leaned so much during the design process to create other models utilizing this technique. Some of them don't need to use the 123D sculpt method and can be designed right inside of Illustrator. And using the extrude technique I can construct isometric study models to view before getting a model cut. This saves so much time. I hope Illustrator can improve this technique and allow for rotation of an assembled model in the near future.

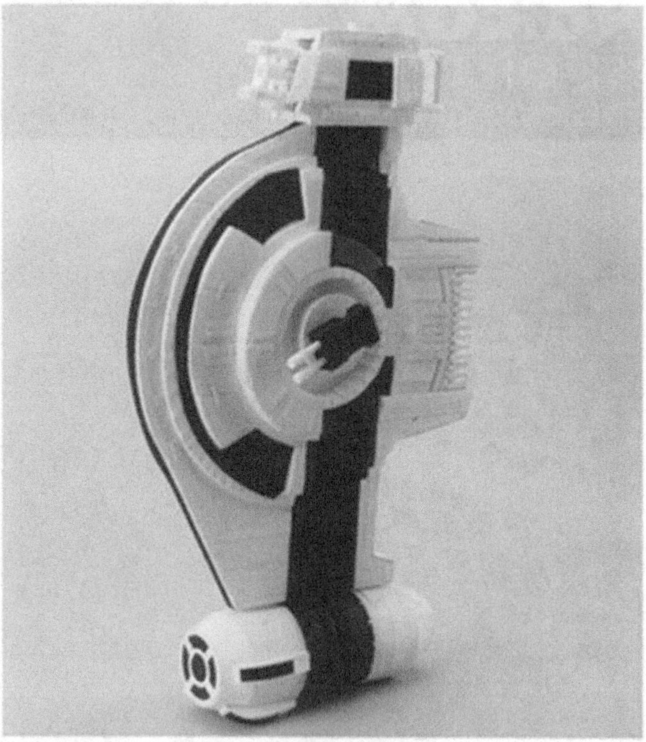

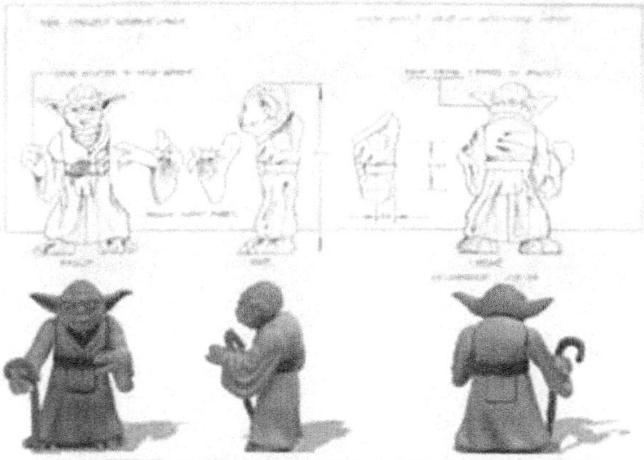

#14953 DASH RENDAR'S OUTRIDER
PHOTOGRAPH © 1996 HASBRO, INC.

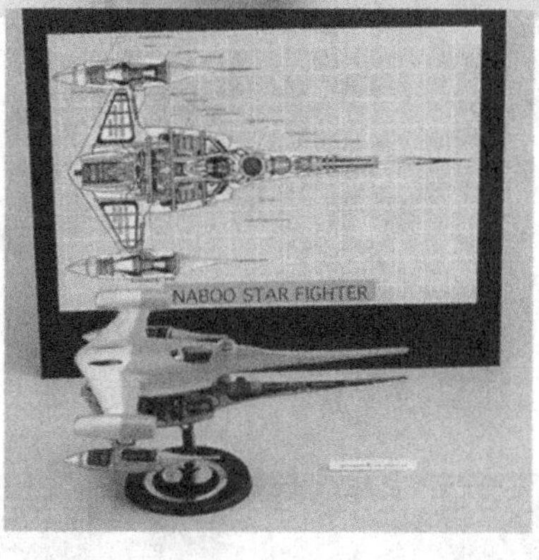
NABOO STAR FIGHTER

Grandma
You
Grandpa
Gran
Nanna
Poppy
Nanny
Ma
Mommy
Mama
Mom
Papa
Mummy
Nan
Daddy
Mum
Dad

# Designers Corner: David Newman & Jerry Rutherford

In Volume 3 of my book, I introduced two designs from David (davidcncking.com); the first, his Bobcat, has since evolved greatly. From his designs, I made a CNC Table Router version that's even bigger and further refined. Going back to an old design, as I have done on several of the ones seen in this book, edition after edition, shows you that a design is never really done. Designing is a lifelong process, not a one-time event though I do try very hard to follow my advice to design forward (new) not backward (old).

Primarily running an engraving business, David also specializes in laser cutting parts up to 2500 mm x 1300 mm in timber, acrylic and metal. In his spare time, he has a passion for designing and cutting small toy kits that he then donates to children's hospitals.

Jerry is interested in robotics, CNC machining, high power rockets and woodworking. In his spare time, he also enjoys writing web pages, teaching electronics, and of course design wooden toys on his laser engraver. He is self-taught and reads constantly trying to learn new things. You can find him at askjerry.info.

The design styles of both David and Jerry are very different as I build primarily in the virtual world while they use the physical one... the end result is a design that uses no more parts than it's needed and looks great at any angle.

What I really like about David's Bobcat is how refined the construction system for it has become with each version, from parts that only fit ONE WAY (good thinking!) to a stepped axle so that you don't have to think about the order of the wheels... there is a lot more thought in his designs and attention to detail than at first sight. He's designed a lot of really neat projects, well over 20 at this point - and I have learned a lot looking at his design and building them that have helped me tremendously for my own projects.

What I really like about Jerry's helicopter, which is why I asked for permission to include it in this book, is how each piece has a specific purpose - there is no filler, no pieces that don't have reason for being. I really like the concept of stacking pieces together to make the cockpit then applying the same methedology for the top to create the spinning rotor. I really enjoyed making a CNC table router version of his original laser design, it was by far the easiest design I've ever had to modify to fit the new parameters of the router.

I want to thank both of them for allowing me to make a CNC table router version of their designs in this book, you can find and buy many more of their designs on CNCKing.com.

# David Newman's MK3 Bobcat (CNC Laser)

Completed 35 piece model dimensions @ 6 mm L: 11.5 cm W: 5.5 cm H: 8.5 cm
L: 4-17/32 W: 2-11/64 H: 3-11/32 (in)

## Product Renders

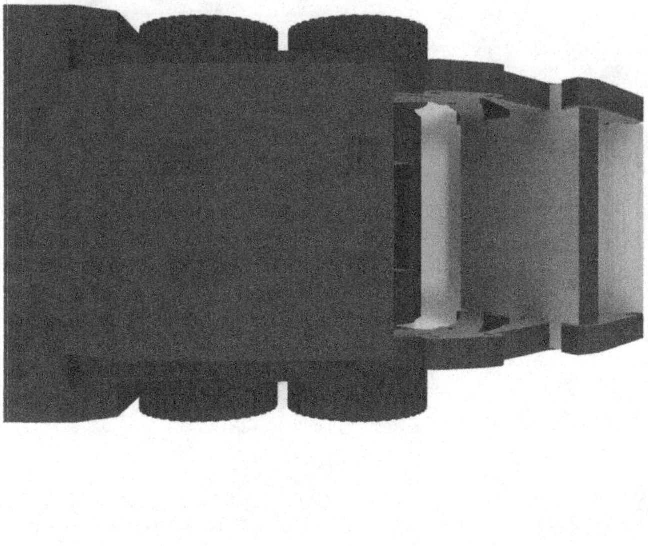

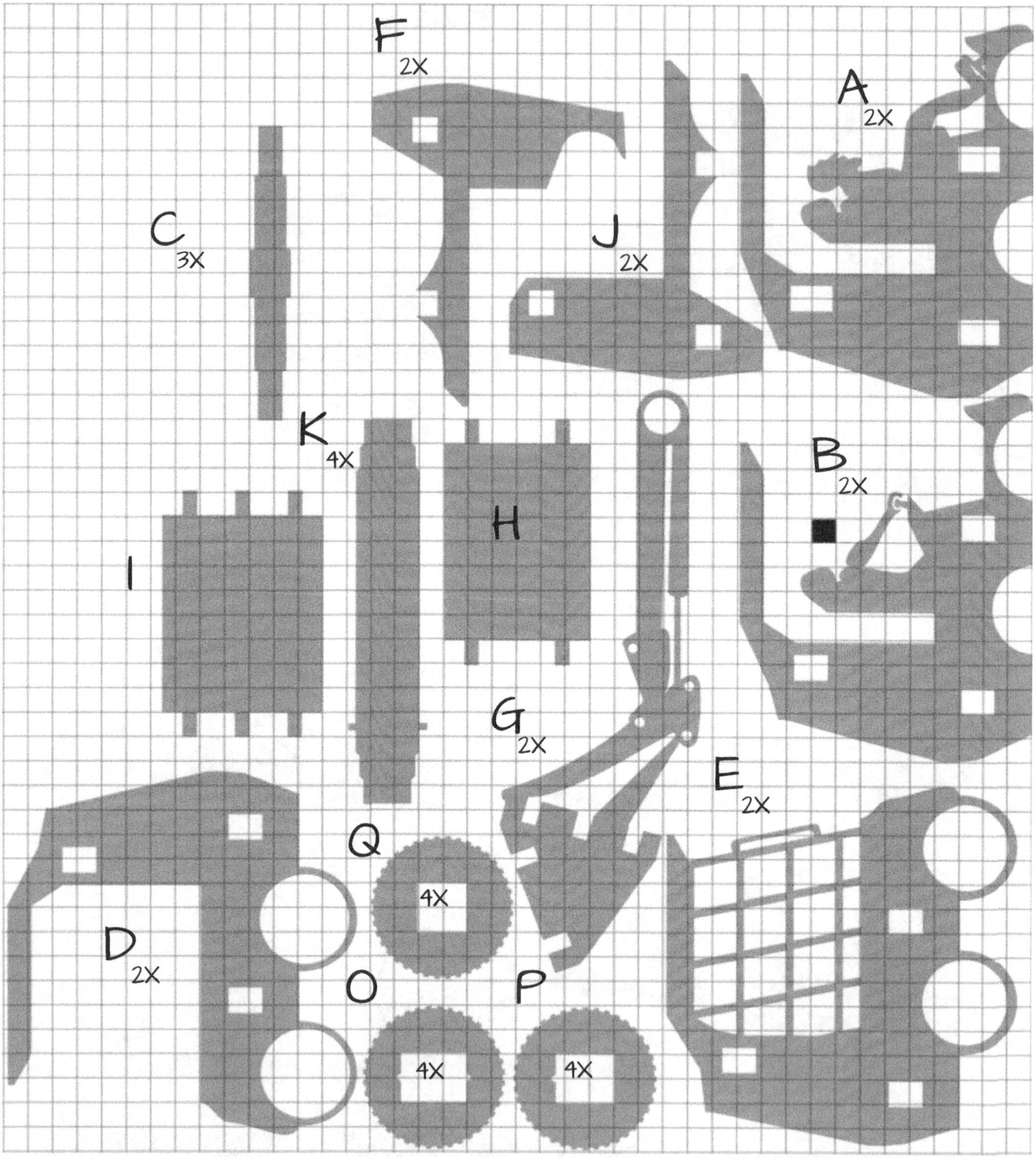
F 2X
A 2X
C 3X
J 2X
K 4X
B 2X
H
I
G 2X
E 2X
Q
4X
D 2X
O
P
4X
4X

# Step-by-Step Assembly Instructions

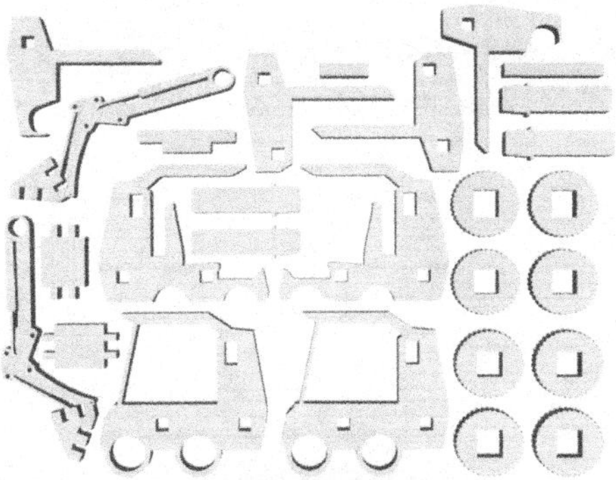

Step 1: Label and check inventory
Once you have all the pieces in front of you, the process of putting them together is no different from assembling pieces of Lego.

Label, using a pencil, the various parts and double-check your inventory. Take your time and enjoy the building process! Before jumping to the glue, be sure to do a dry run.

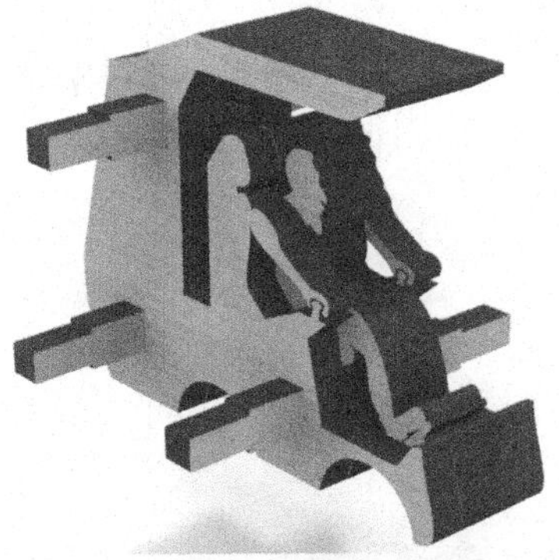

Step 2: Build the base
Glue together the part As (man with head) and Bs (man with shoulders) intersected by all three part Cs.

Step 3: Add some sides
On both sides, slide and glue into place the part Ds (cockpit without a cage) followed by Es (cockpit with cage).

Step 4: Add the bucket
Slide and glue into place on both sides the part Es (cockpit side) and Fs (bucket arm) followed by the buckets (parts H and I).

Step 5: Add axles

Slide and glue into place the outer cab ledge (part J) on both sides followed by the axles (part K) in alternate directions.

Step 6: Add wheels

You can now glue into place the wheels, starting with part Os followed by part Ps, and Qs for the final edges.

Step 7: Perform some finishing touches

There is no need to apply any preservatives or paint; feel free to sand any sharp edges down. ENJOY!

# David Newman's MK3 Bobcat (CNC Table Router)

Completed 44 piece model dimensions @ 6 mm L: 30 cm W: 12 cm H: 19 cm
L: 11-13/16 W: 4-23/32 H: 7-31/64 (in)

Product Renders

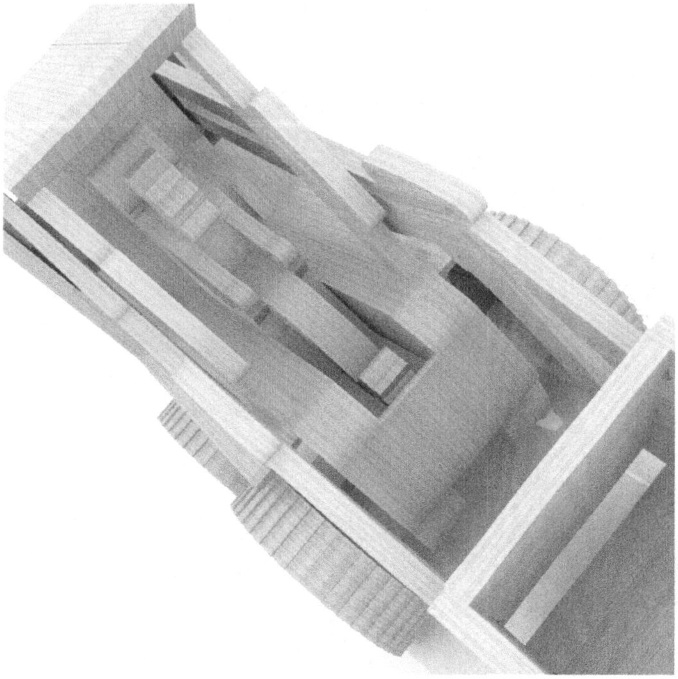

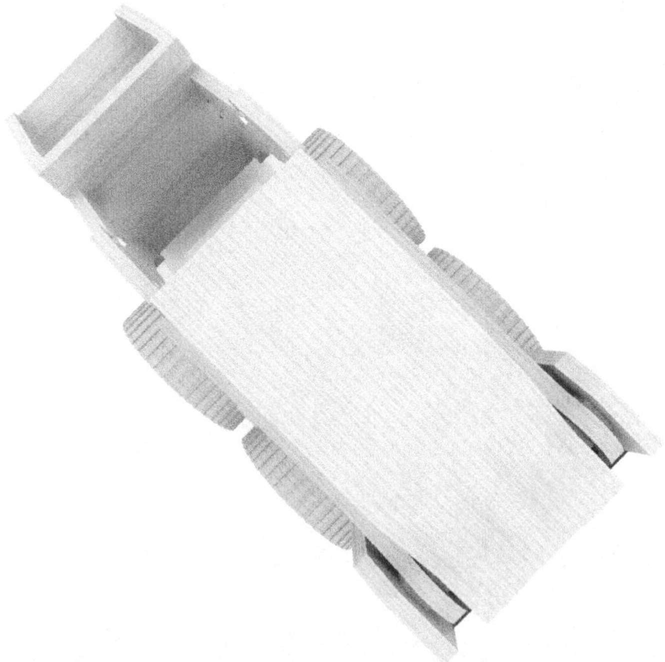

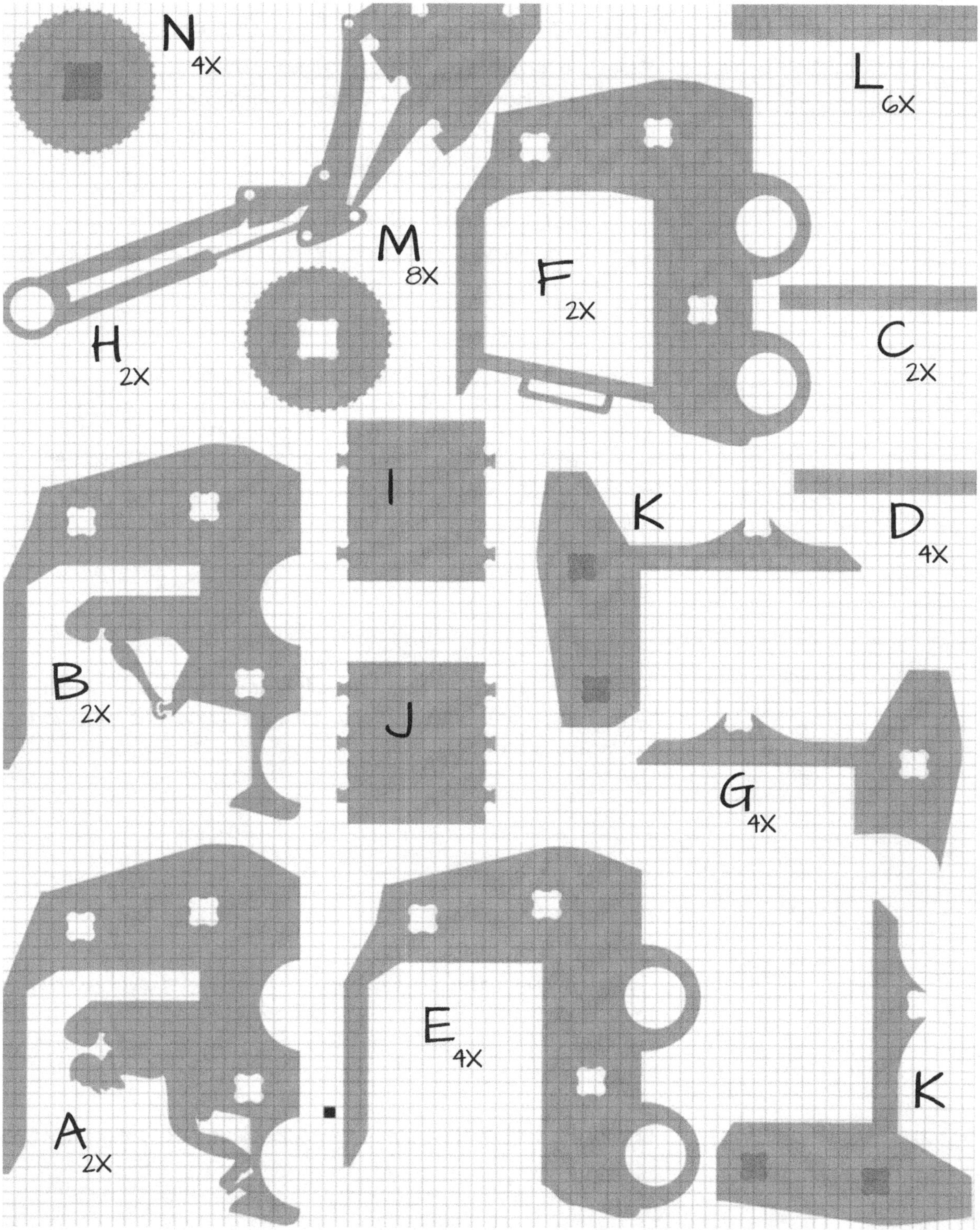
N 4X
L 6X
M 8X
F 2X
H 2X
C 2X
I
K
D 4X
B 2X
J
G 4X
E 4X
K
A 2X

# Step-by-Step Assembly Instructions

Step 1: Label and check inventory

Once you have all the pieces in front of you, the process of putting them together is no different from assembling pieces of Lego.

Label, using a pencil, the various parts and double-check your inventory. Take your time and enjoy the building process! Before jumping to the glue, be sure to do a dry run.

Step 2: Start the cockpit

Slide and glue into place the part As (man with head) followed by a part B (man without head) on both sides. In the back, slide into place doubled up part Cs (short guides) and, under the man, doubled up part Ds (long guides).

Step 3: Complete the cockpit

Slide and glue together doubled up part Es (cockpit without cage) followed by part Fs (cockpit with cage) on both sides.

Step 4: Add the shovel

Starting with the shovel, add a part H (shovel arms) to both sides into the upper doubled up part Cs on the top of the bobcat. To form the shovel, glue and slide into place parts I and J.

Under the shovel arms, glue and slide into place doubled up part Gs.

Step 5: Lock the shovel in

Starting from the sides, slide and glue part K on both sides.

At this time, glue tripled up part Ls to form the axles.

Step 11: Perform some finishing touches

Glue on the axles, on both sides, and doubled up part Ms, followed by a part N to cap the wheels.

There is no need to apply any preservatives or paint; feel free to sand any sharp edges down. ENJOY!

# Jerry Rutherford's Helicopter (CNC Laser)

Completed 23 piece model dimensions @ 6 mm L: 28 cm W: 23 cm H: 9 cm
L: 11-1/32 W: 9-1/16 H: 3-35/64 (in)

Product Renders

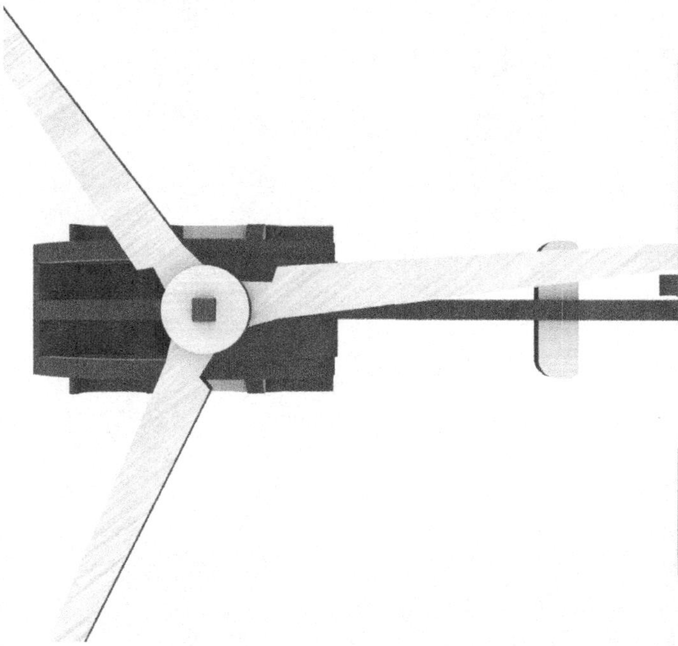

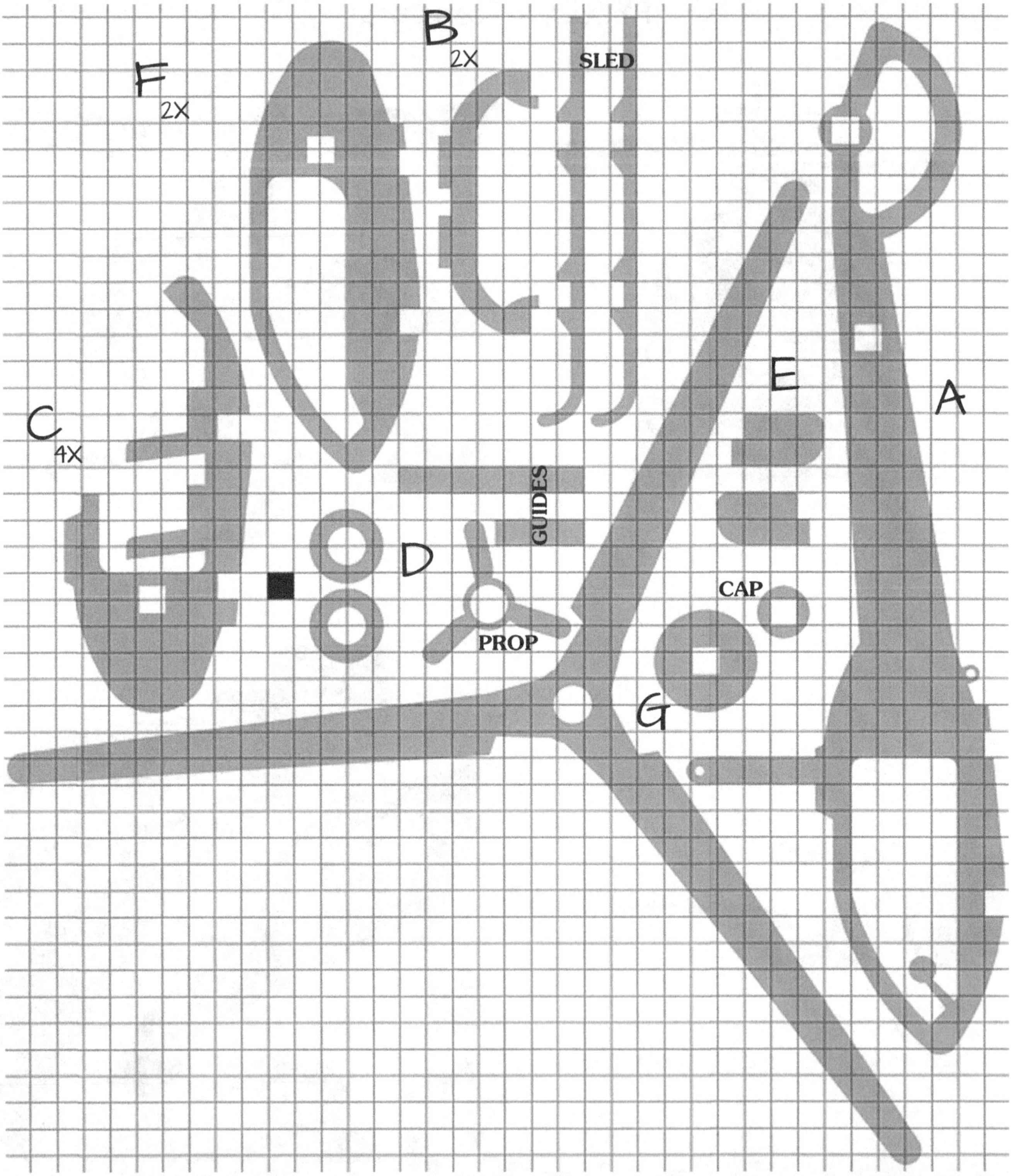
B
2X
SLED
F
2X
E
A
C
4X
GUIDES
D
CAP
PROP
G

# Step-by-Step Assembly Instructions

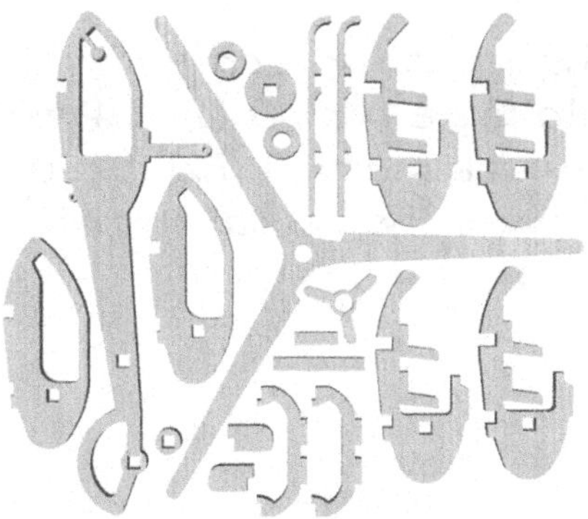

Step 1: Label and check inventory

Once you have all the pieces in front of you, the process of putting them all together is no different than assembling pieces of Lego.

Label, using a pencil, the various parts and double check your inventory. Take your time and enjoy the building process!

Step 2: Assemble the foundation

First, put both part Bs onto the sleds, followed by part A and the two guides through it. Apply glue where the various parts intersect.

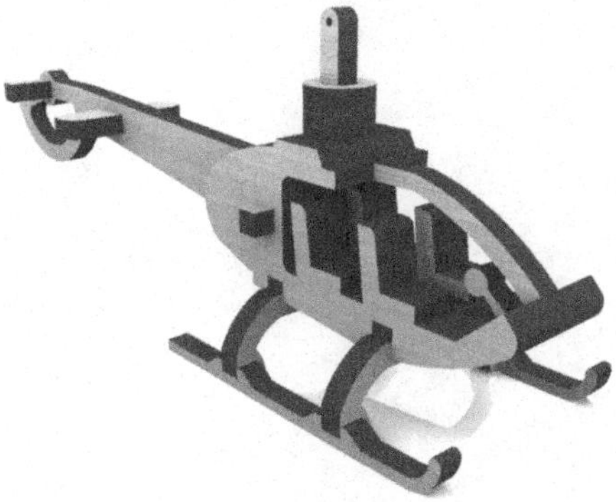

Step 3: Add the wing

Glue into place the part Cs, followed by part D and the two small wings (part Es). Allow the entire assembly to dry.

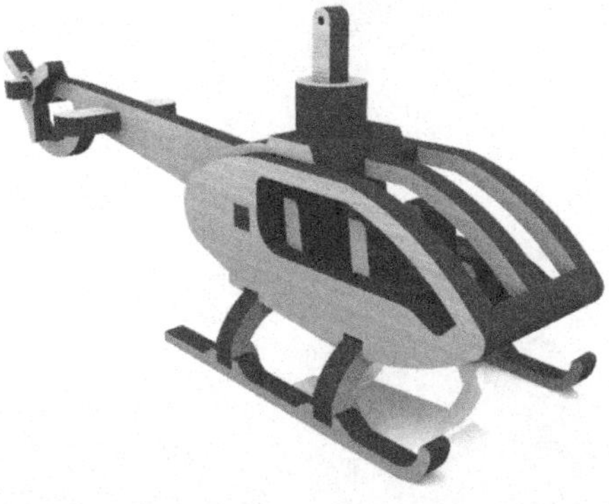

Step 4: Finish the cockpit

Glue into place the part Fs on the front and the prop on the tail end. Do not apply glue to the prop unless you do not want it to spin.

Step 5: Cap the rotors

You can now slide into place the top rotor. Leave a little space just above it and glue into place part G (feel free to use spacers to keep it from sliding down). You can use the same process to glue into place the cap at the tail end as well.

Step 6: Finishing Touches

You may now sand any protruding parts flush, along with the outer edges if you so wish.

There is no need to apply any preservatives or paint. ENJOY!

# Jerry Rutherford's Helicopter (CNC Table Router)

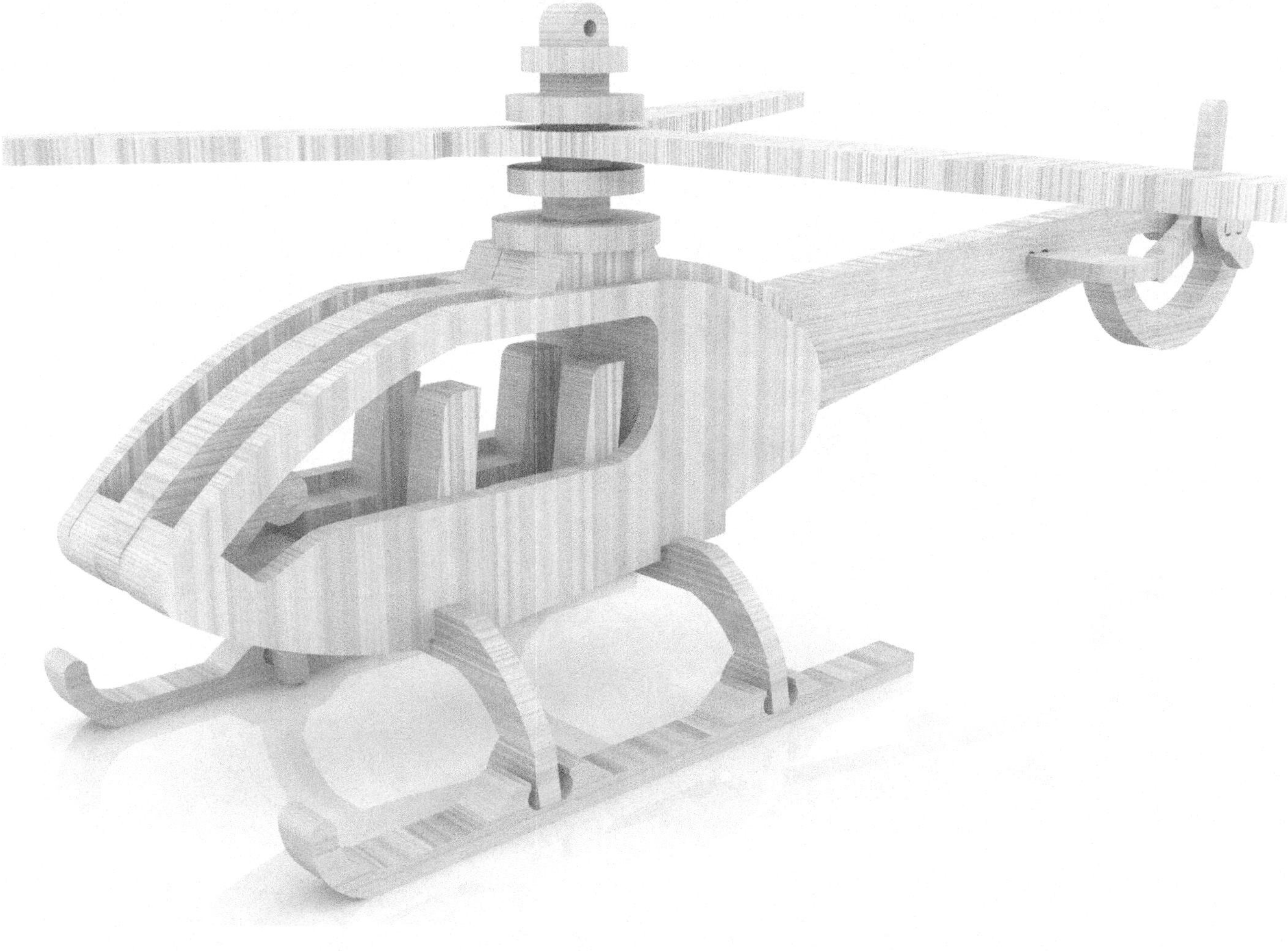

Completed 30 piece model dimensions @ 6 mm L: 56 cm W: 36 cm H: 18 cm
L: 22-3/64 W: 14-11/64 H: 7-3/32 (in)

## Product Renders

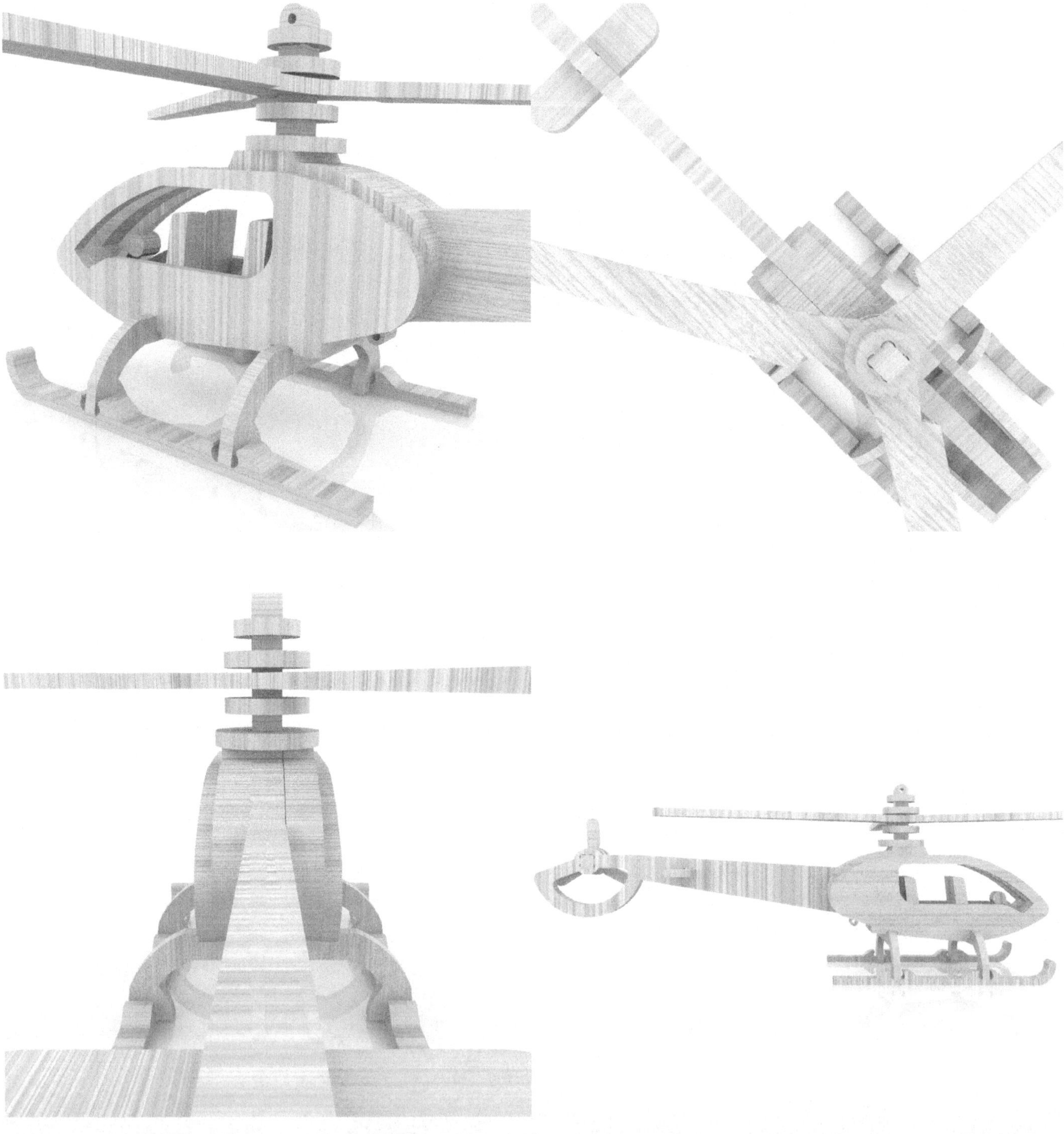

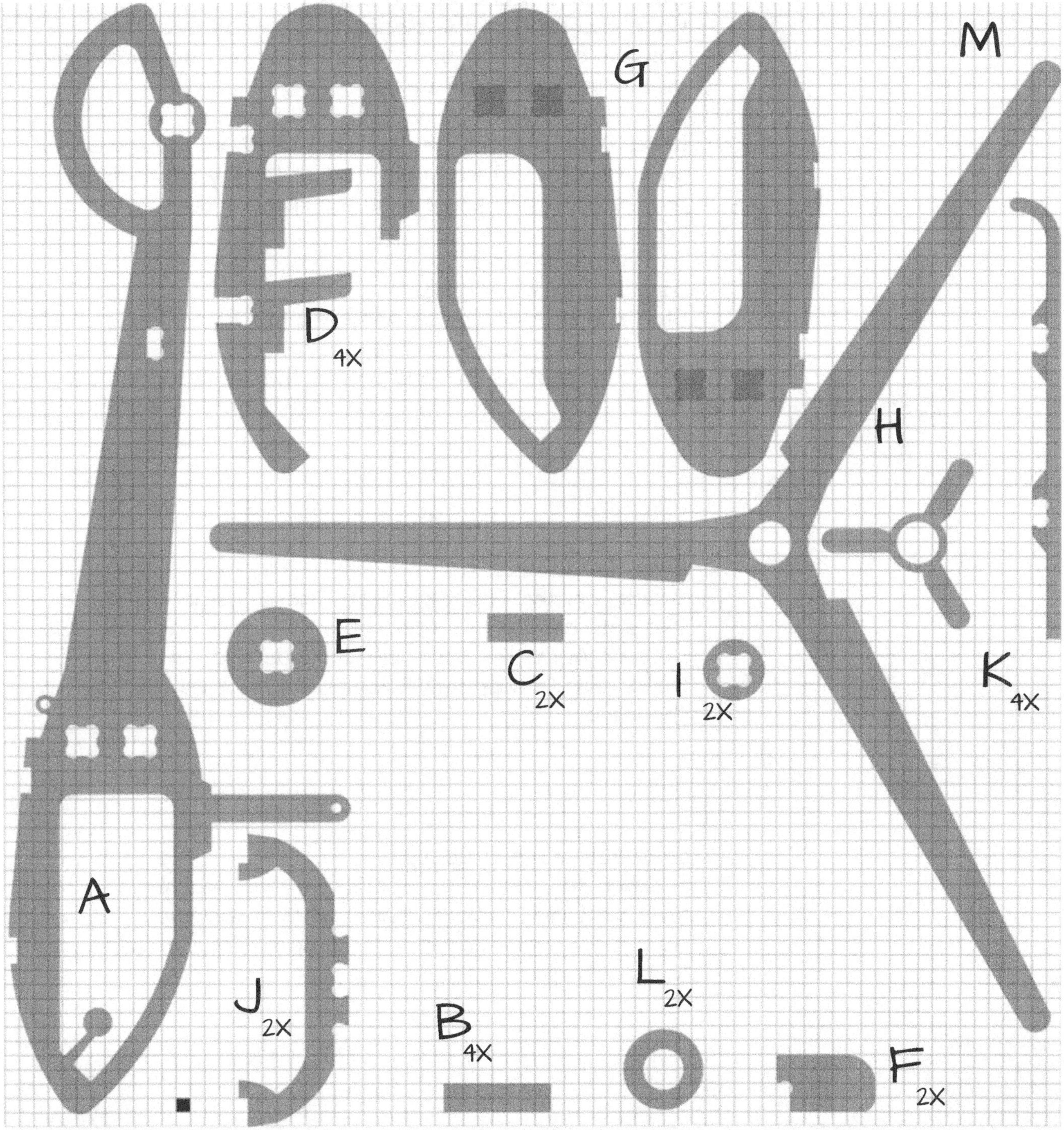
M
G
D 4X
H
E
C 2X
I 2X
K 4X
A
L 2X
J 2X
B 4X
F 2X

# Step-by-Step Assembly Instructions

Step 1: Label and check inventory

Once you have all the pieces in front of you, the process of putting them together is no different from assembling pieces of Lego.

Label, using a pencil, the various parts and double-check your inventory. Take your time and enjoy the building process! Before jumping to the glue, be sure to do a dry run.

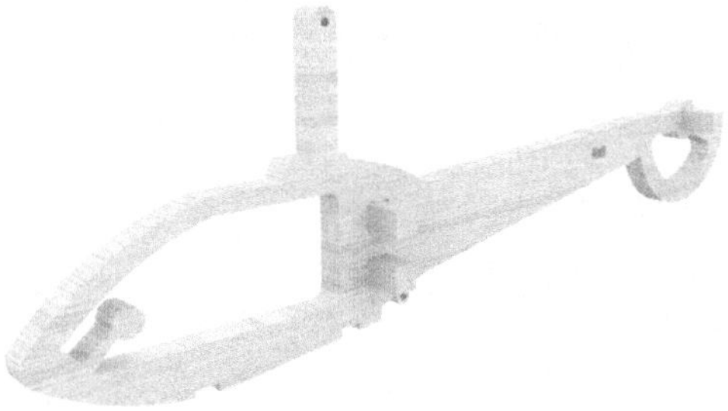

Step 2: Start the body

Slide and glue into place part A (large body) using all four part Bs (long guides) in the cockpit area and part Cs (short guides) at the end where the tail rotor will be.

Step 3: Add chairs

Starting in the front cockpit, slide and glue into place two sets of part Ds (cockpit with chairs) on both sides of part A. On the top, slide and glue into place part E (rotor spindle support) followed by part Fs (tail fins) in the back.

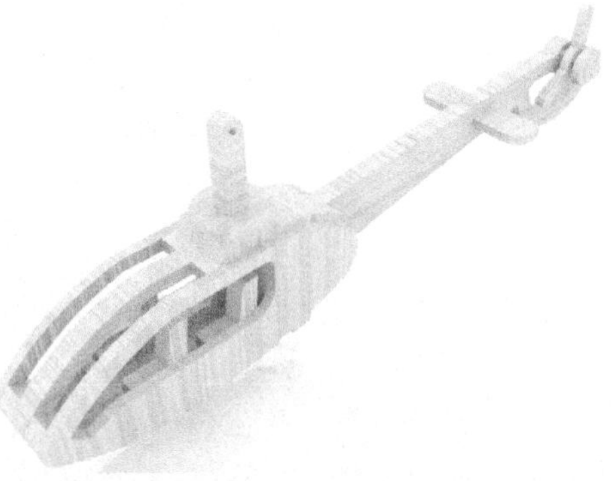

Step 4: Finish the cockpit

In order to complete the cockpit, slide and glue into place both part Gs (cockpit doors) while in the back, slide into place part H (tail rotor; without glue) followed by a part I (cap; with glue).

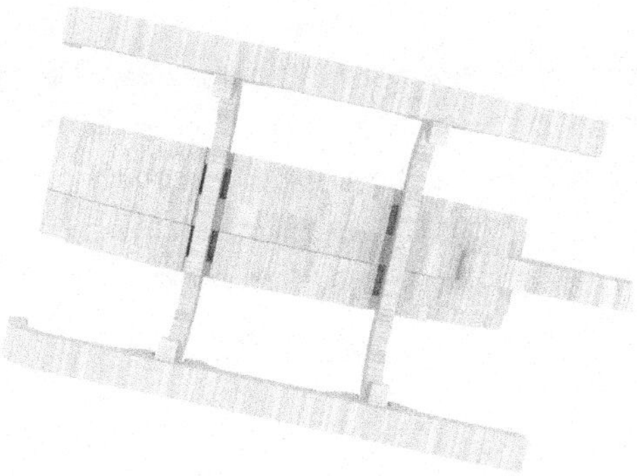

Step 5: Add legs

Underneath, add both part Js (legs) and Ks (feet) for the helicopter, using glue and clamps to keep everything together until dry.

Step 6: Add the main rotor

On the top of the helicopter, slide into place -- without glue -- part L (rotor support,) then part M (rotor,) followed by another part L and I (cap) with glue.

This allows you to easily adjust the height of the rotor.

Step 7: Perform some finishing touches

Allow everything to dry overnight.

There is no need to apply any preservatives or paint. Feel free to sand any sharp edges down. ENJOY!

# 3D Printing Technology Overview

3D printers have come a long way since the 1980s when the working technology was first developed. There are two types of 3D physical manufacturing processes out there. The first one developed was subtractive, which involves removing material to make everything from engine blocks that start as one solid piece. This process is as old as mankind, but the second, called additive manufacturing -- the focus of this article -- is the most recent and is more exciting in my opinion.

## Why is it called additive manufacturing?

3D printers go layer by layer, adding a little material after each pass. Depending on the technology, curing may occur in addition to other processes such as adding color (via an integrated inkjet printer) or switching between various materials at the same time using different percentages of each for different rapid prototyping needs (one material that bends and is soft while the other is inflexible and hard). Here is a list of the additive technologies out there now (in alphabetical order).

- Chocolate Additive Layer Manufacturing (ChocALM)
- Direct metal laser sintering (DMLS)
- Electron beam free form fabrication (EBF3)
- Electron beam melting (EBM)
- Fused deposition modeling (FDM)
- Laminated object manufacturing (LOM)
- Plaster-based 3D printing (PP)
- Selective laser sintering (SLS)
- Stereolithography (SLA)

So, let's review them so you can understand the differences between them. First, what they generally have in common and technological advantages.

## General 3D printing workflow

In my case, I use Autodesk 3DS Max to generate the 3D model, but you can use Blender, SolidWorks or a host of other CAD type programs to do the same. That is the fun and most creative part in my opinion. After this is done, you export it as a STL format into proprietary software that may or may not come with your machine, which then slices the model and orients it. Depending on the technology, it may require the addition of support material. Wait anything from a few minutes to a few days, and it's done!

**General advantages of 3D printing**

3D printing has been used for manufacturing over the past decades because it requires no tooling and parts can be built in hours (days) which otherwise would be impossible or too complex.

Chocolate Additive Layer Manufacturing (ChocALM)

Developed by Choc Edge (http://www.chocedge.com), ChocALM uses yummy chocolate, added layer by layer by a heated extruder, to make your stomach happy!

Material

chocolate ... chocolate and more chocolate!

Advantages of ChocALM

source material is both easy to find and edible
instantly makes you the most popular person in town
can make multiple treats in one shot

Disadvantages of ChocALM

resolution isn't that great
can't make moving parts
can make multiple treats in one shot (leading to waist expansion)

Direct Metal Laser Sintering (DMLS)

Developed by EOS Electro Optical Systems (http://www.eos.info), DMLS machines use a 200 watt laser that fuses metal powder laid onto the build platform from the dispensing platform by transforming the powder into a solid by the process of melting.

Material

any metal alloy powder (not cheap!)

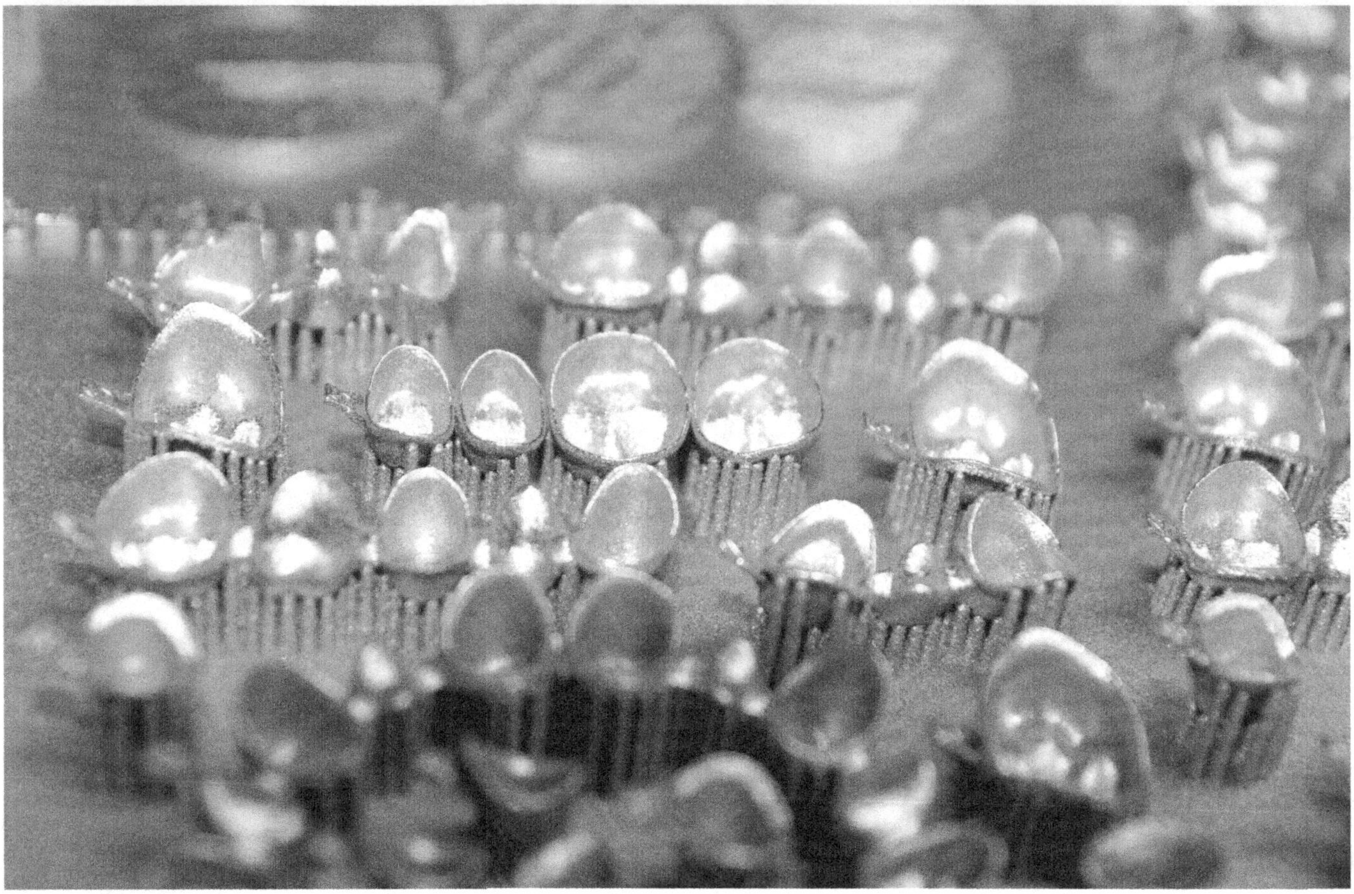

Advantages of DMLS

- models are of high accuracy with great detail due to each layer being about 20 micrometers thick
- can have mechanical properties (meaning you can have moving parts)
- the metal used during the additive process is the same metal that can be used during the general manufacturing process
- prototypes are strong (model depending) and can be tested within their intended application

Disadvantages of DMLS

- models are small (build envelope is "only" 250 x 250 x 185 mm)
- surfaces need to be polished (depending on application, you may need a grinder!)
- removing metal support structures and thermal post-processing is time-consuming (you can't have the supports in a different material than the part)

Electron beam freeform fabrication (EBF3)

Showcased by NASA (http://nasa.gov) about a year ago (it was invented with funding from DARPA and Lockheed Martin), electron beam freeform fabrication (EBF3) uses an electron beam and wire to fabricate metallic structures. This allows you to basically build anything within a 3D environment as basically solid wire is fed and, by using an electron beam, fused into place.

Material

wire

Advantages of EBF3

can be done in a vacuum (which makes it great for welding in outer space)
incredibly strong and smooth results
works in a 3D environment, not a closed box like the other additive technologies
possible to change chemistry and incorporate sensors while the part is being built

Disadvantages of EBF3

the most expensive of all the additive technologies
requires a large capital investment for machines, material and qualified staff

Electron beam melting (EBM)

Developed by Arcam (http://www.arcam.com), EBM melts titanium alloy powder, layer by layer, using an electron beam to bind them together within a vacuum.

Material

Titanium alloys

Advantages of EBM

titanium is safe to use in the body for implants and heavily used in high-end aerospace manufacturing
requires no oxygen (done in a vacuum)

ready to go right after it's built as there is no filler to remove or support material
can have moving parts
layer thickness of 0.05 mm and fast build rates of 80 $cm^3/h$

Disadvantages of EBM

requires a high operating temperature (700 and 1 000° C)

Fused Deposition Modeling (FDM)

Invented in 1989 by Scott Crump, who is also the co-founder of Stratasys Inc. (http://www.stratasys.com) along with his wife. FDM use a plastic filament or metal wire that is unwound from a coil which goes through a heated extrusion nozzle. Of all the additive technologies, this one is the simplest to build and operate though the resolution isn't as high.

Material

ABS
ABSi
polyphenylsulfone (PPSF)
polycarbonate (PC)
Ultem 9085
(other types of thermoplastics)

Advantages of Extrusion

cheap and easy to operate
plastic filament easily sourced
some plastics, like Ulterm 9085, are fire retardant
can use multiple extrusion heads to print support materials using a different material from the intended model
can build multiple models at once
generally no post processing required (depending on model and intended use)

Disadvantages of Extrusion

resolution is relatively low (compared to the other additive technologies)
is widely used so profit margins are slim as are barriers to entry
no post processing required (depending on model and intended use)
material shrinkage, warping and increasing Z-layer error rate as model builds up
can't really have moving parts (unless they are HUGE)
one model at a time (even with large build platform) meaning unlike PP for example, 3D printers, you can't optimize build volume with models at all angles, shapes and sizes

Laminated object manufacturing (LOM)

Originally invented by Helisys Inc. (now Cubic Technologies http://www.cubictechnologies.com), LOM is a relatively simple process of adding successive layers of adhesive-coated paper, plastic or metal laminate which is then cut with a knife or laser cutter (cross-hatches occur with area outside of build to facilitate removal).

Material

sheets of paper, plastic, metal or other material laminates

Advantages of LOM

cheap and abundant material
characteristics of paper model are similar to wood
can build HUGE models at a very affordable price
can build multiple models at once

Disadvantages of LOM

can't have moving parts (unless they are HUGE)
resolution limited by material used (layer wise with 0.2 mm (x,y); 0.3 mm (z) accuracy)

Plaster-based 3D printing (PP)

Invented and developed by Z Corporation (acquired by 3D Systems Inc. http://www.3DSystems.com), PP involves a head moving over successive layers of powder (similar in consistency to corn starch or flour) with an inkjet-like head that deposits liquid binding material. This technology is amazing to see in person and is the technology I hope to use here on 3DMarvels.com to demonstrate and test products!

Material

white powder

This 3D Systems ZPrinter® 850 is my dream machine, plaster based 3D printing has so many advantages from full color printing and large model sizes to full material recycling. It's a perfect machine for affordable static 3D printed model testing.

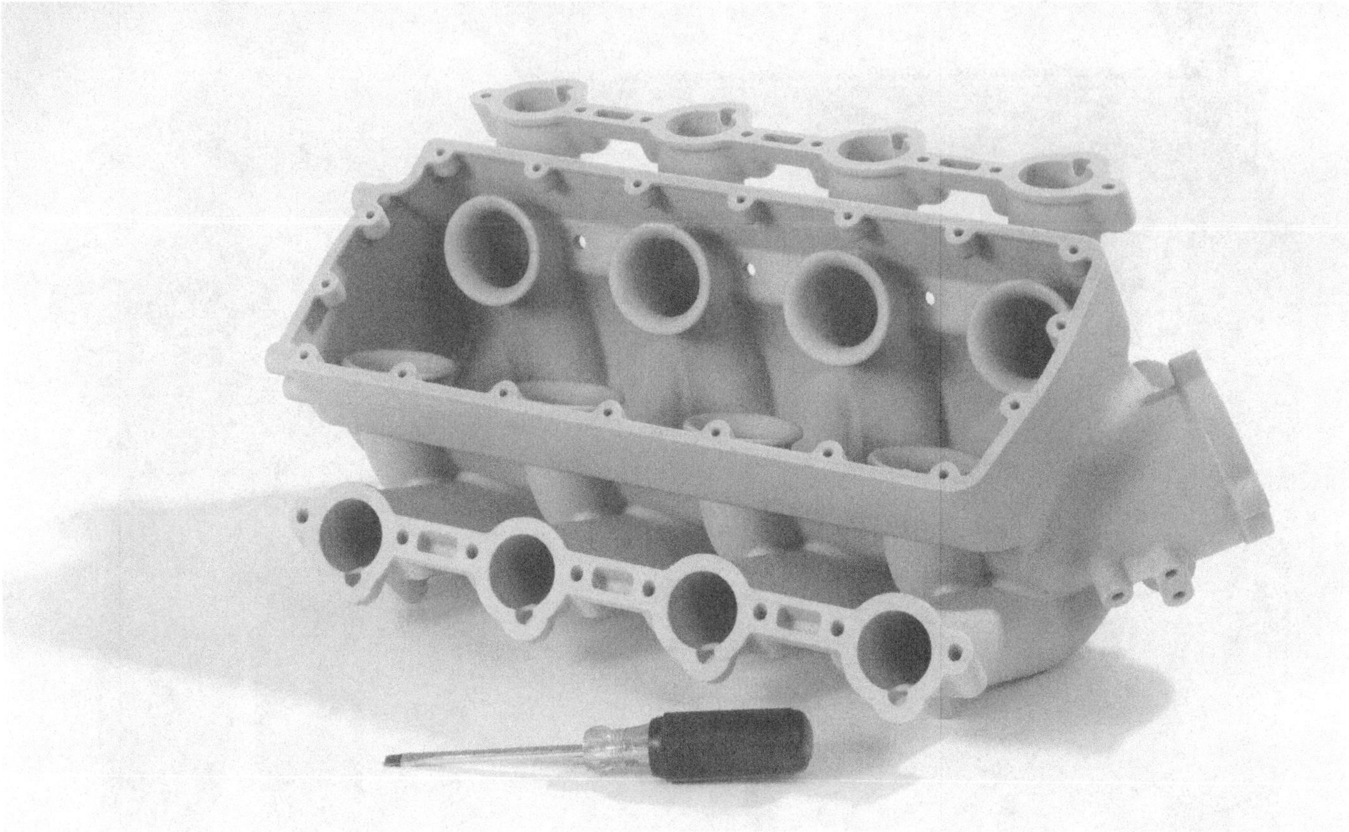

Advantages of PP

- capable of color printing each successive layer using normal HP inkjet cartridges
- prints about an inch per hour
- internal material recycling system (no waste!)
- no need for support materials
- can be used for casting fine jewelry
- very nice finish
- can build multiple models at once

Disadvantages of PP

- requires post-processing (basically spray on a layer of crazy glue)
- finishing (air blasting) required once the model is picked out of the machine

Selective Laser Sintering (SLS)

Patented by Dr. Carl Deckard at the University of Texas in the '80s thanks to some DARPA funding, though technologically similar to another patent held by R. F. Housholder in 1979. SLS is very similar to PP except instead of an inkjet nozzle squirting some liquid, it uses a laser similar to DMLS to fuse particles of plastic, metal (DMLS), ceramic or glass powder one layer at a time over a bed of powder. The powder is also heated so that the laser doesn't have a long way to go for the melting point to be achieved, unlike DMLS.

I saw these massive machines on the production floor of 3D Systems during the tour and was very impressed with the build quality they produce. These wonderful heaters produce incredible details across all three axis with the ability to 3D print in glass and metal alloys (something Bathsheba would love).

Materials

metal alloys (DMLS)
ceramic
composites
glass
polymers (nylon)
polystyrene
green sand

Advantages of SLS

variety of materials can be used, each with their own distinctive material traits
multiple physical processes such as full melting, partial melting or liquid-phase sitering
strong parts with 100% density (strong and resistant) can be built
can build multiple models at once
support materials not required
large build area (machine dependent)

Disadvantages of SLS

cool down period required

Stereolithography (SL/SLA)

SL uses a liquid-filled vat full of UV curable photopolymer (resin) that works in conjunction with a UV laser which builds the parts layer by layer by curing the resin on the surface. The layer thickness is between 0.05 mm and 0.15 mm.

Material

Ultra-violet curable photopolymer (resin)

Advantages of SL

can build relatively large parts with a build envelope of 210x70x80 cm
relatively fast compared to other additive technologies
strong enough for injection molding, blow molding, metal casting and thermoforming molding

Disadvantages of SL

parts need to be cleaned in a chemical bath and cured in a UV oven
requires support materials
resin costs between $80 and $210 USD per liter

# CNC Opportunities: Develop Strategic Partnerships

Once you get into the CNC community, you'll discover that there is a certain level of respect and pride that goes with having a machine. People love comparing machines, finding out each other's capabilities, and sharing ideas. It's a rather tight-knit group! Take advantage of this; join local manufacturing associations and to go industry events.

At a more local level, you will sometimes get the opportunity to take on extremely large projects, like a new hospital that needs thousands of widgets or a hotel that is undergoing major renovations and has hundreds of doors needing some engraving. These jobs sometimes fall out in the blue, and a RFQ (request for quote) might be very tight to generate if you don't know the capabilities of other CNC operators in your area.

If you are smart, you will have developed relationships with others in the industry so that, if needed, you can easily scale and subcontract some of your overflow. Don't see these other businesses as competition, see them as friends who can help you out in a bind!

Imagine for a moment that you have a large job, and you come into the shop early in the morning to discover your machine isn't going to be able to do the job and you'll miss the client's deadline. You have three options:

1) Call the customer and tell them the situation and explain to them why you won't be able to make the deadline. Sometimes this may be all right, but what you are really telling your customer is that when the shit hits the fan, you will take the easy way out and let them bear the consequences. This causes a ripple effect that may cost your business in the future, especially if the customer was REALLY counting on you to get the job done.

2) Subcontract the job to an unknown competing shop in a hurry, knowing that if it comes back at a lower quality that you are willing to accept, you won't have enough time to remedy the situation -- and at the end of the day, that competitor has your client's information -- you may lose your customer to them.

3) Call your local competitors, take them out for a beer, and tell them that although you are gunning for similar jobs, that you'd like to develop an understanding that you can count on them to do jobs for you (and you will do jobs for them) should any issues arise. This takes lots of courage as it takes time to develop trust, but it will help you weed out the bad apples in the bunch as you can visit one another's shops, get an idea of quality, and, more importantly, scale far beyond what your financial resources may allow in a moment's notice. At the end of the day, you are in business to make money, whether it's you doing the job or a subcontractor.

What I would suggest is that you also go one step further, if you can, and develop a relationship with your sales representatives! They know the machine industry and may even send you some leads. How? Not everybody will be able to buy the biggest and most efficient machine the sales representative has to offer. Tell him or her that if they have clients interested in similar machines, you are always open to tours and to answer any questions their potential clients may have.

Here's a secret, though: most of the people visiting your shop won't have or purchase the capabilities you have, but your sales representative is a great way to send you strategic partnership opportunities. Most people pick the first shop they find. If that shop has similar equipment to you but their bed size or manpower is smaller than yours, you can offer them a way to accept jobs that normally, they'd have to turn down.

Your first fear might be that if you quote a job to this smaller shop and then get a call from the same customer to quote on as well, you will be competing against yourself. Don't worry, if you win the job (easier for you as you can show the customer your ACTUAL capabilities with a quick tour) then you are in the green. If the competing shop wins the job and sends it to you, you STILL WIN! The biggest money pit in any CNC shop is a machine that isn't producing product. Sure, your margins may be lower working for your competitor, but at least that CNC machine isn't sitting there collecting dust, it's collecting some coin that you wouldn't have gotten otherwise. Of course, if you are too busy, you can always quote a very high price to a competing shop so you don't need to bother with the hassle, but treat them fairly with respect and you'll get the same in kind.

This is where it gets fun though. One shop going back and forth is nice, but imagine if you had several smaller shops feeding you jobs that they don't have the time or resources to complete! This can quickly become a good profit center for you without having any of the client issues. You may think that you aren't building a name for yourself, but most business owners are rather lazy. If something works and they can just make straight cash from the transaction, they'll keep repeating that cycle instead of taking a chance to grow their operations. After enough time, some of these other shops may close or downsize, rendering you all the more profitable. What you know and they do not is that an idle machine doesn't pay. Use their jobs to keep building yourself up -- let them find the small projects so you can have the basics covered while you gun for the larger jobs that they have no hope in attracting.

Success in business has a very large "luck factor." The more opportunities you can bring your way and the more diversified they are, the more protected you will be from market upheavals, and you'll be able to invest in better equipment and staff.

# Interview with Cathy Lewis VP of Global Marketing at 3D Systems

I was invited down to 3D Systems HQ in Rock Hill earlier last year (2012) to spend two days learning first-hand about 3D printing from the industry professionals at 3D Systems. What I liked about 3D Systems was how diversified they were across hardware, software and as a global solutions provider -- whether it's machines, supplies or having models printed, they can do it all and then some!

During my visit, I was able to spend a lot of time with Cathy Lewis, talking about the future of 3D printers. She is extremely well known within the 3D printing world -- she also happens to be the VP of global marketing for 3D Systems and lead Desktop Factory towards creating one of the first consumer-friendly desktop 3D printers before being acquired by 3D Systems.

**Can you tell me a little about 3D Systems?**

Today's 3D product design revolution -- additive manufacturing or 3D printing -- was started over 26 years ago by 3D Systems. In 1986, Charles 'Chuck' Hull patented Stereolithography and founded 3D Systems.

When Stereolithography, SLA®, was developed there was no such thing as rapid prototyping. Creating a concept model, let alone a functional prototype, took months and cost thousands of dollars. And while designers, model makers and engineers were using computer aided design and manufacturing software, there was no method for that software to communicate to what became the very first Rapid Prototyping System, SLA.

So Chuck and 3D Systems also developed the .stl file format, still in use today, to complete the electronic 'handshake' from the CAD software and transmit files for the printing of 3D objects.

Fast forward to 2012 and all of that has changed. From medical prosthetics, car parts and unmanned drones to iPhone cases, these are just a few examples of the tangible items that manufacturers large and small are making with 3D printing technology.

With over 1,100 patents, over 100 specialty engineering materials and 7 print engines, our years of expertise and myriad materials can be brought to bear on any application challenge.

**How does 3D Systems help model makers?**

We strive to support the professional model maker by providing what we believe are the best solutions on the market in terms of choice, price and performance. As an example, in the ideation step of early product development a client benefits from concept models to help them iterate and gain consensus on a proof of concept. Depending upon the size and sophistication of the company or institution, it is important to offer a couple of options to create that first concept model.

In some cases cost may be one of the most important factors, so our Plastic Jet Printing with a RapMan kit or the beautifully evolved 3DTouch provide a durable color model at a very reasonable price. If a smoother surface finish is desired, then a desktop solution like the ProJet 1000 or the 1500 with a choice of colors may be the better alternative. And if extremely fine detail and smooth surface finish is desired, there is nothing that compares to the ProJet 3500 series, including the variety of casting alternatives.

If the company demands a realistic model in the ideation phase there is nothing more powerful than full color 3D printing to express every nuance of the concept and allow everyone to 'see' the idea come to life.

Zprinting is the only available alternative and with access to a full spectrum color palette of over 390,000 shades. This solution is now being adopted for applications as diverse as shoe and apparel design, architectural modeling, new consumer electronics products, the arts and entertainment categories.

If the use of in-house 3D printing is not justified or the need arises for special materials or the demand increases -- the professional model maker can take advantage of our on-demand parts services. We find that through Quickparts.com the model maker can access all the technology choices that will either complement their internal efforts or provide the perfect model just in time, at a great price.

**What sets 3D Systems apart from your competitors?**

We have found that no single solution offers the answer to every product design or development challenge. There is a complete imagination to production cycle with numerous steps and stages where 3D printing brings substantial value in compressing time and reducing wasted resources, financial and human. This is why we now offer seven unique printing technologies and over 100 specialty engineering materials.

To further support our users, 3D Systems offers a complete ondemand custom parts service with a global footprint. So no matter where you are in the world, whether you use in-house 3D printing or not, we can support your needs for concept models, functional prototypes, tooling, metal parts and assemblies - complete design to manufacturing services, 24x7.

Our Alibre CAD design software allows almost any user to begin to create their models and concepts affordably while our newest brands, like FreshFiber and Bespoke Innovations, provide access to customized electronics accessories and customized prosthetics. When you add to that our new Cubify.com meeting place and marketing place that is bringing 3D printing to the non-expert and consumer markets, you have a companlike no other in our space.

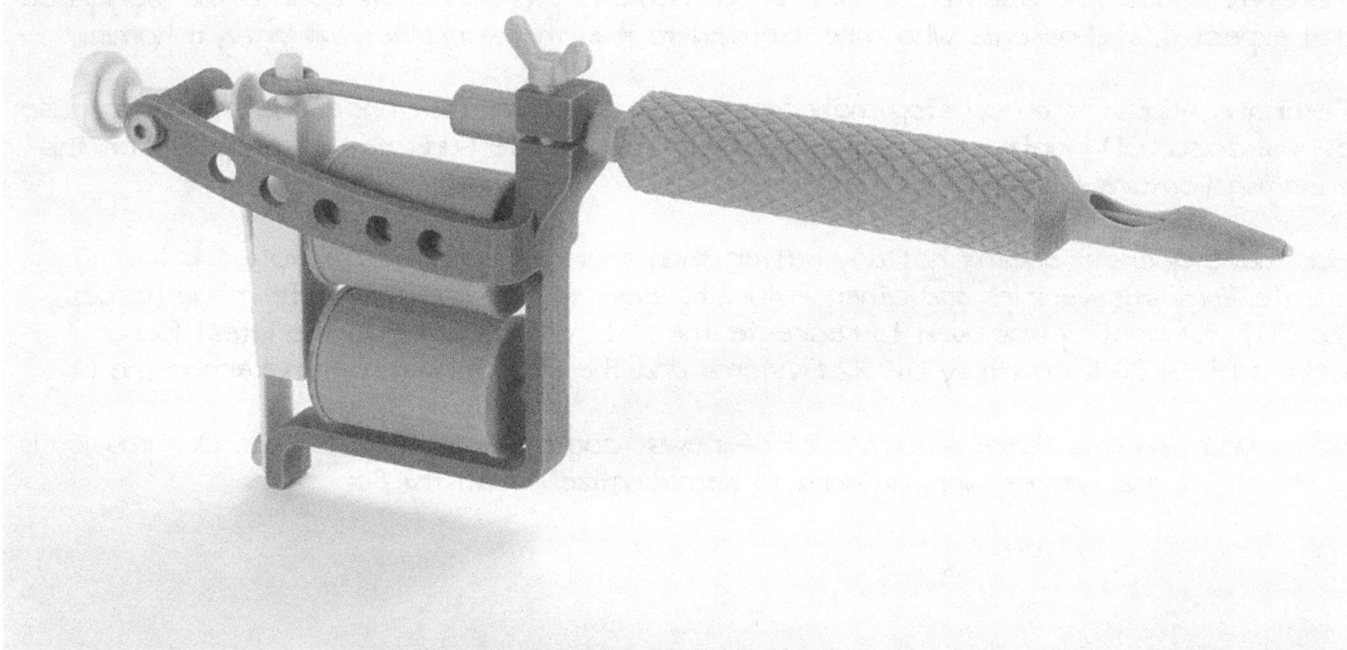

**What's going on inside 3D Systems right now?**

Today there is an explosion in custom consumer products from lighting, furniture and accessories for the home to jewelry and other wearables for the individual. In addition, a growing number of global companies and governmental agencies are beginning to re-localize manufacturing for core applications in transportation, aerospace and defense.

Another great example is medical and other healthcare applications. Prosthetics, orthopaedics, dental restorations, hearing aids, anything related to the body is organically shaped and customized per the individual; so, there is enormous opportunity. Computer-aided surgery, for example planning surgical procedures based on skin data, etc., is a fascinating and growing field.

Some companies and institutions are even doing research on how to print body parts, replicating everything from bones to organs to tissues. This is a field that's ripe for growth and where 3D Systems will continue to innovate.

This May, 3D Systems opened up this world of creation to everyday consumers with the Cube, a personal 3D printer that the company promises even an 8-year-old can master. This $1299 USD desktop printer pulls heated plastic material from a single-colour cartridge, then applies layer upon layer with a jet print nozzle to create a tangible item. We believe that Cube complements and amplifies our professional solutions by converting more fans. Kids who discover 3D printing at home and in school will go into the workplace and expect it, and parents who are exposed to it in the workplace will bring it home.

Coming in August is a new stop motion movie, ParaNorman, that has made an amazing use of full colour 3D printing technology to design and create hundreds of puppets for this engaging feature film.

For movie goers it doesn't get any better than seeing 3D printing put to work to create fantastic vehicles and other futuristic devices for our favourite movie heroes, like 007. 3D printing was used to recreate the Aston Martin DB5 in the latest Bond installment, Skyfall, courtesy of 3D Systems and the PropShop model makers in the UK.

So as you can see, there is quite a lot of news coming from 3D Systems. Our passion is all things 3D and we are working hard to democratize creativity for all.

# Extrusion to Stereolithography to Plaster: My Roadmap

There is a very wide range of "consumer level" 3D printers out there, but for my needs, I've focused on where I see myself heading as a designer on CNCKing.com. 3D printers are rapidly evolving in the marketplace, so some of this information is likely to change by the time this book is published and in print.

**Tier I: "Cheap" Extrusion**

The advantages of extrusion, as I outlined in the Technology Overview, are numerous -- such as abundant material supply, relatively simple technology (platform and head moving along XYZ axis with a heating head) and decent durability of the finished models. Within extrusion 3D printer ranges, you do have different levels relating to the end quality of the finished print.

At the low end of the spectrum of extrusion 3D printing, you have the Cube/CubeX by 3D Systems and the Replicator by MakerBot Industries, which range from just over a thousand to several thousand dollars depending on the options you wish to have. These are very capable machines with upgrades including multi-head extruders (nice for making support material in a weaker dissolvable material to that of the model), relatively large print areas and increasing precision with speed to match.

At the higher end of the spectrum you have the Mojo by Stratasys which sells for just under ten thousand dollars and is capable of producing incredibly detailed 3D prints! There is very little in between; both 3D Systems and MakerBot Industries are the market leaders at the "low end" as of this print.

The quality is better as you move up, leading to higher resolution models when you are done printing. Generally speaking, these affordable extrusion 3D printers have a print resolution of about 200 microns (0.2 mm / 0.008 inches per layer) with a build envelope of about five inches square.

As of this printing, 3D Systems and MakerBot have since released higher-end 3D printing machines which feature larger printing envelopes and other neat features. I suspect that extrusion will dominate the consumer marketplace for at least the next year or two until SLA or PP become more affordable.

Over the long haul though, extrusion based 3D printers will be a dime a dozen like inkjets are now -- you'll get a machine for a few hundred dollars and the companies will make money off their "proprietary" materials much like current inkjet cartridges.

There are major drawbacks to this technology, though, the biggest in my opinion being inability to print COLOR (equivalent to an inkjet in range) and inability to make 3D moving parts without requiring any type of extra assembly. For instance, you can't make a toy car moveable unless you make each major part independently and THEN put it together. I think the issues with material warp, shrink and errors will go down dramatically -- being a non-issue by then, though this is a common issue that I have to deal with printing thicker material especially at the model's base.

Regardless of the material you use to 3D print, one very useful feature of my Cube is something called "rafting" which is the mesh you see going all around the model. The purpose of this is to keep the model stuck to the platform as the 3D printing progresses, but it still takes trial and error to get things to stay STUCK to a platform.

**Tier 2: Quality Stereolithography (SL)**

There are many advantages to SL, including up to 10x better resolution than what is possible with current generation extrusion machines, the ability to make moving parts, and superior build quality; with the major drawback being that you need to do some post-processing before you can play around with the part to remove support material.

Depending on the model, you may need to do this anyways with extrusion and other 3D printing technologies, but the advantage of SL is that generally, that support material can be dissolved away in a vat instead of having to go in with pliers to manually remove excess material.

This technology shows the most promise for affordable 3D printers looking at build quality over cost. For instance, the B9Creator will retail for about the same as the MakerBot with material costs of 10 cents per gram, while the extrusion material used for the MakerBot costs (at $90 USD per kg) about 9 cents per gram. But since SLA is a liquid, there is a lot less waste. On the higher end of things, you are looking at either a 3D Systems Projet or Objet Polyjet -- both start at about the same price as mid-range car and go up almost exponentially as you increase the build envelope.

The one major reason I don't think these systems will be huge in the consumer marketplace is that although the end material easily accepts paint, the general public doesn't want to learn how to paint any more than they want to learn how to build 3D models. 3D printing companies are starting to realize this and coming up with apps that make the process of generating a 3D print "easy," but it's still rather limited. Ultimately, nothing can replace the flexibility that comes with having true 3D modeling experience under your belt.

As of this writing, there is a new 3D printing company on the block, but its release date will happen after the publishing of this book. It's called FormLabs Form 1, and if it catches on as I suspect it will, it will quickly become the 3D printer of choice for those currently looking at the higher-end extrusion 3D printers. The reason is simple: with a price of just over $3000 USD, you get all the benefits of SL at the cost of an extrusion: moving parts, paintable surface, and incredible detail. Although I love my 3D Systems Cube, this would make a great complement to test designs that are simply impossible to print with an extrusion printer. The process is slow, but for my purposes, if the reviews are positive, this will probably be the next 3D printer I get before moving on to a zPrinter. You can buy a low-end SL printer for about $3000-$8000 USD.

The image above could easily be printed using even an entry level SL 3D printer, but it is simply impossible using an extrusion 3D printer. The reasons are several, including moving parts, incredibly tight tolerances, and negative spaces inside the model which would be incredibly difficult to access using anything but a chemical bath to remove support material. This is why I'm looking at this space very carefully, though the costs of resin are far more nose bleedingly high than extrusion.

Although the price of resin varies significantly, as does PLA/ABS for extrusion 3D printers, 3D printing companies are smart to learn from the mistakes of their inkjet cousins -- they are developing proprietary blends that only work with their machine, which technically can't be used or found elsewhere, lest you void your machine's warranty. Most 3D companies make money two ways: one, the machine, two, material!

**Tier 3: Plaster-based 3D printing (PP)**

The major advantages over extrusion and higher end Stereolithography are that plaster-based 3D printing (PP) allows you to print in color while the model is being made, and it has material recycling capabilities, meaning anything not used on your model can be used for the next one. The resolution you get out of it is about 100 microns per layer.

Why can a system that prints twice that of current generation extruders but 1/4 that of SL based systems be able to have overhanging and moving parts? It's all about the support material! You see, with extruders, you print and gravity pulls down on the model so if you have an overhang, it goes DOWN. With SL you are using a liquid so again, gravity can take over (though nowhere near as severe). Plaster-based 3D printing, on the other hand, builds layer by layer across the entire build envelope -- like placing sheets of paper one at a time and applying resin only where you want it (this is how Mcor Technologies works).

If you have a part that's "floating" in space, it will "fall," but only after you remove the supporting material -- not during the actual printing process. You do have to spray the model with super-glue as plaster falls apart, so there are some disadvantages, but it's a small price to pay in my opinion -- you can just print another one!

Cost-wise, a plaster-based 3D printer starts at about the same cost of a new entry-level car and goes up from there, with the most notable difference being a larger build envelope. Material-wise, it's MUCH CHEAPER than professional level SL binding material, though you can't have transparent or translucent parts. Generally, you have the advantages of SL without the disadvantages of extrusion regarding build options at a fraction of the cost.

For toy making, this is really the Holy Grail as far as 3D printing. So, you have the resolution, speed, material (which is non-toxic by the way) and color printing capabilities all in one machine. Once I saw this machine in action at 3D Systems HQ in Rock Hill, I realized that this was the way to go to test models out. I REALLY liked the texture of the model afterwards, too. Extrusion is rough even visually while SL is incredibly smooth, but PP is just perfect in my opinion for easy grabbing, and you don't see the material layers.

I call this a real 3D printer from a general consumer point of view as it most closely mimics the replicator in Star Trek! Edible, functional but with a limitation of flexibility which for me, is an easy compromise to make.

**Special Mention: Mcor Technologies**

I briefly mentioned Mcor Technologies technology before, but I think it deserves a lot more attention than it's currently getting both in the media and in general design forums. The major advantage of this technology is that you have the full color printing capabilities of Plaster-based 3D printing (PP) with the incredibly cheap source material that's available just about anywhere -- paper!

The cost of operating this machine is very cheap, as is the source material, though the machines themselves aren't as competitive as high-end extrusion 3D printers as of yet.

The advantage of using paper is that it readily accepts (absorbs) paints, glues and anything else you can think of but at a major cost to the manufacturers themselves. You see, 3D Systems makes good money selling you the base material, so they can afford to make less money on the machine knowing you will be buying material for the life of the machine. The same holds true for MakerBot's extrusion machines and other manufacturers.

I've been intriged by Mcor Technologies for quite a while, which is why I interviewed them for this book. I also think an Mcor 3D printer would suit my needs wonderfully for models that don't require too much movement.

All the advantages of PP minus material costs, though that technology also has limitations but could be easily designed around -- as with any CNC technology, design for the machine for optimal results!

# Choosing PLA over ABS

I bought a 3D Systems Cube, and when it came to material choices, after I did some research, I decided that for my needs, PLA was a better fit for me than ABS. Although the Cube is capable of using both materials, I prefer to keep things simple and as straightforward as possible, and I wanted to share my reasons behind my decision.

**More Environmentally Friendly**

3D printing has many material options but for the 3D printer that I bought, I had the choice of either man-made plastic (ABS) or natural corn starch/equivalent (PLA, or Polylactic acid or polylactide).

Yes, plastic ABS can be thrown in the recycling box and to a certain extent, converted back to the oil from which it is derived, but PLA can be thrown into the garden or even eaten (though I wouldn't recommend it). Environmentally speaking, PLA is light years ahead -- we already have far too much plastic on this planet; if there is anything I can do to support something that doesn't add to this mountain of waste found in our oceans, I'm all for it! PLA is biodegradable while ABS is recyclable.

**Great 3D Printing Properties**

When printing with a 3D printer, what would you rather smell: burned plastic or potato chips ... CHIPS! That sounds healthier to me! PLA also has less warping and requires less power to produce a 3D FDM model, with the added benefit of being all the easier to remove from the heated table after 3D printing has completed. ABS is more durable, flexible, and heat resistant while PLA is more brittle. For my purposes, it doesn't matter.

**Easy Disposal After Filming**

My purpose with the 3D Systems Cube is to test models which I want to design and sell on CNCKing.com -- like all my previous models that I filmed with my ShopBot Desktop CNC table router... I'll be doing the same thing! The scraps from the ShopBot were great firewood while for the PLA, after a model has been proven, thrown into the compost heap. What I use for testing only needs to last as long as the film is rolling. Afterwards, it has served its purpose and needs to find a new home.

I could have thrown the ABS into the recycling bin, but then a truck has to pick it up, process it, etc. ... PLA for me just seems more sensible.

# Tale of three Pyramids: Virtual to Physical

For those unfamiliar with 3D modeling, I want to show you how easy or hard it is to create a simple object like the 3D printable pyramid I designed to be extrusion friendly. As you are now aware, there are a lot of different types of 3D printers in the market, each with its own mechanical and material quirks.

I enjoy challening myself and in this case, we have the tail of three pyramids. The first will be a science experiment that's done at the elementary school level around the world to demonstrate what happens when dried yiest and hydrogen peroxide mix. The second is the more traditional view of an Egyptian pyramid and the third from Mayan/Aztek origins.

The combinations are truly endless and I hope this glimpse can ignite your curiosity to explore the world of 3D, as I have for almost 20 years, to develop the skills required to make your very own projects. The learning curve is steep but the results are rewarding, especially when you have access to a 3D printer like my Cube to test your ideas!

What I find most fascinating about the 3D modeling process is that of time, you can spend 10 minutes or 10 days on the same model and the model that took you 10 minutes may look better and be more functional! It's so easy to go down the rabbit hole where you, eroneously, believe spending more time on a model produces a better result. 3D modeling isn't about time, it's about being SMART. The three projects outlined here can be done in endless permutations, if you want, you could easily create each side of a pyramid and stitch it together if you were so inclined, it would take more time but the results would be the same.

I believe the biggest issue 3D printing enthusiasts will have is that of modeling, it's not easy, there are no shortcuts and up until now, only possible on high-end machines. Rendering a model can take from a few seconds to a few days depending on the resolution, mesh topology, surface complexity and "special f/x" you add to a scene. If you are interested in getting into this field CNC design, I highly encourage you to learn at least one 3D modeling package. CNCKing.com wouldn't exist if it wasn't for my passion for 3D as the visualization processes behing creating a model to be cut with a CNC is absolutely mind blowing beyond a few pieces - it's beyond any human's intellectual abilities.

Use a computer to help you visualize a 3D model and focus on making things that are interesting. Start slow but start NOW... it's one of the leading high-demand skills that will be increasingly required if you wish to have a prosperous design career.

# Erupting Pyramid

Before you model a project, you have to think in advance the purpose behind it: it will open the door towards solving many problems before they occur throughout the modeling process. Creating an erupting pyramid is pretty easy, first we need to create a pyramid, then we need to create the inner chamber that will hold both the chemical mix producing the pressure required to make the eruption happen.

I want to produce a "shooting" reaction instead of a "flowing" action in this pyramid, so right away, I know I need to create the inner chamber as large as possible with the hold in the top relatively small. The wider I make the hole on top of the pyramid, the less pressure will build-up and the less "shooting" I'll get. This chemical reaction happens fast so I still need the hole large enough to easily put in the base ingredients.

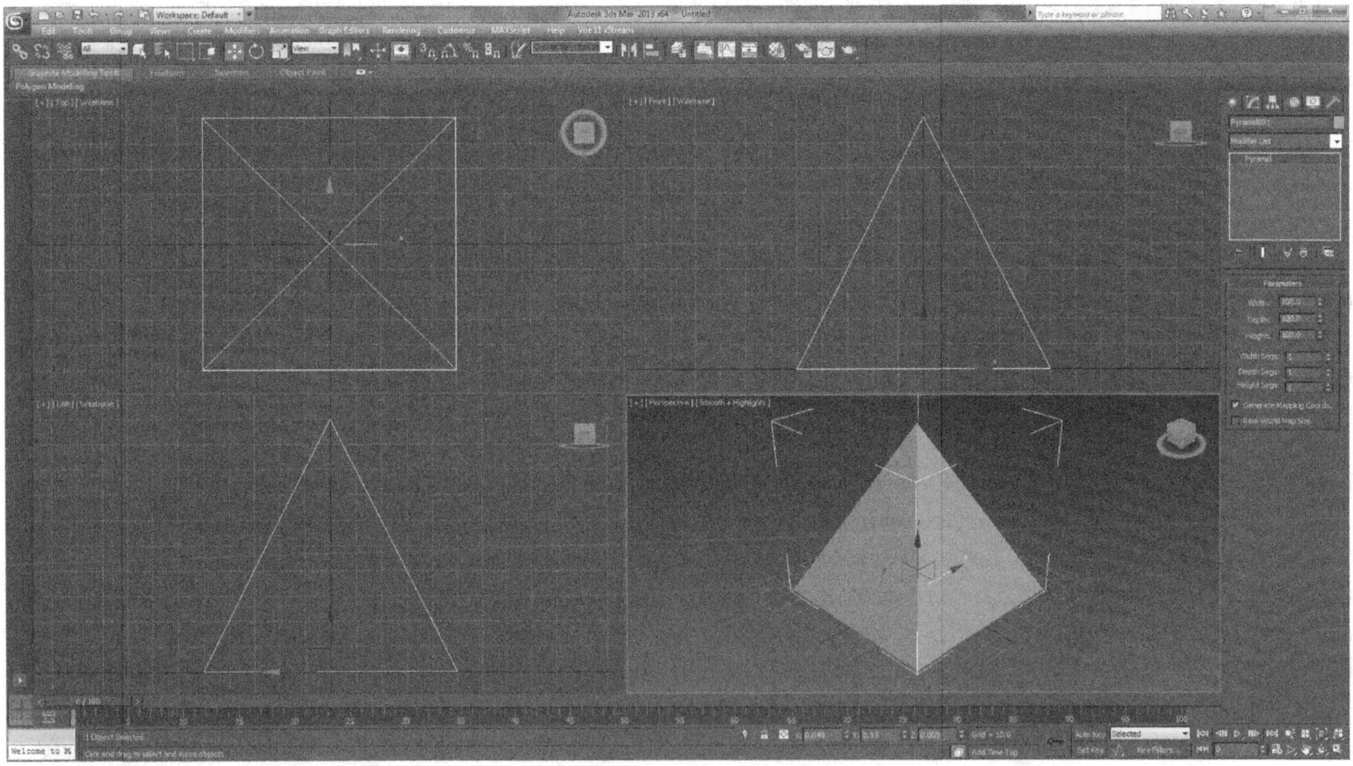

Step 1: Create a pyramid

I opened my 3DS Max 2013, clicked the little ball on the right, and chose a standard primitive of a pyramid that's 100 units (mm, inches, feet or kms ... doesn't matter as these are arbitrary units) in size along the X, Y and Z axis. Easy! That took a whole of five seconds. Could you 3D print this? Absolutely. It's boring, but this just goes to show you how easy creating a primitive pyramid is in most 3D modeling programs.

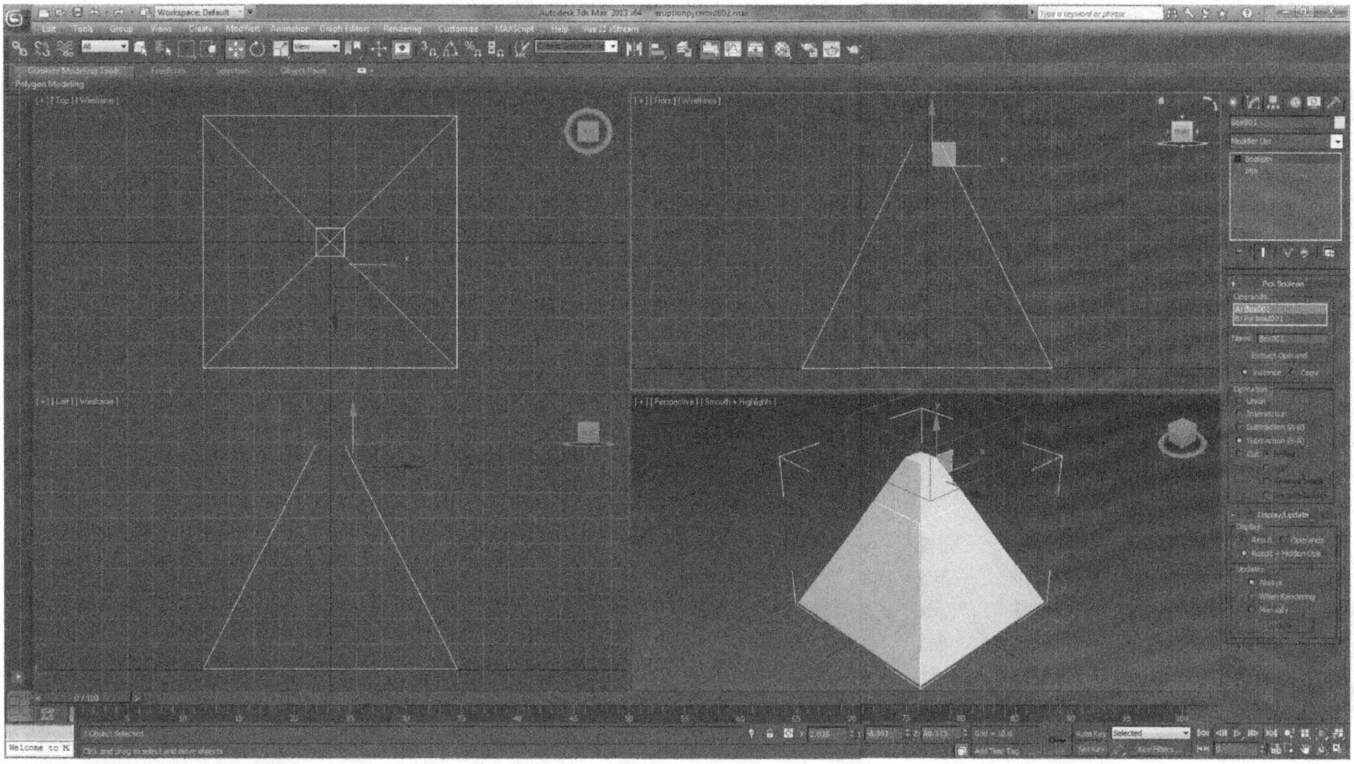

Step 2: Cut and Extrude

Due to the fact that I decided to use a pyramid instead of a cone to produce this project, I have to lop off the top of it to produce a flat surface. Why? Because in 3DS Max, a cone is created with a flat surface on the top by default, it allows you to adjust the angle of the final shape. A cone can then be easily extruded inwards after adding a cap segment to produce the inner chamber so that I have both a ledge and inner mesh. If you are interested, I made another version of this same idea by creating an actual volcano further in this book - the construction is very different.

In this case, I needed to first increase the cap segments to 2 (if I were to build this using a cone or cylinder) but this primative pyramid doesn't have that option, so what I'll do is create a box (doesn't matter where I put it as long as it covers the area I want cut), at the height I want then use a boolean to cut it off. This will produce a solid cap instead of just using a "edit mesh" which produce a hollow mesh that later needs to be filled or shelled. What you see above is the result plus the hidden operands. I have a nice cap!

I do forsee a slight problem here though, this model will ultimately take-up a space that's 4 inches cube so I fear by the time I add a ledge, the hole in the top will be too small to easily push in baking soda much less effectively clean the model after the experiment.

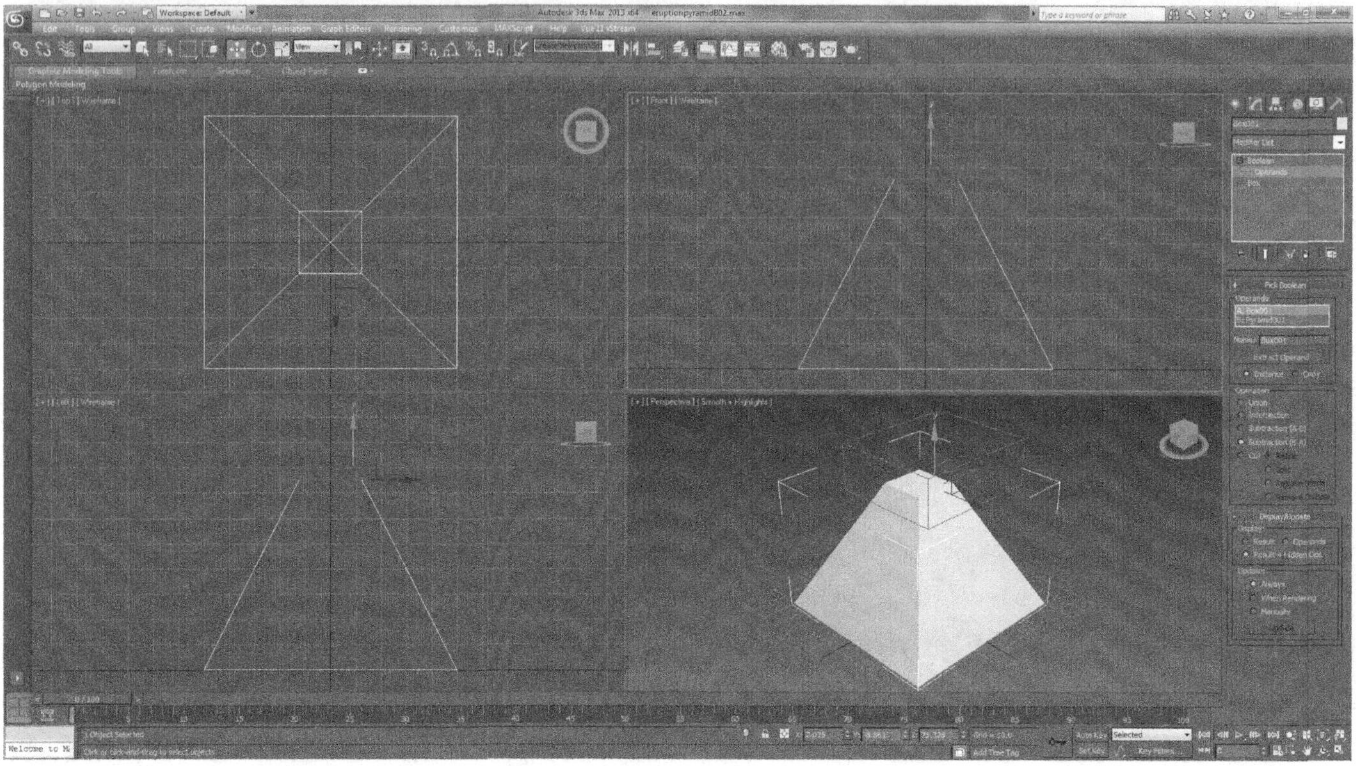

Step 4: Create a larger cap

That's much better, I just clicked on the Box operand and moved it down to cut away more of the pyramid... in hindsight, I should have made the box a bit higher so I could go further down but in this case, I'm fine so let's continue!

Step 5: Create Inner Chamber (next page top)

Originally, I was going to make a round chamber inside the pyramid but looking at the model, it's easier to simply make another pyramid that's 10% smaller than the original, center it and perform a subtractive boolean operand.

Step 6: STL Check (next page bottom)

Although I don't normally do this as most of my meshes are rather complicated and I KNOW there won't have any errors in the mesh, I decided to do an STL check and this assumption is proven.

It's normally incredibly difficult to create a mesh that has zero errors, regardless of the software I've played with, they all produce mesh errors, sometimes it's nothing major and easy to fix but othertimes, you need third party applications to fix this (as you'll see for the Egyptian Pyramid). For 3D extrusion printers, you need the surface to be "water tight."

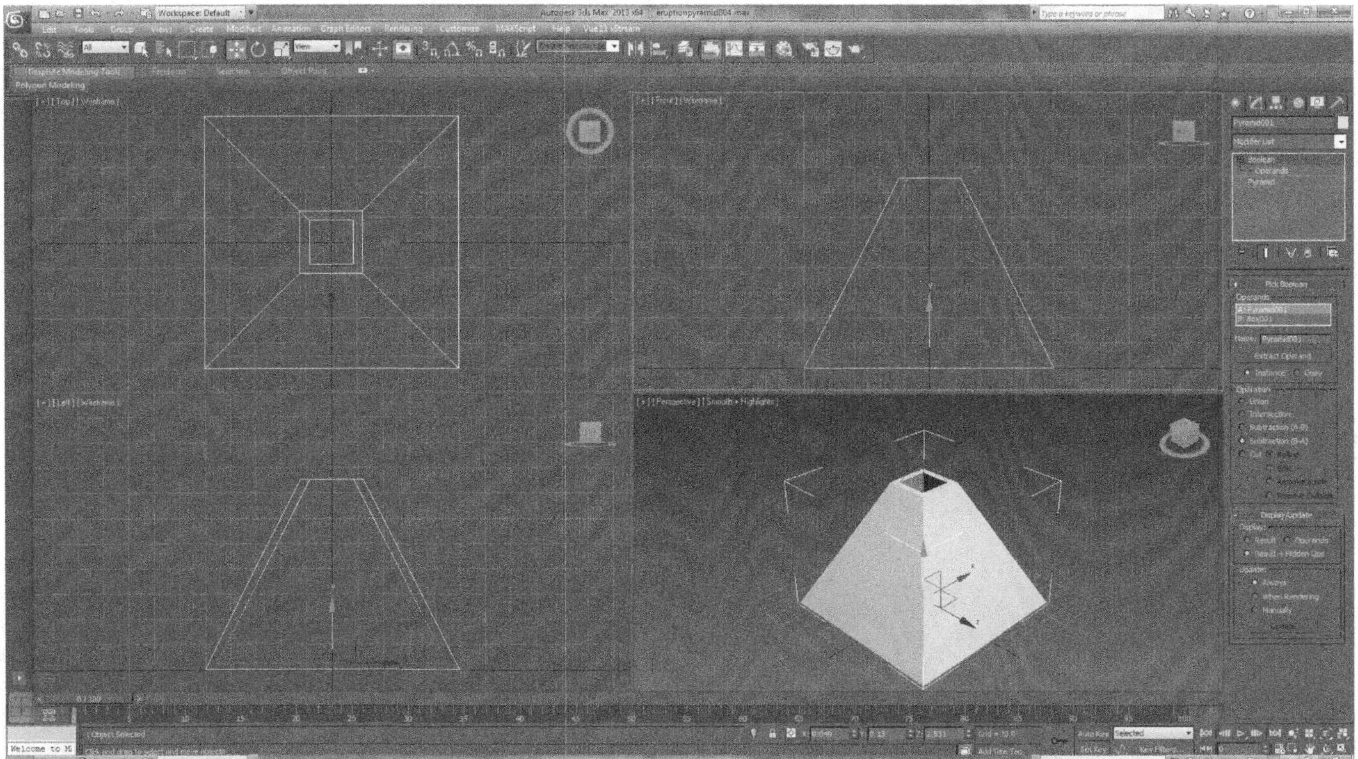

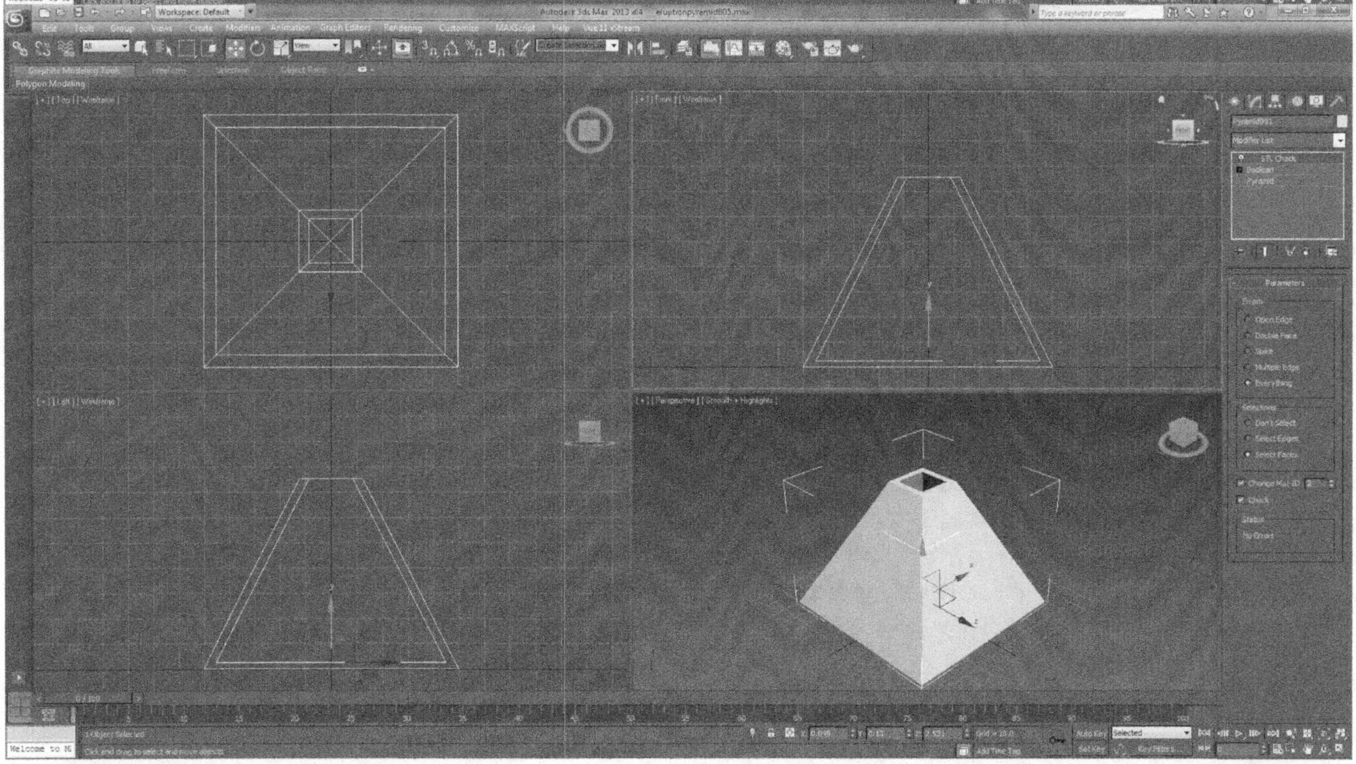

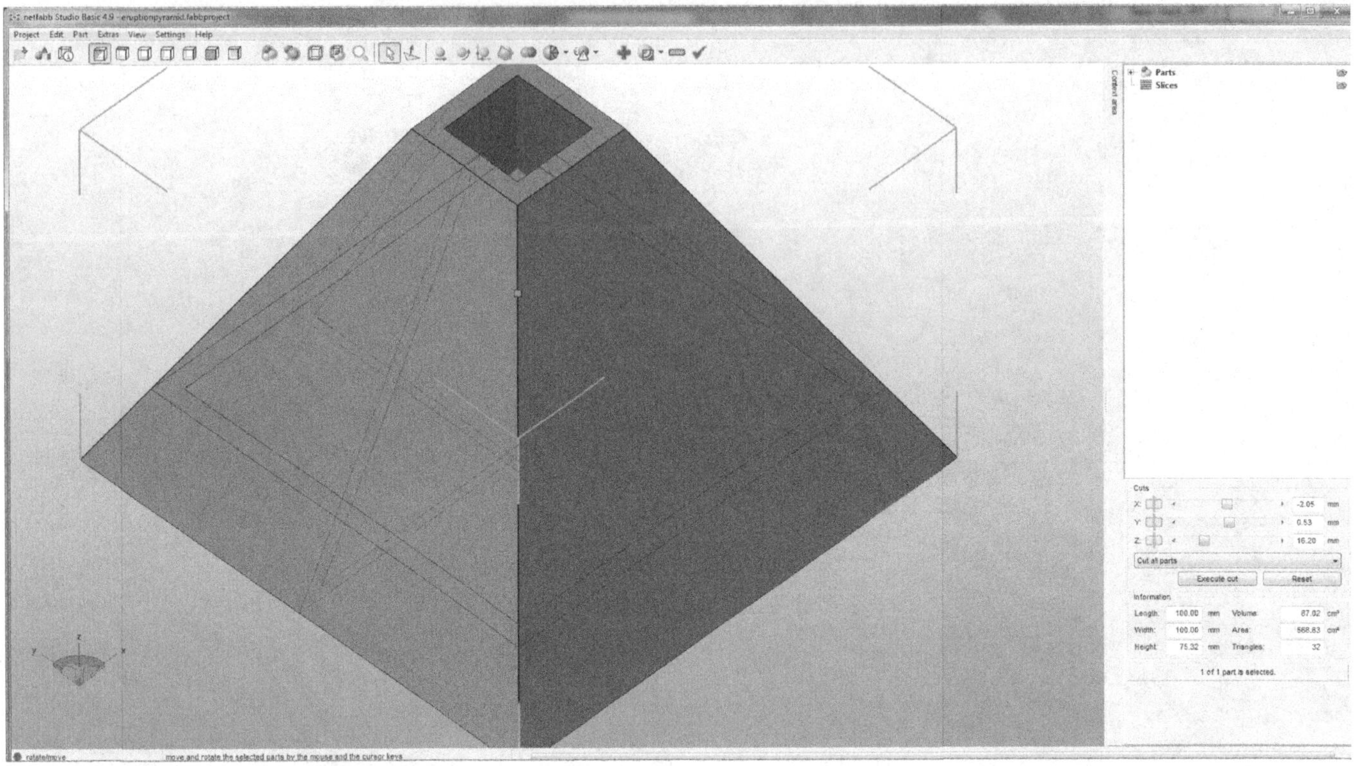


Step 7: Resize the Mesh

Looking at the mesh inside of netfabb, the mesh looks perfect, just like it did in 3DS Max with the added advantage within netfabb of allowing me to review the wall width all around, which is easily seen without having to do boolean. As luck would have it, 100 units from 3DS Max works perfectly to roughly 100 mm, the bed size on my Cube is 150 mm cube, so I'm fine. Should be a quick build!

Step 8: Cubify my Creation (next page top)

I can now import my original STL file directly into my 3D Systems Cubify software and generate the build file. Almost 5 hours to print this model... not bad considering most of my tests take half days to print!

Step 9: Print my Creation! (next page bottom)

As you can see, the printing process is very straight forward. I use PLA which is great but I know it will dissolve after a bit of usage in a wet environment, especially when I start adding chemical reactions. What did this model produce? I took a picture of the results two pages from this one - this model was a great success. I designed, built and tested this model within the comfort of my own home! You can also get this model for FREE on CNCKing.com and give it a try in your own house as well.

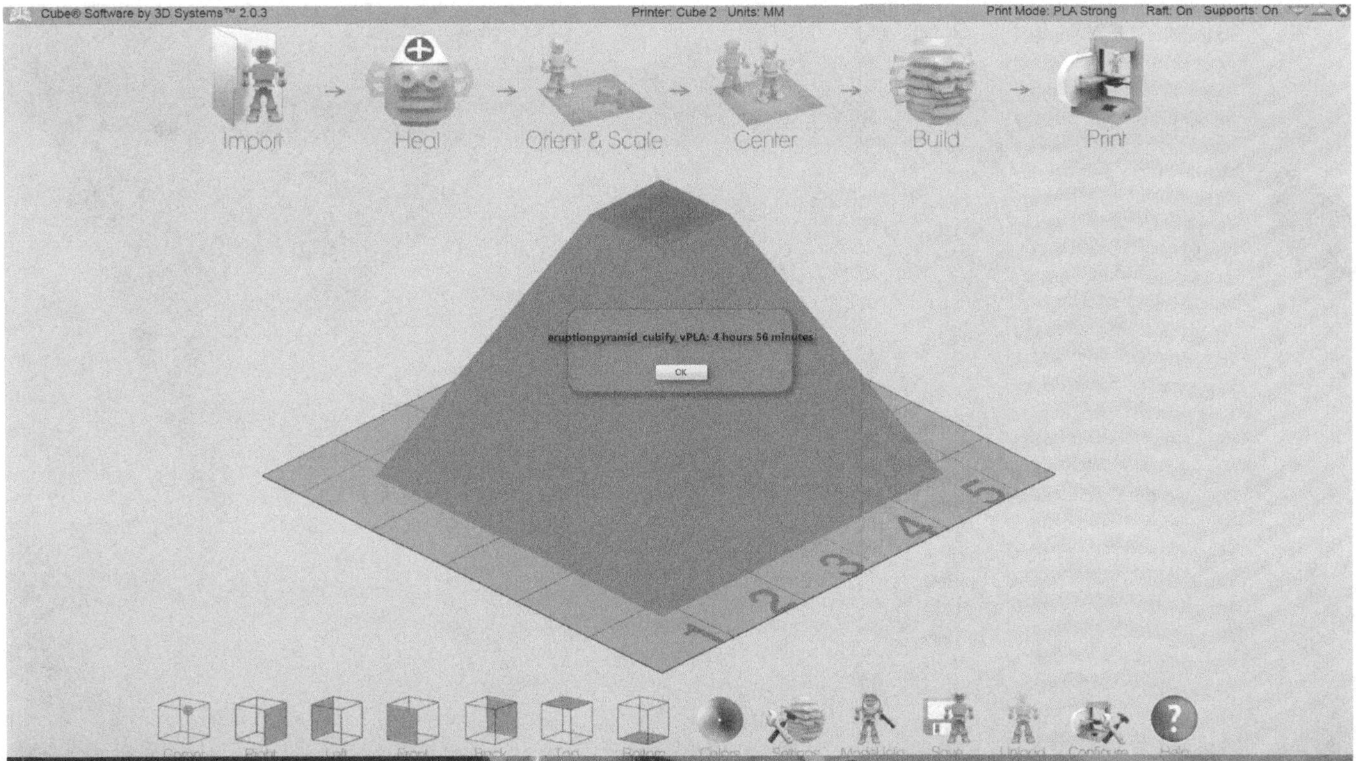
Cube® Software by 3D Systems™ 2.0.3
Printer: Cube 2 Units: MM
Print Mode: PLA Strong
Raft: On
Supports: On
Import
Heal
Orient & Scale
Center
Build
Print
eruptionpyramid_cubify_vPLA: 4 hours 56 minutes
OK
1
2
3
4
5

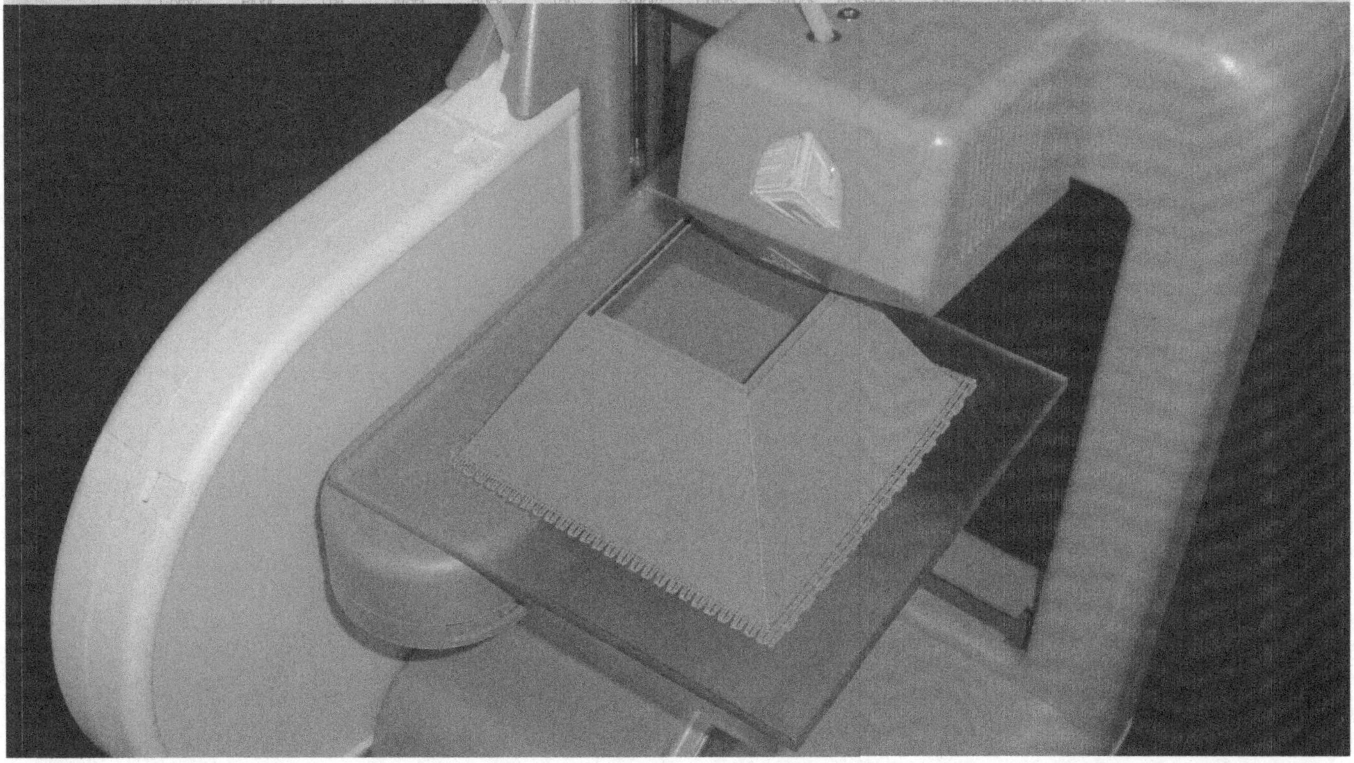

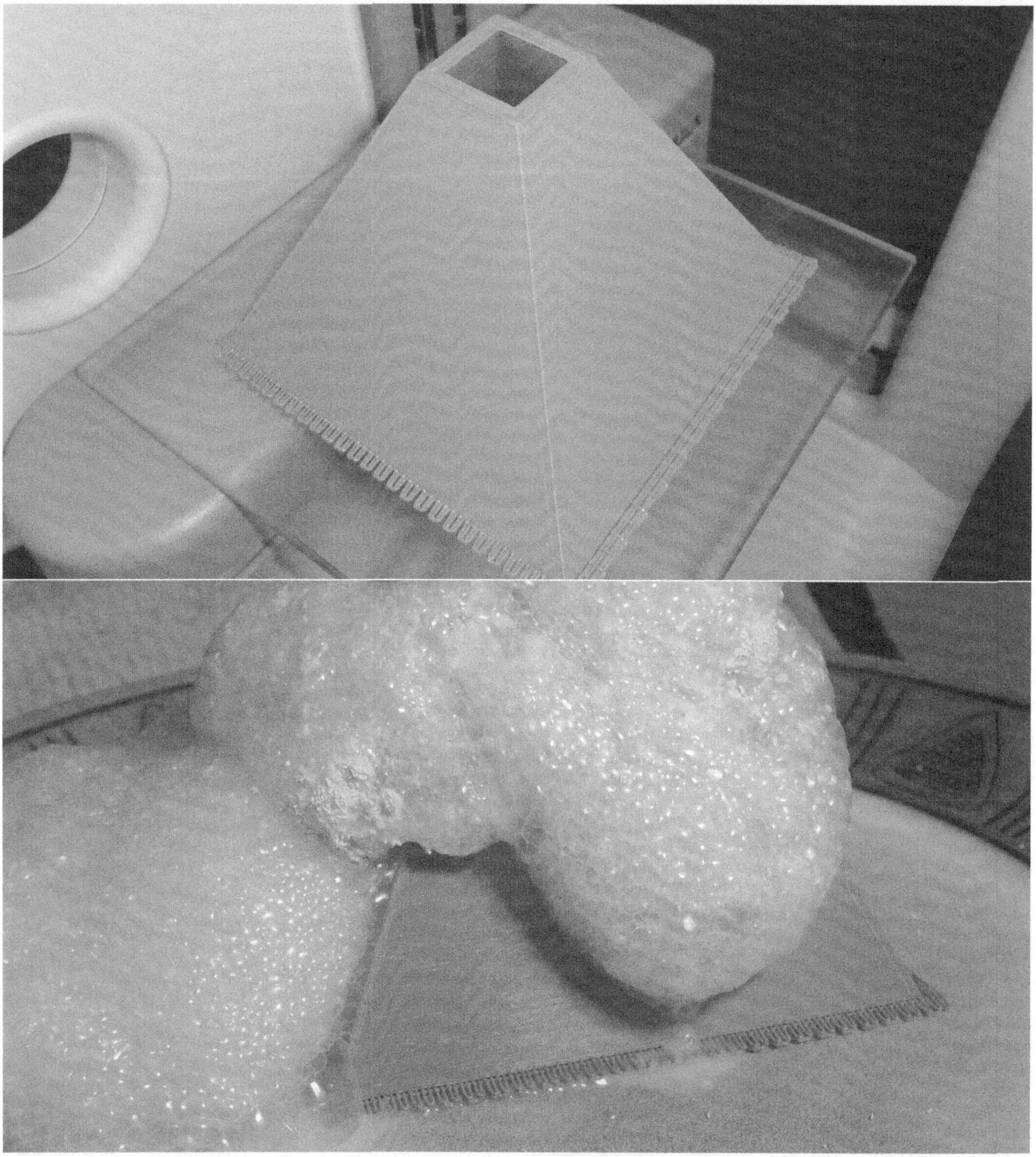

# Egyptian Pyramid

This Egyptian Pyramid was designed with a lot of detail but, because each level is smaller than the previous one along the z-axis (vertical), it can be efficiently printed using my 3D Systems Cube as the gantry is above the platform. I could technically make it inverted using support material but supports are something to be avoided with a single-head extrusion 3D printer. As you see in the close-ups photographs compared to the 3D model, you lose lots of detail due to resolution loss across all 3 axis with extrusion but the information is there for a better print using a zPrinter or other high-end 3D printer.

This project has endless applications and deviations so I tried to show you each step I took to create this Egyptian Pyramid but within the scope of the purpose behind this book.

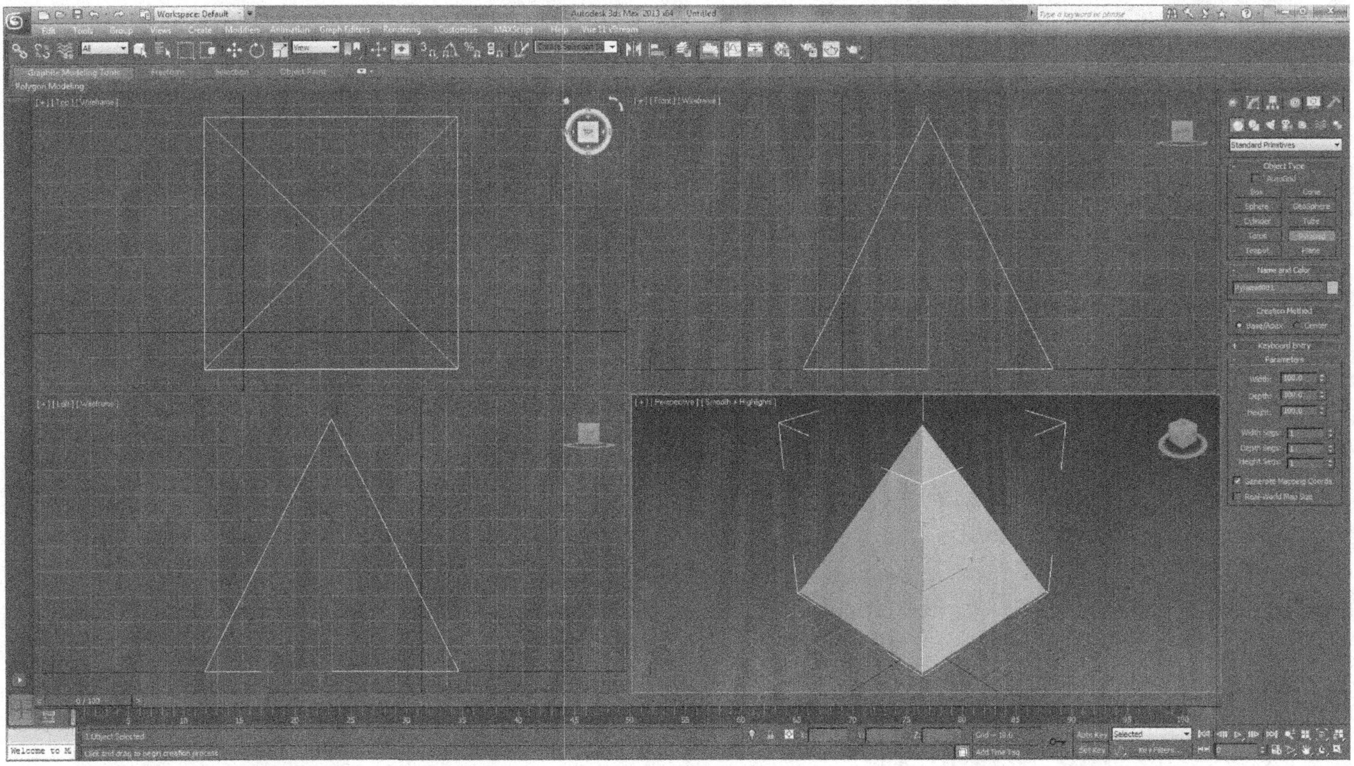

Step 1: Create the same pyramid

I opened my 3DS Max, clicked the little ball on the right and choose the same settings and process I used to create the erupting pyramid but, unlike the previous project, this pyramid will be used as a reference and scaffold, not the finished model I'm looking at creating then printing.

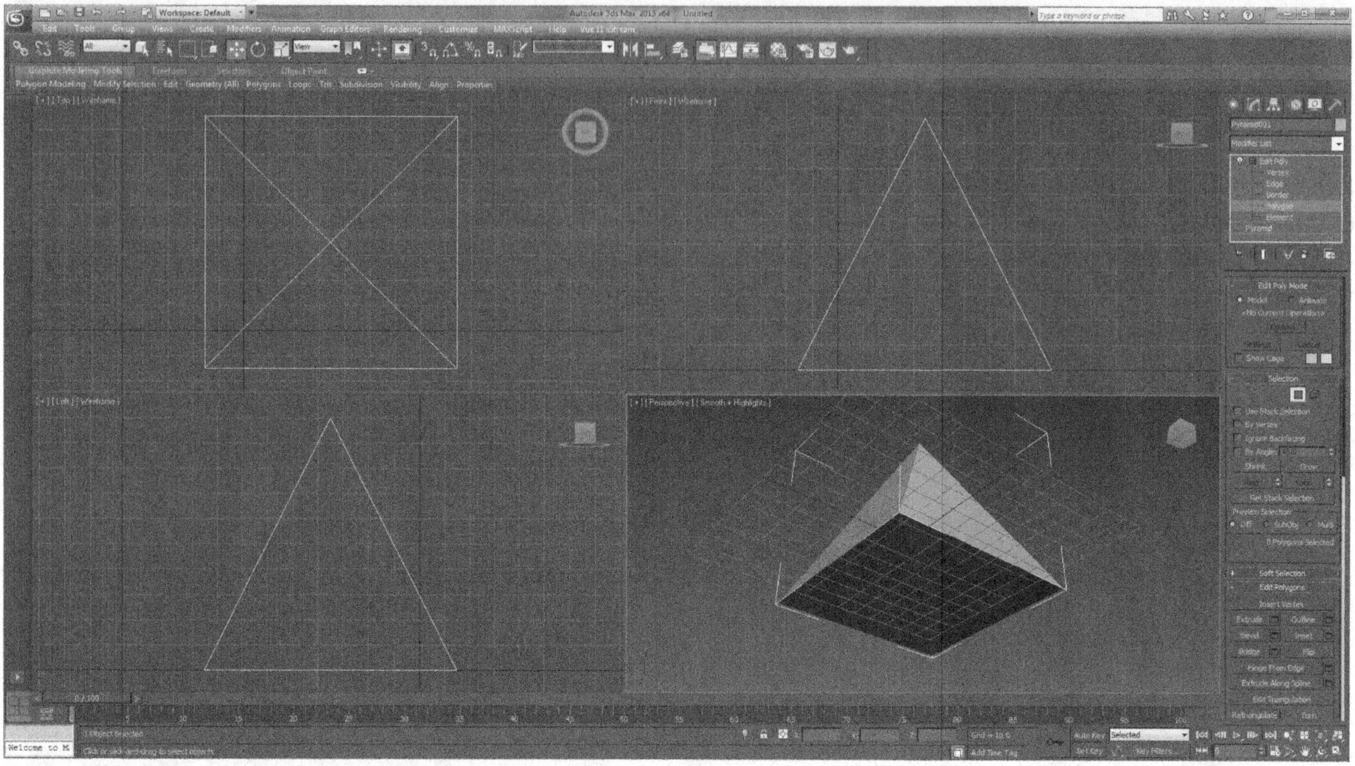

Step 2: Remove polygons

My goal is to create stone blocks that go all the way up this pyramid; as such, I need to automatically generate them within 3DS Max. Sure, I could do this manually by placing one stone after the other until I have a pyramid, but I want this model to be finished within the next few minutes, not HOURS.

What I did was change the initial shape into an editable polygon, click the bottom of the pyramid, and manually delete the polygons there. I don't want the stone blocks to show up under my model, just from the sides. They'd also use up more 3D printing material and not bring anything "new" to the model, as they are underneath.

Step 3: Create a block (next page top)

I go back into my primitives and create a small box, roughly equal to what I'm looking for my pyramid to have, with dimensions of 5" square.

Step 4: Create a sick pyramid (next page bottom)

What I did here was select the block I made, select the round ball, and this time select Compound Objects --> Scatter. Then click the pyramid and tell the computer I want this little block scattered onto the pyramid shape 494 times. Looks awful, so let's fix it!

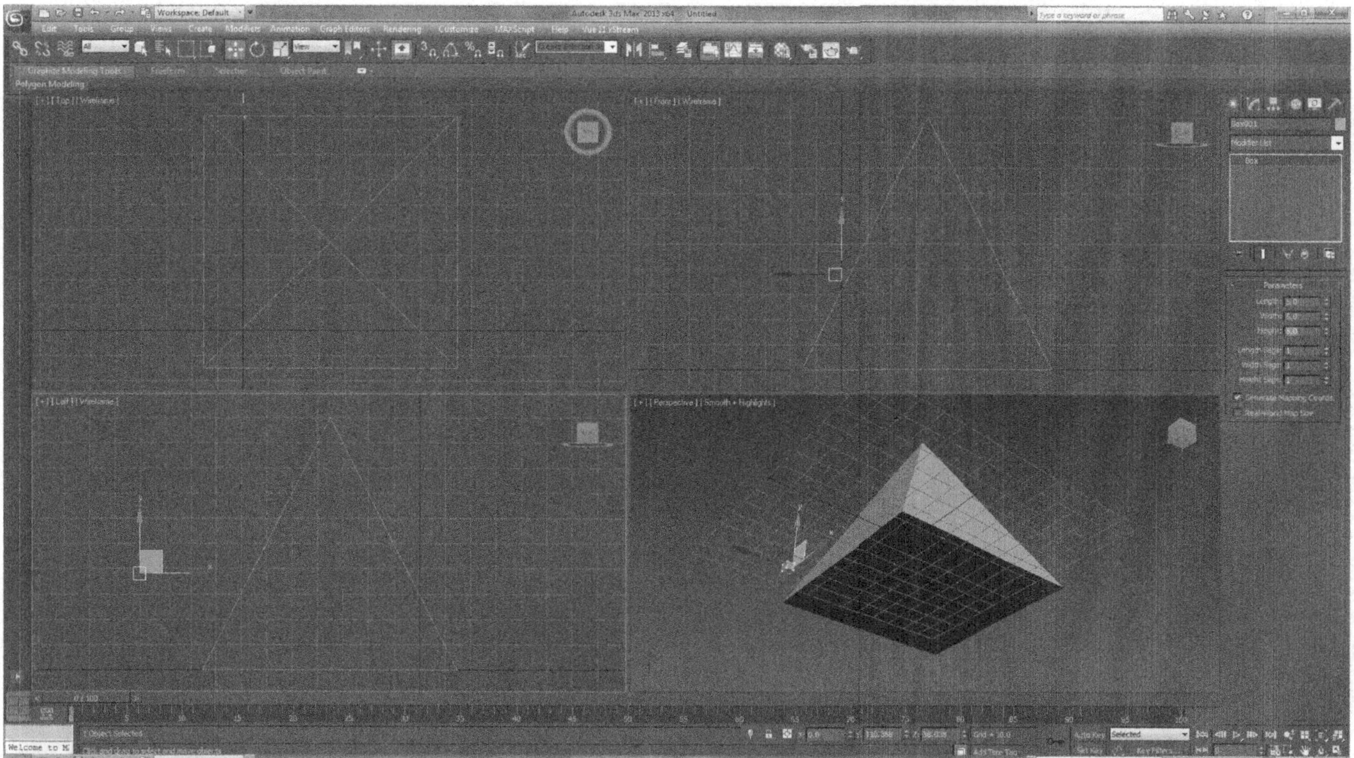

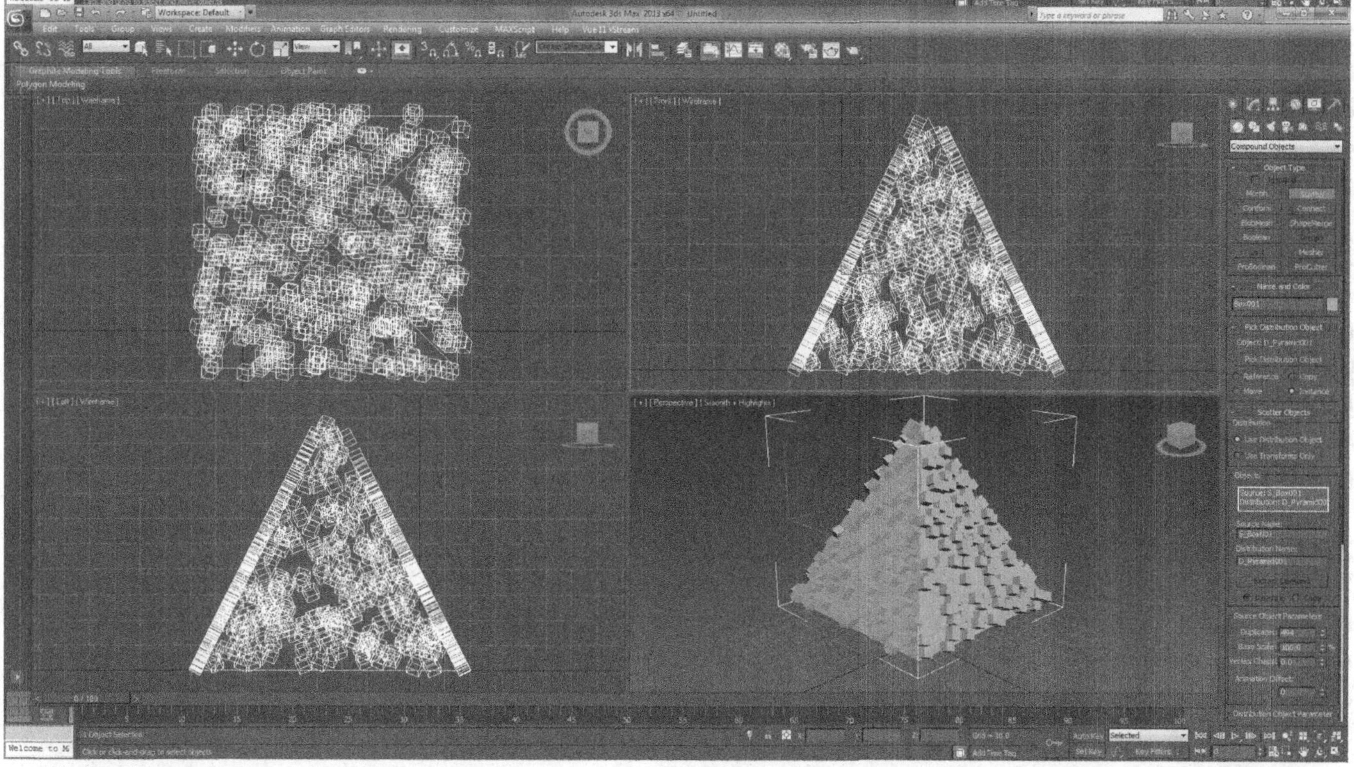

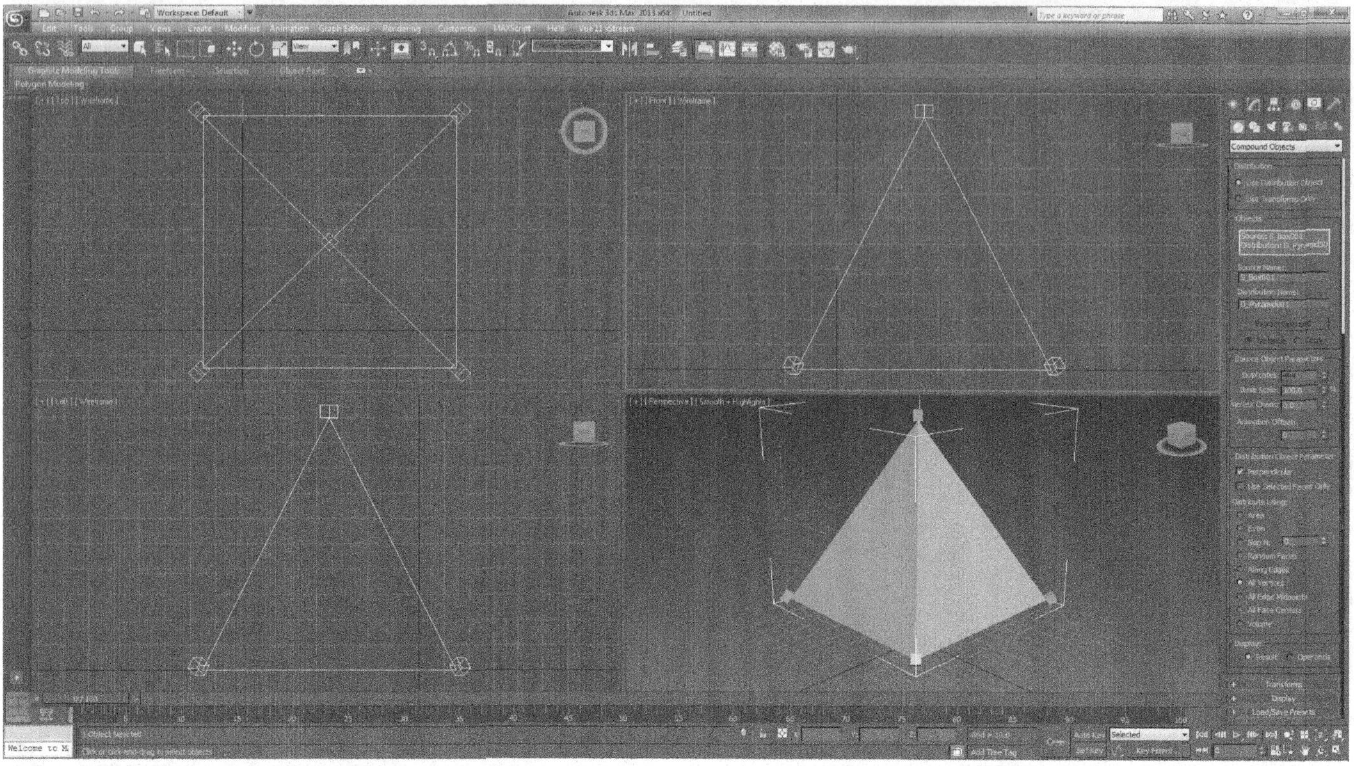

Step 5: Source of the error

I scrolled down the menu of options on the right under scatter and selected All Vertices, and we see the problem: the pyramid shape doesn't have enough of them ... not even close! See the little blocks on all corners -- this tells us we need to create more vertices on this model ... A LOT MORE!

Step 6: Subdivide and conquer! (next page top)

As luck would have it, the shape is no longer an editable polygon, so I have to click undo (something you do a lot of in 3D modeling) to undo the step so I end up back with only a pyramid and a square as seen at the top of the previous page. Then under the Modifier List, I selected Subdivide. Now that's looking better! Where before I had five vertices, I now have exponentially more! I choose a size of 3 under subdivide -- why? It made the mesh look about right for the block, as I want it to overlap.

Step 7: Create another sick pyramid (next page bottom)

What I did now was the exact same step I did in Step 4. Looks like the same result, right? Not exactly. Following the same step I did in Step 5, let's see what happens this time around. In this case, I wanted to make a pyramid, but all these steps remain the same for any other shape you want to create. How do I know these things? Practice and over a decade's worth and endless tutorials.

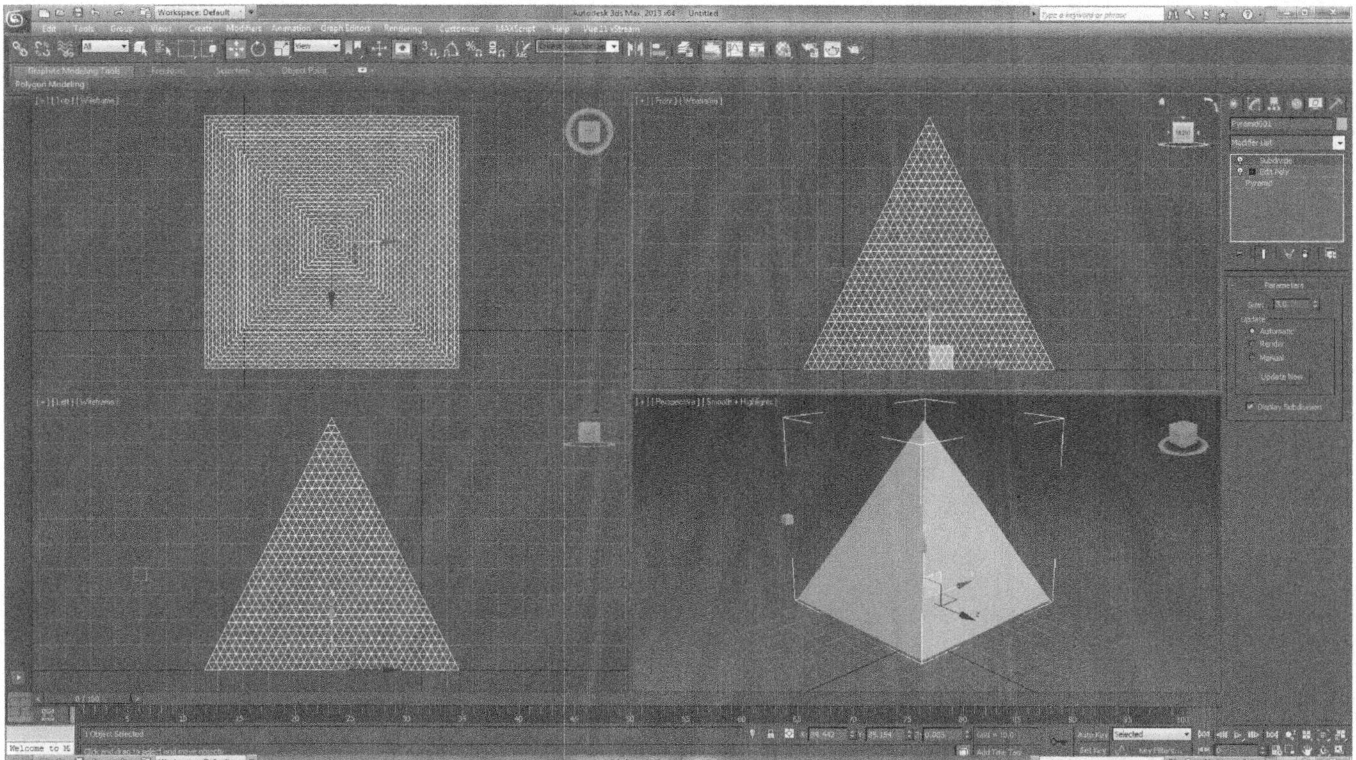

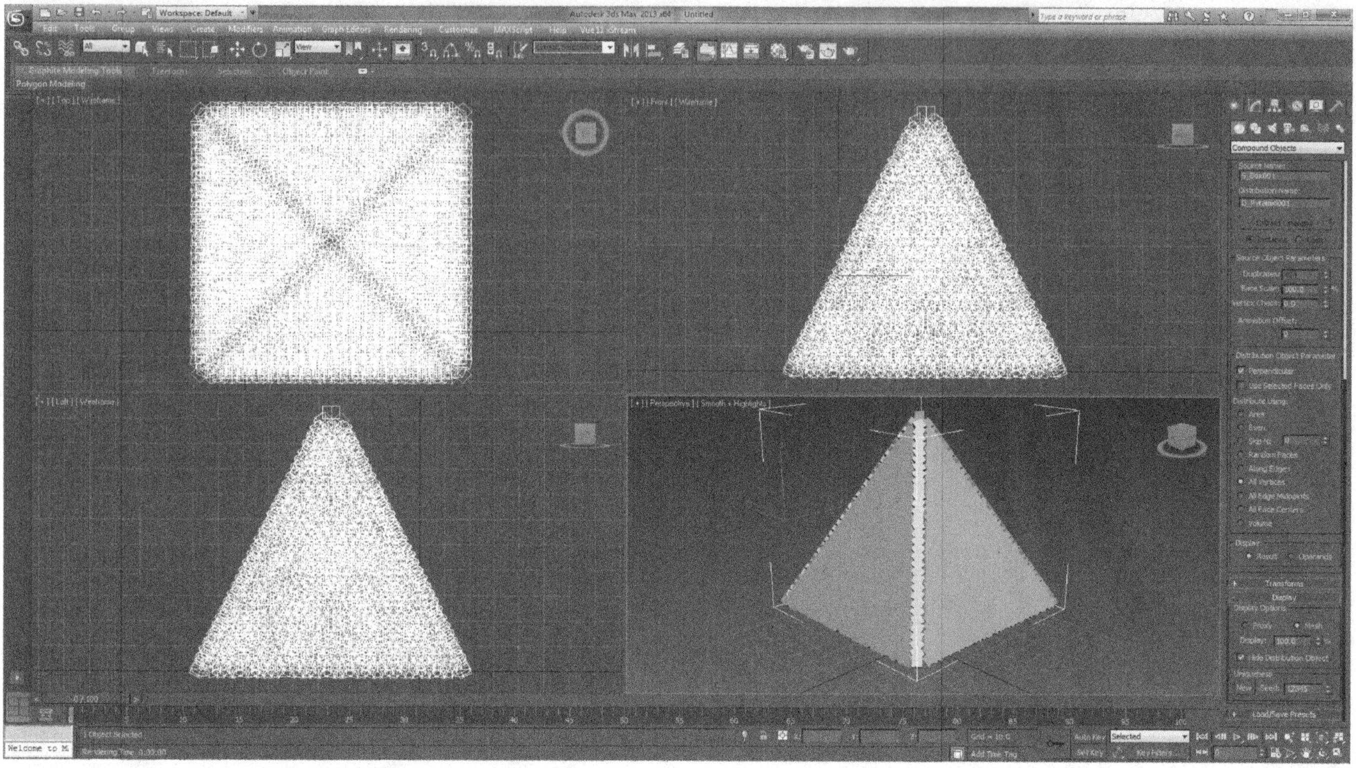

Step 8: Perpendicular blocks

The results are very different. Under display I decided to hide the distribution object, and well, we still got our pyramid but it seems like we have way too many square blocks! We also don't want to make the blocks go up perpendicular to the sides of the pyramid so let's turn that off in the options list as well.

Step 9: Finally, some steps! (next page top)

This mesh has issues. It's too dense and the blocks don't look like stone at all, but we have some progress -- skipping through all the boring details to make this NICE -- after half an hour of playing around (a lot of 3D modeling is playing around and trying endless options), I generated my pyramid as you see it now in Step 10.

Step 10: End result (next page bottom)

What I did here was add lots of detail to my initial block, add lots of extra vertices, and deform it, then scatter it across the pyramid. The results might look chaotic at best, but we now have a mesh. Let's export it as an STL file and open it within netfabb, a great free program for fixing mesh up! What do you do to create another object such as the beautiful sculptures that Bathsheba makes? Each model can be generated countless ways; this is the beauty and challenge of 3D modeling -- it's a lock with endless combinations and no one way is really better than the other once you know the basics.

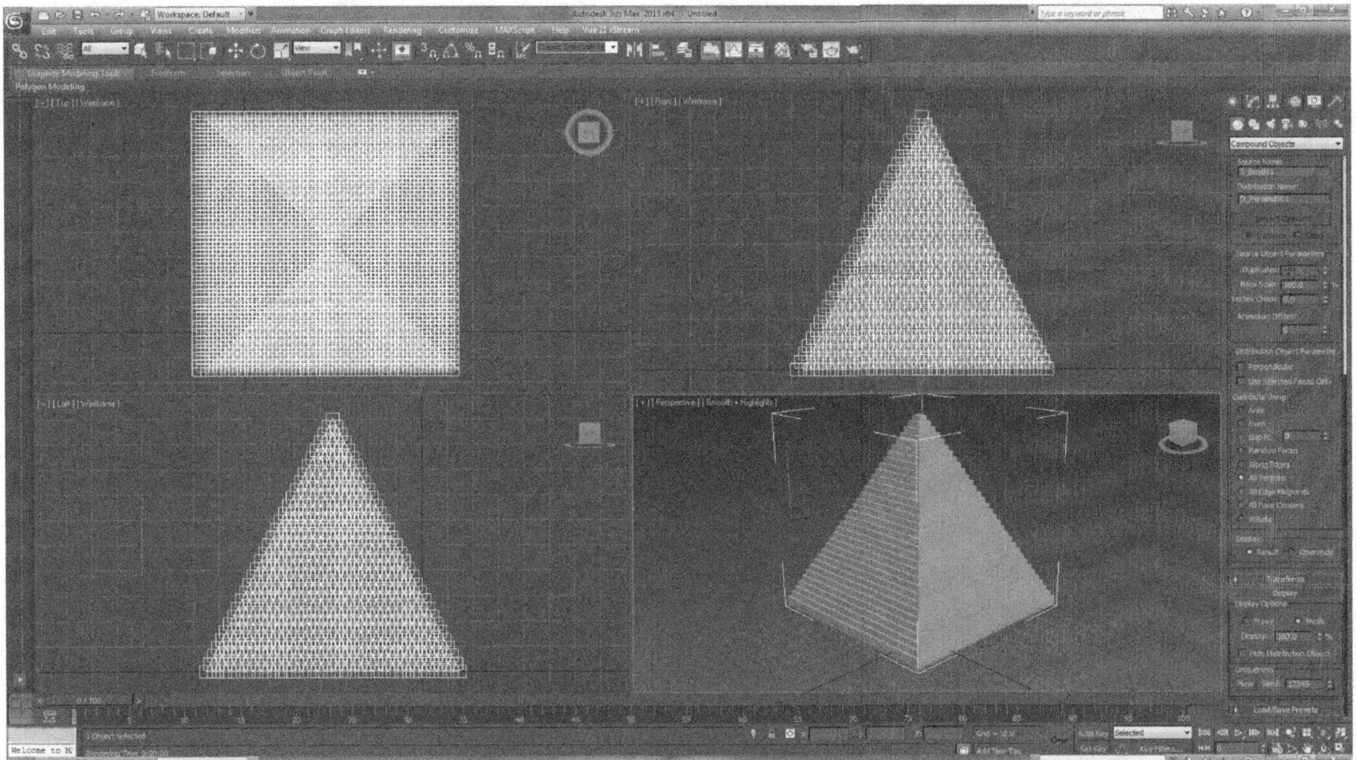
Workspace: Default
Compound Objects

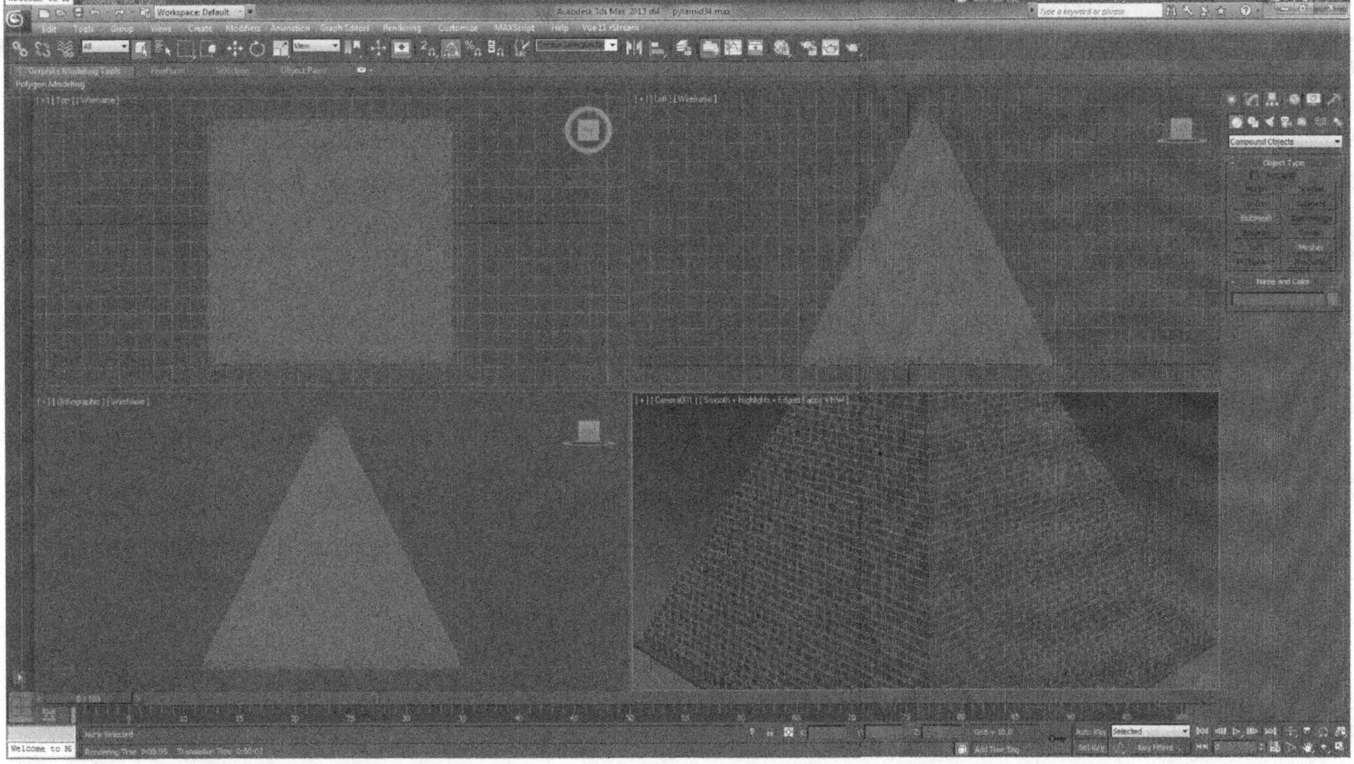
Workspace: Default
Compound Objects

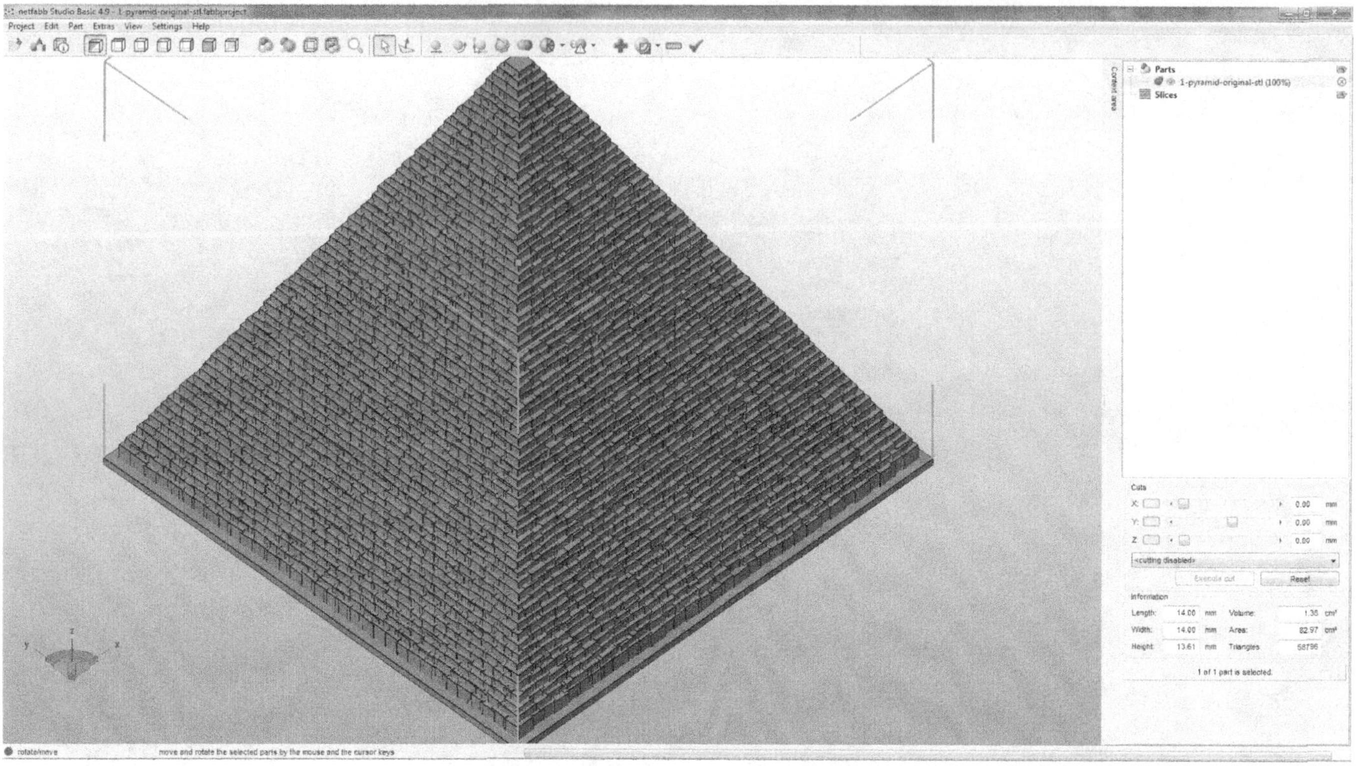


Step 11: Netfabb to the rescue!

In netfabb, you can see I have my scales all wrong, and I'm sure the mesh I generated has more than its fair share of problems -- netfabb allows me to fix these issues and a whole bunch more. The slicing feature is fantastic for looking through the model to find more errors! Sometimes you find so many problems it's quicker to go back into 3DS Max and regenerate a fresh mesh based on what you see. Once that's done, I export the STL and import it within my Cube 3D printing program.

Step 12: Cubify my pyramid (next page top)

As you can see, the Cube software is incredibly bare-bones -- hence why I prefer fixing things in netfabb, then exporting it to 3D Systems software.

Step 13: 3D Pyramid printing (next page bottom)

You can now finally print your 3D model ... only after all this software work can you finally see if what you imagine actually works! In this case, the model printed out great -- there was some mesh bowing on the bottom (you'll see it go unstuck in the bottom right corner) but luckily it didn't completely unglue from the platform.

With 3D printing or any other CNC machine, you can do everything perfectly but still have problems -- this is why it's vital to test your stuff out or buy proven meshes!

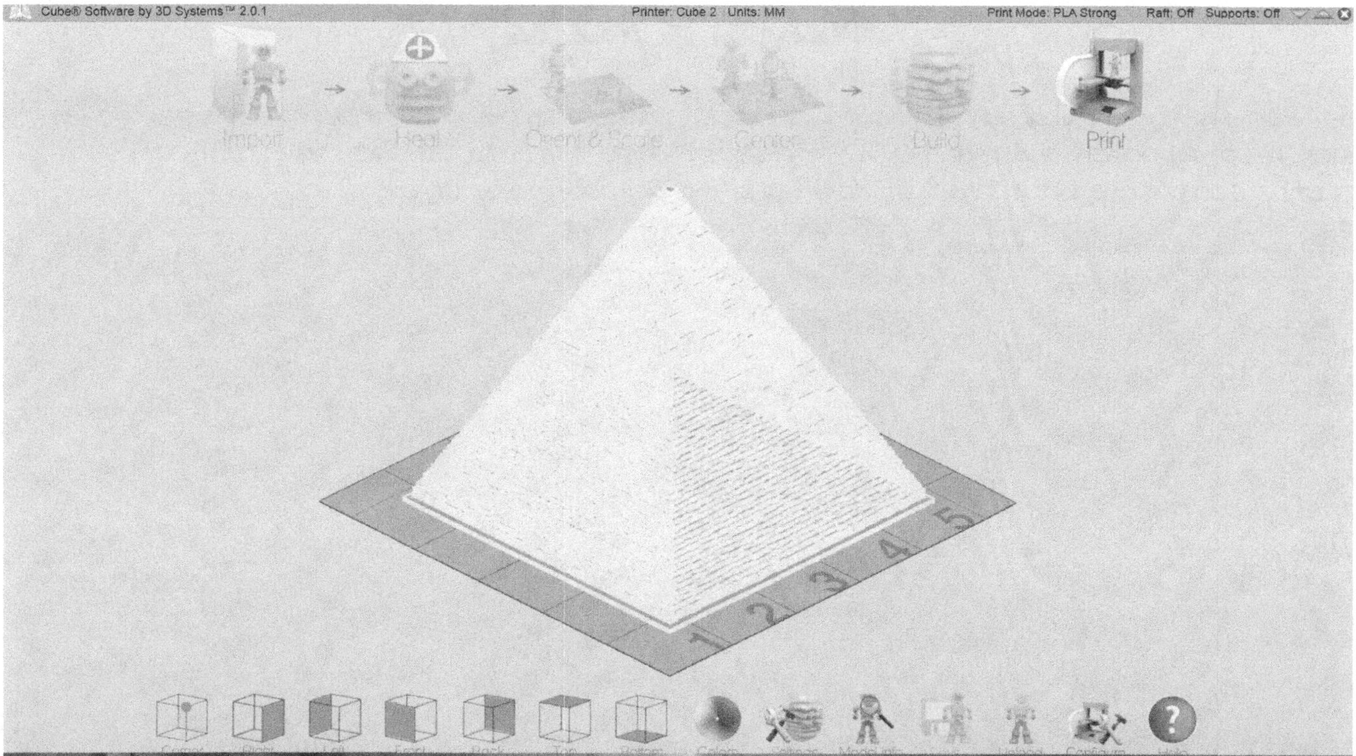
Cube® Software by 3D Systems™ 2.0.1
Printer: Cube 2 Units: MM
Print Mode: PLA Strong
Raft: Off Supports: Off
Import
Heal
Orient & Scale
Center
Build
Print
1
2
3
4
5

# Mayan Pyramid

So far, the two "pyramid" projects have been rather simple and straight forward, so let's kick it up a knotch and make a 3D printable Mayan Pyramid. At first glance, it looks pretty complicated but break it down and it's actually very simple!

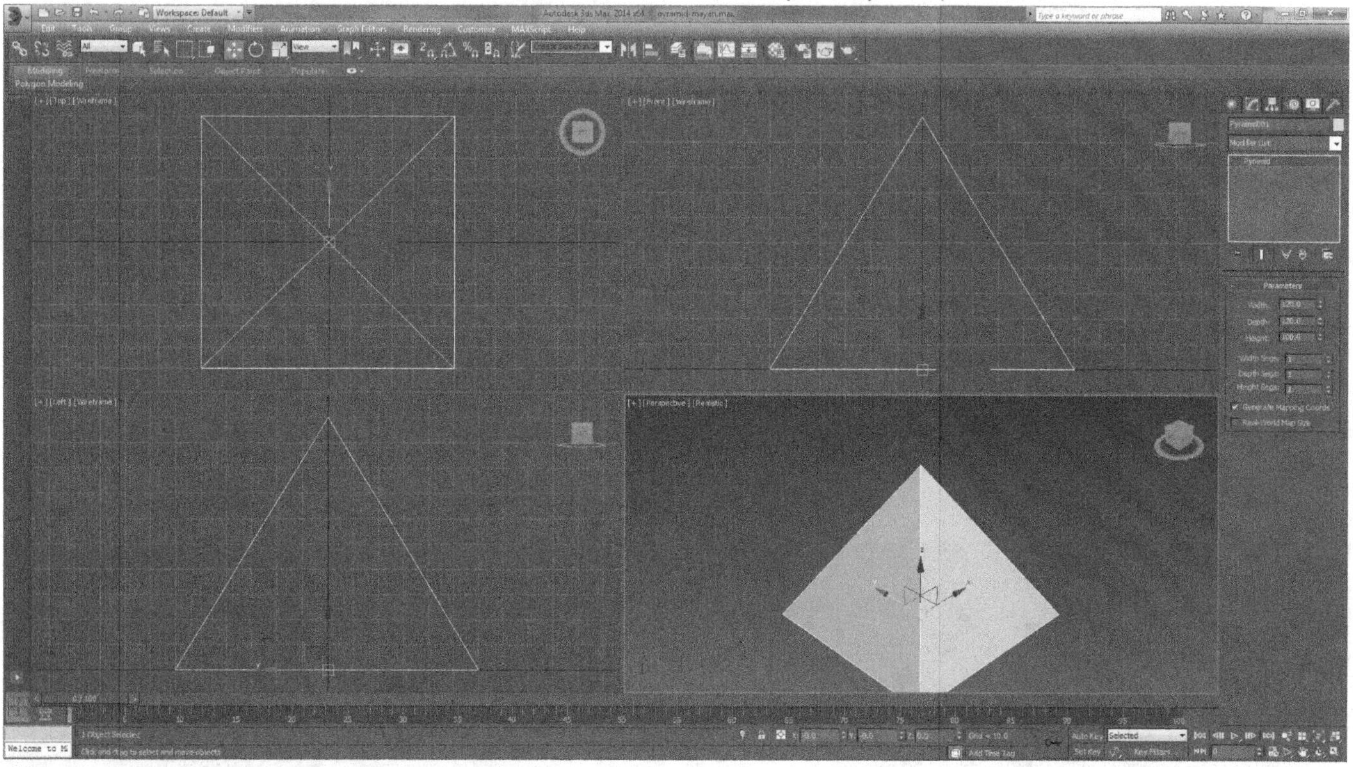

Step 1: Create a pyramid

Like the previous two designs, I started with a primative pyramid but in this case, I'm only going to use it for reference. This is for an extrusion 3D printer, as such, this shape is great because each stage is smaller than the previous one, dimension wise. I'll give mine 8 large stages with 4x stages for the stairs along the front. I've made it slightly wider.

Step 2: Step splines (next page top)

What I've done here is create a spline of steps which we'll then "lathe."

Step 3: Initial shape established (next page bottom)

A pyramid has four sides, thus applying a lathe modifier to a 2D spline creates the same shape when applied four segments. This method is far faster than building this up, one square at a time (producing 7 pieces instead of one) or playing around with the pyramid geometry, one side at a time.

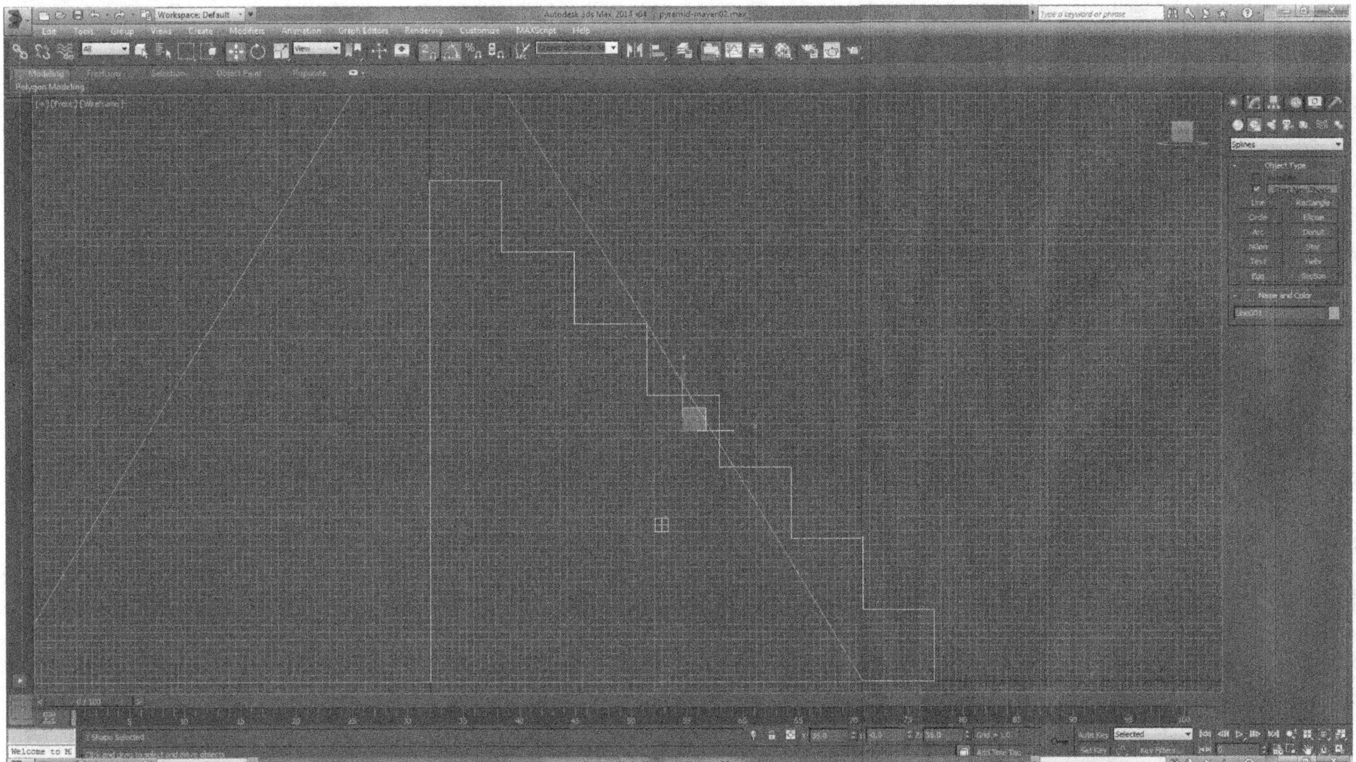

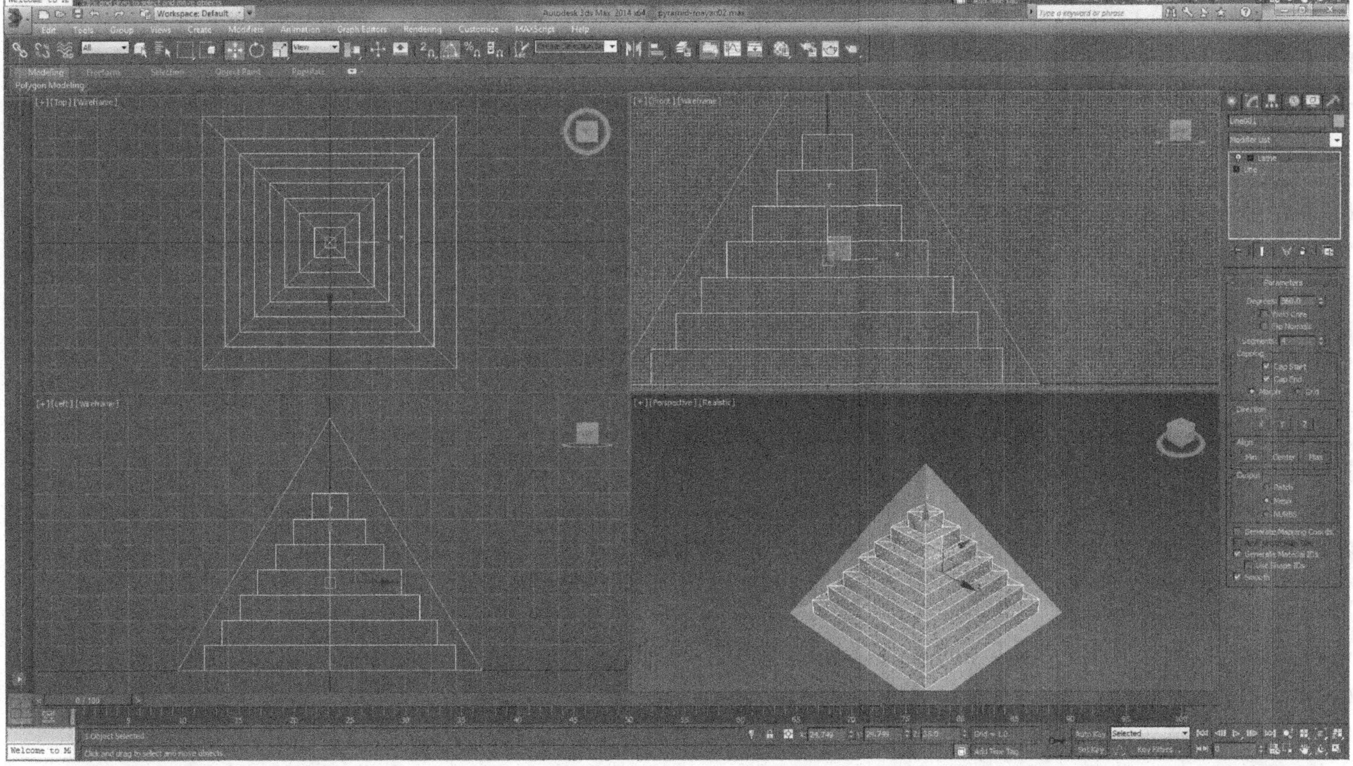

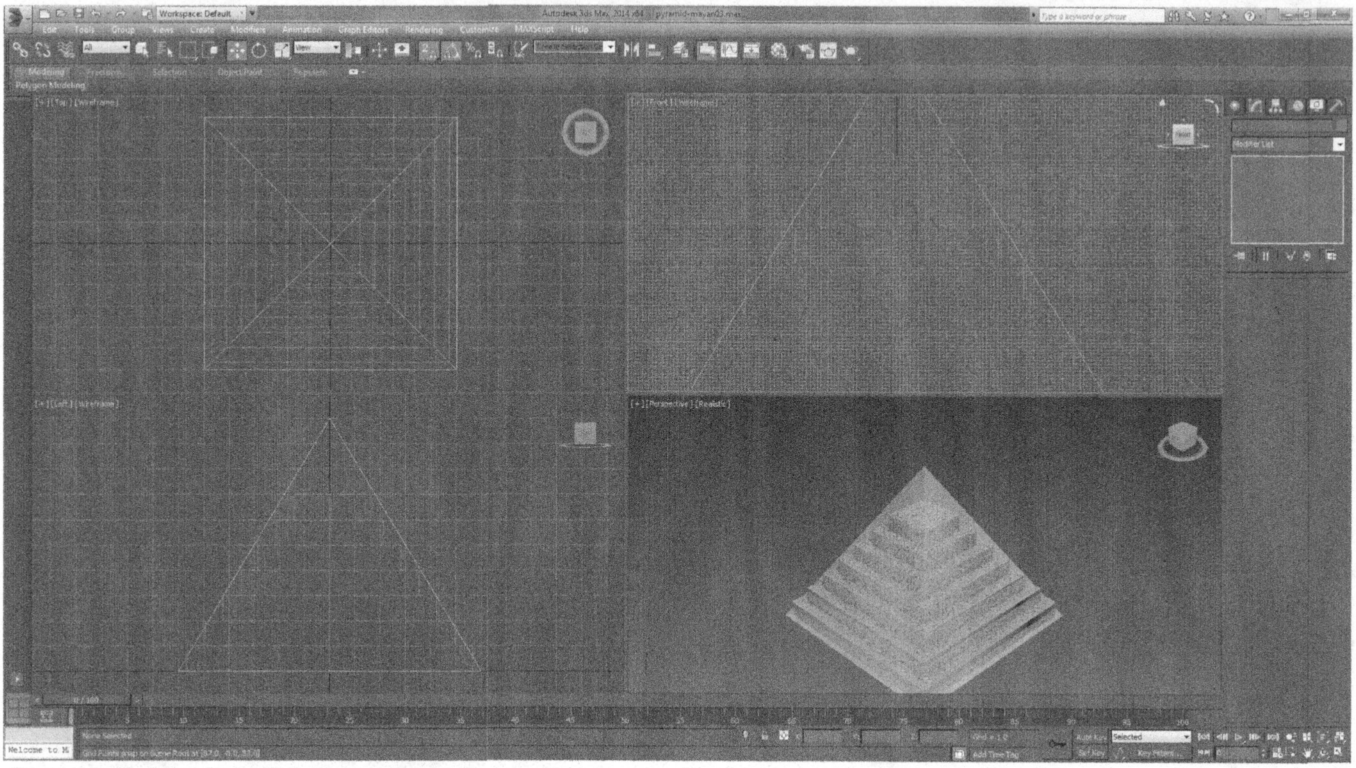

Step 4: Widening the pyramid
I want to make my pyramid wider so I went back a step and extended the step a little further and re-applied the lathe. I now have the shape I was after.

Step 5: Create a ramp (next page top)
Unlike the Egyptian pyramid you saw earlier, Mayan pyramids have a ramp going up to the top with small little steps. We'll first establish the angle of the ramp we want and extrude it on both sides. I've also deleted my reference pyramid as I no longer need it. At this point, I already see a few minor issues I'll have to resolve as having square steps going all the way up will go at a 45 degree angle, but we'll deal with it later.

Step 6: Details to match resolution and context (next page bottom)
Now that I have the steps done, I just extruded a 2D spline that went up like the steps I did in Step 2 but smaller, it's important to discuss resolution. I'm not designing this Mayan pyramid to be in a Hollywood movie; I'm designing it to be made with my 3D Systems Cube where having extra resolution doesn't produce a better model while teaching people who probably have no background in 3D modeling to easily follow these steps. If I were making this model for a more capable industrial 3D printer, this build process would probably take over a hundred pages, not just the few seen here. I generally try to design for the machine that I have and the intended audience its for.

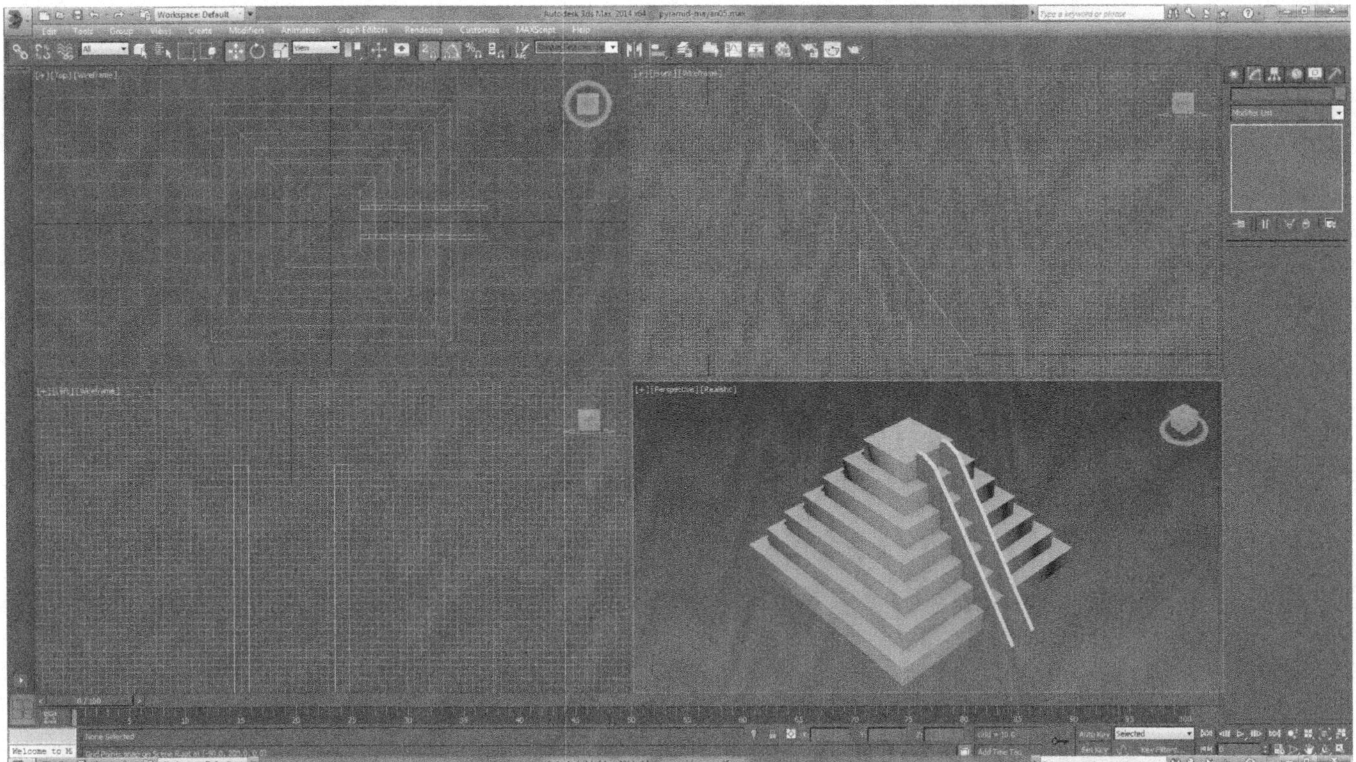

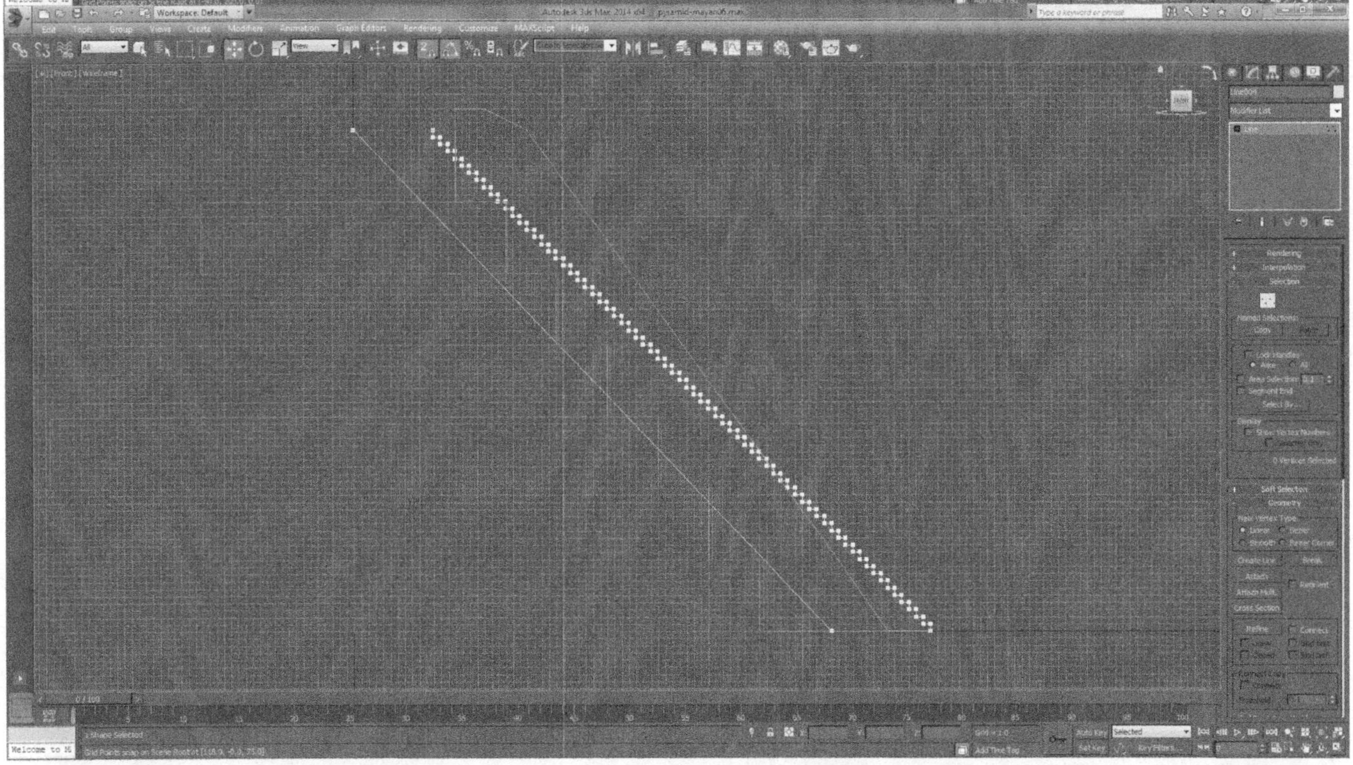

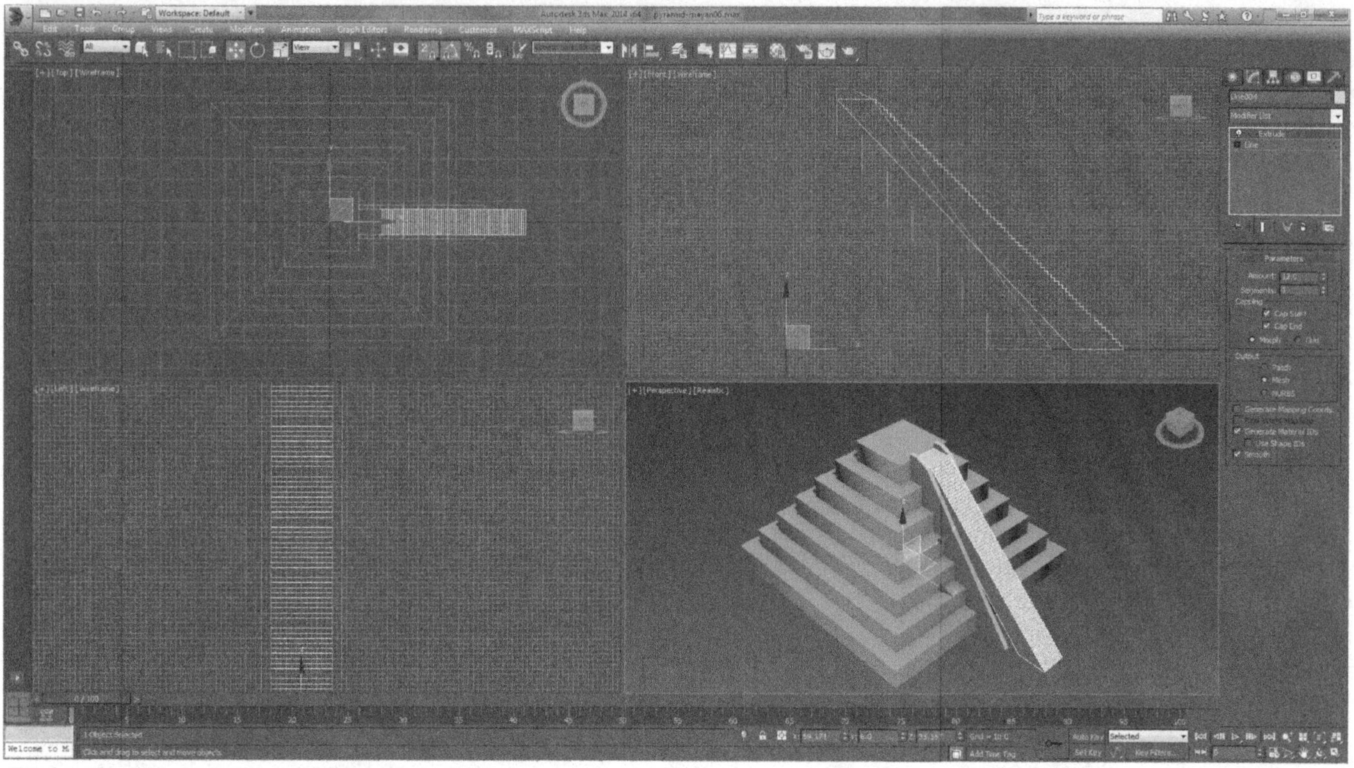

Step 7: Step extrusion

Now that I've extruded the steps, you can clearly see the issues I though I'd have to contend with in Step 5. Let's fix the ramp so that it comes out over the steps, not under them and move the steps closer into the model.

Step 8: Fix a few details (next page top)

That's looking far nicer now but a little too narrow; let's make the steps slightly wider and add a pavilion on the top.

Step 9: Almost done (next page bottom)

It's looking better now - I just wanted to make something nice and easy for my 3D printer to make. The mesh is incredibly clean (unlike the volcano you'll see later in the book) which means it will be an easier print to make without lots of mesh errors.

With an extrusion 3D printer, if you are going to make overhangs, they generally need arch supports, though you can go at right angles if you wish, up to 30 degrees. Although not accurate, I want to make the mesh a little bit more interesting; so lets make the stages all the way up the model slightly more tough looking. Remember that this is a lathed mesh, so it's easy to go back, add details and re-lathe the model.

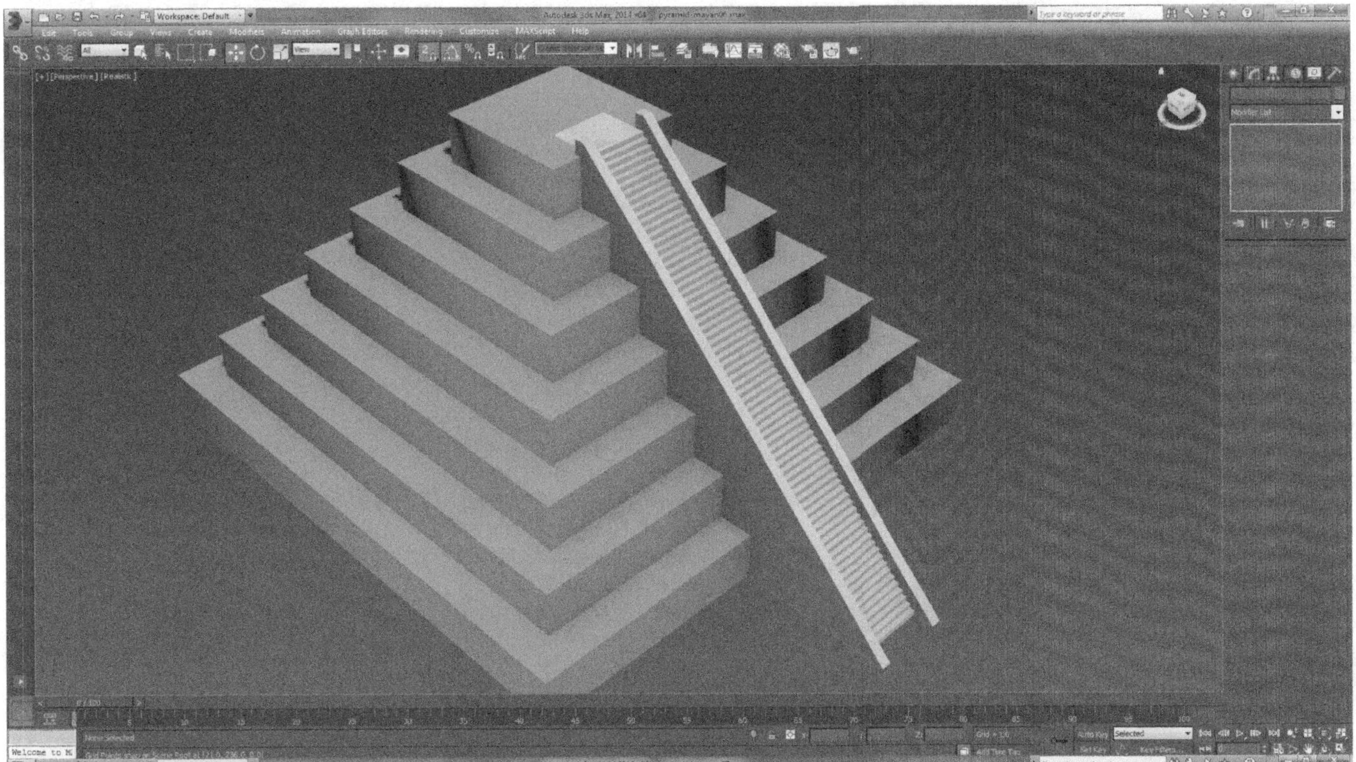

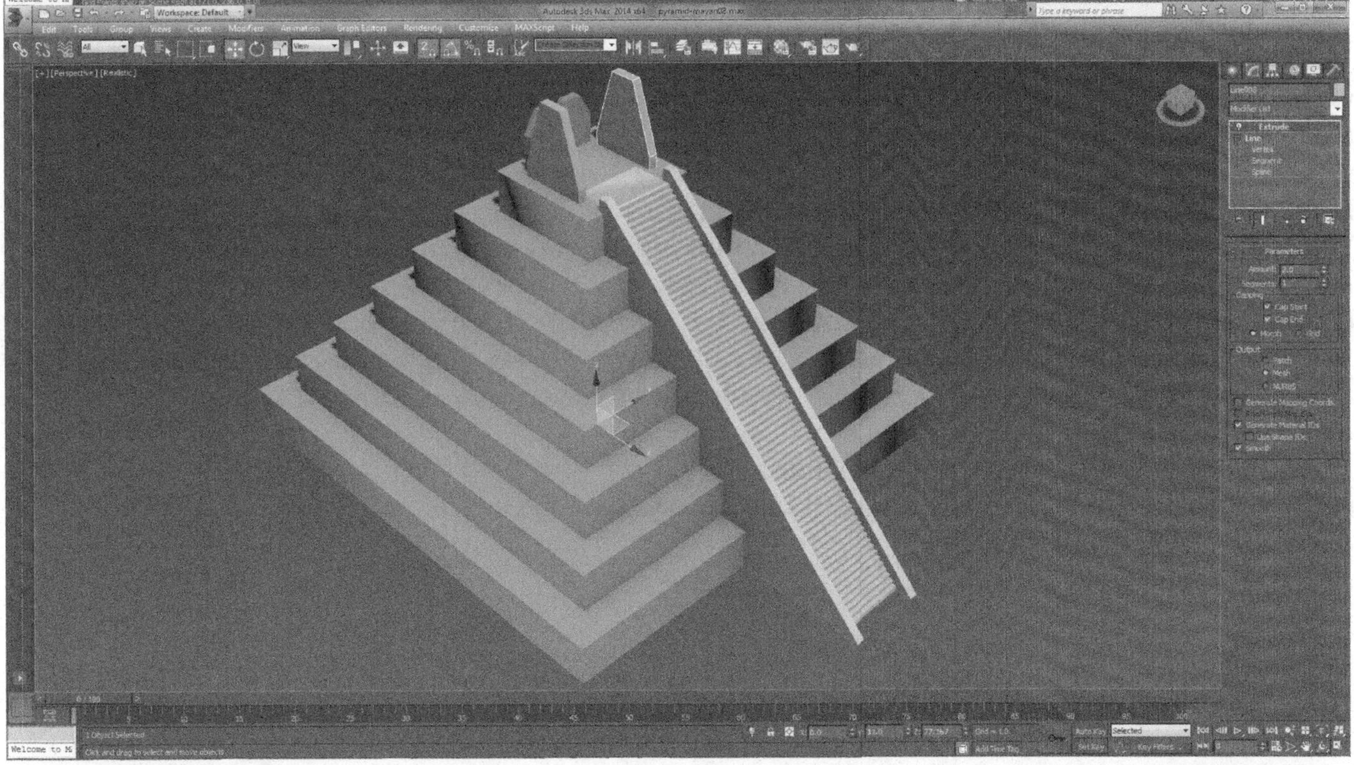

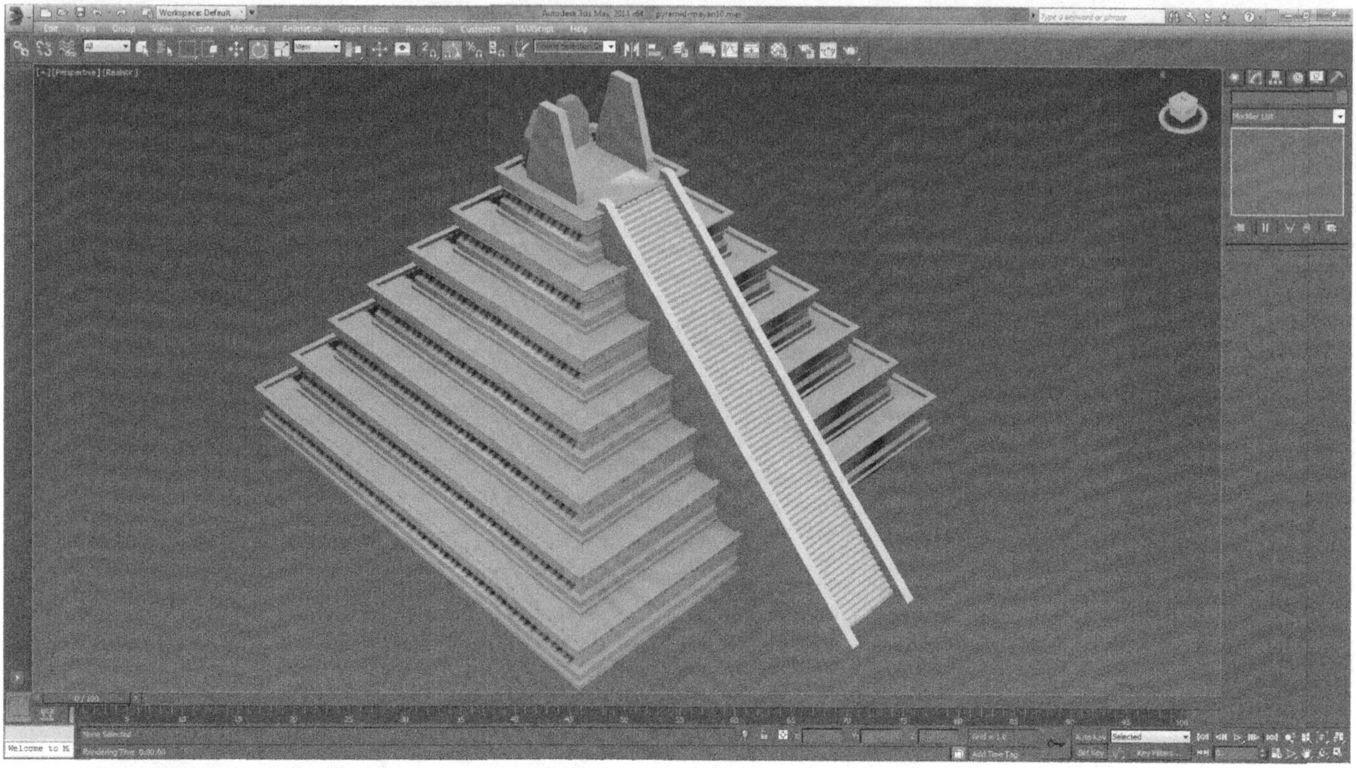

Step 10: Finishing details
That took about a minute and now looks more interesting. I could keep going but I'm happy with this. As we've done before, what we'll do now is hollow-out the inside so that this uses less material and prints faster!

Step 11: Double-check the pyramid (next page top)
So far, my entire model has been built to be "water tight" but once you start cutting away at a model, errors begin to be produced. What I'm looking-out for here is that the edges of the pyramid I'll use to hallow-out the inside of the mesh doesn't take-out anything other than what's intended... looking closely at the steps you can see I'm fine. I've decided to make it take-out some of the top, let's make this a death chamber!

Step 12: Almost done (next page bottom)
Now that we have a finished model that's hallow, we'll export it as an STL and do some checking inside of netfabb.

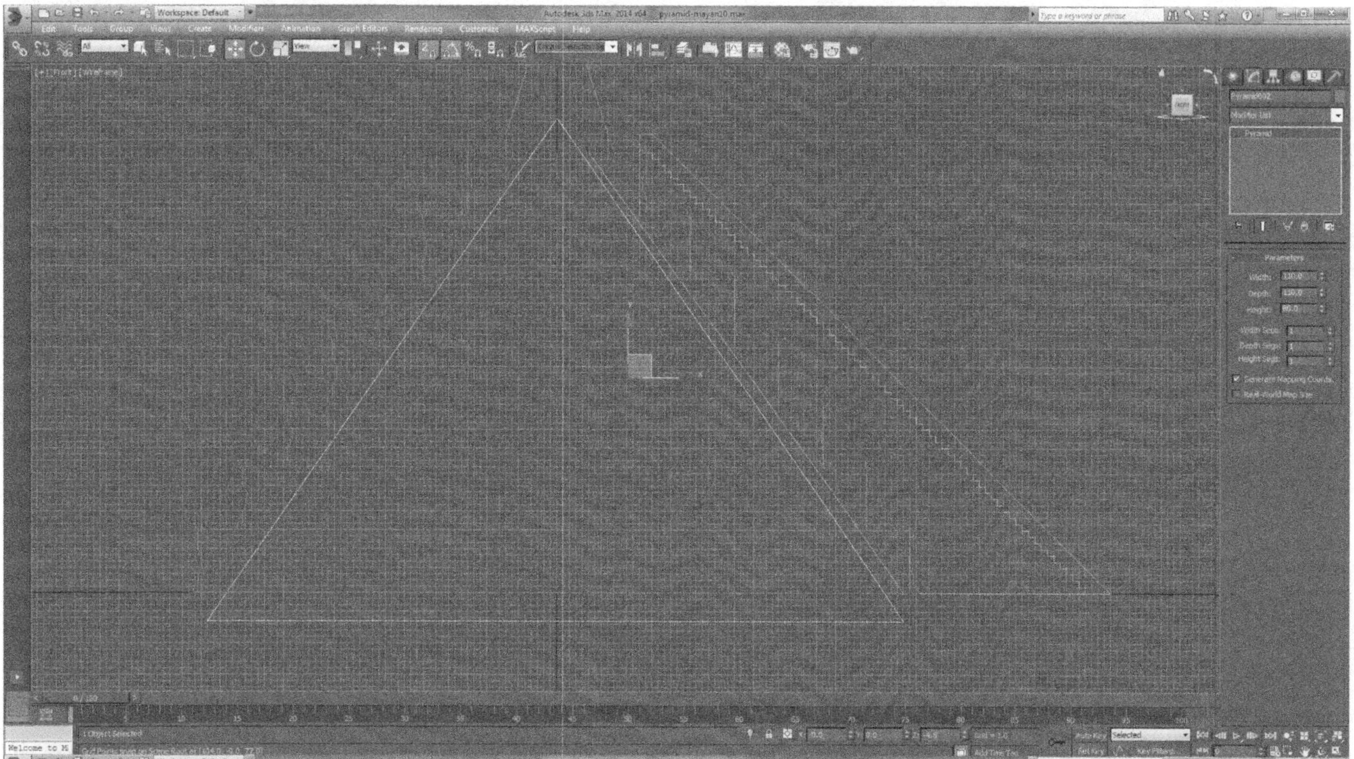

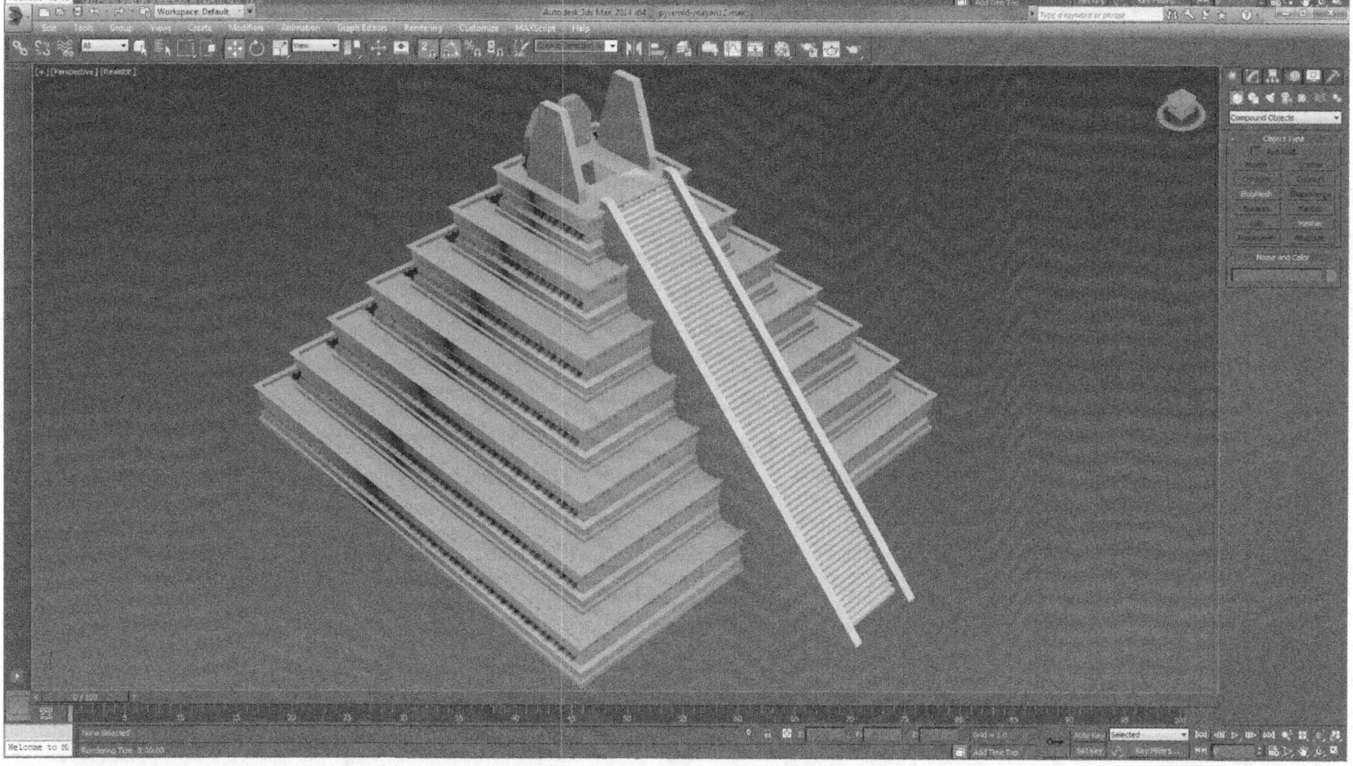

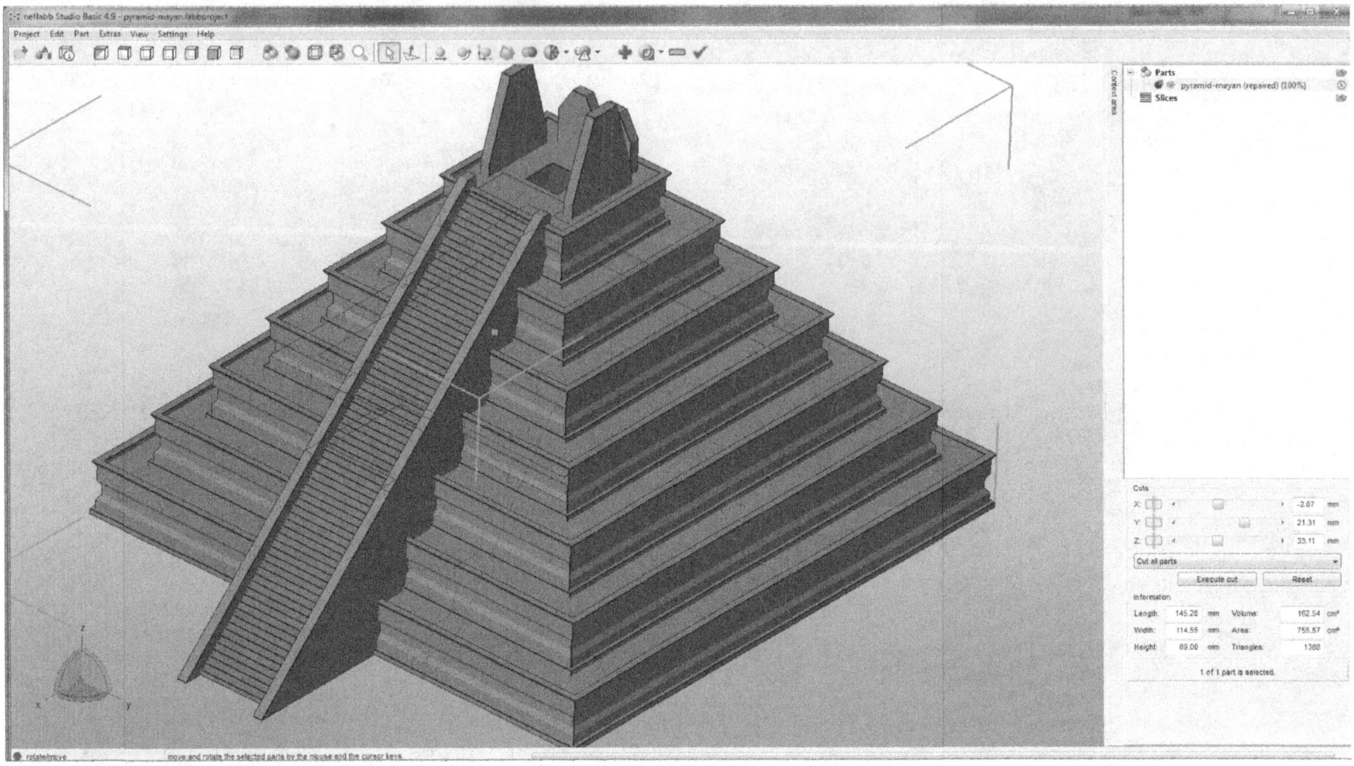


Step 13: Mesh looks great!

Mesh is looking great, nothing to repair though it's slightly too large for my Cube's platform. I can't go over 140 mm so I made it slightly smaller before importing the final mesh into my Cubify software.

Step 14: Quick build time (next page top)

Cubify tells me it's a 6 hour print... pretty neat considering how relatively large it is. I turned supports off but kept rafting to reduce the chance of it getting unstuck to the platform during the printing process. I did a few things to make it print faster, first was hallowing the inside... that's a major win but secondarily, I made the inside stage walls concave which produced a slightly quicker build rather than making them convex which would have required more material. By making the top with a hole with the small details on top minimal, it also allowed me to produce something better looking at a cheaper price.

Keep in mind that 3D printing is still very new. Expect to have software and hardware failures even on the most dedicated and supported 3D printers out there. If you are lucky enough to have a REAL industrial 3D printer at your disposal, many of these issues will simply not exist to the same extent as they do at the consumer level. At the consumer level, cost of entry is a deciding factor behind the engineering that goes-on behind the scenes while at the industrial level, quality and ROI are the primary motivators.

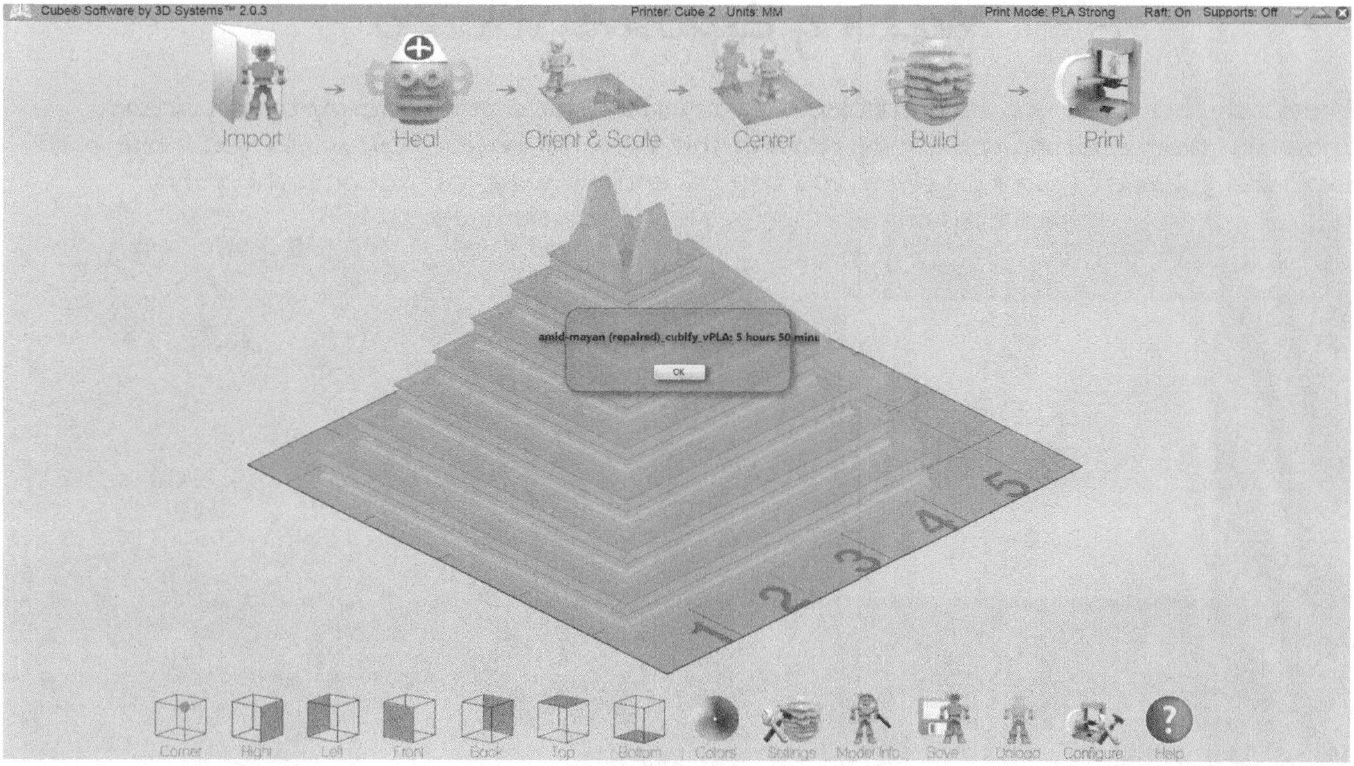


These models are't very functional but I hope that these three "pyramid ideas" have inspired you to realize what you can make with your own 3D printer. You could do the same exercise using a sphere, cylinder or any other combination of shapes... as with other CNC equipment, you will need to practice a lot before something cool is produced.

Generally, I'm very happy with my Cube even with its limitations - 3D printing isn't a panacea where you can make amazing things with an entry level CNC machine but in a few years, it will be. My primary goal with acquiring my Cube was to get myself thinking of the possibilities and challenges I will face once I move onto larger more industrial machines and processes. For what it is, an entry level 3D extrusion printer - the 3D Systems Cube is the best entry level extrusion 3D printer out there. You have lots of pretty colors to choose from (even more so with multiple heads in the CubeX range), it uses cartridges for easy loading and unloading and it's about as simple a machine I've ever used. Load-up the model, press "print" and a few hours later, you got your model.

The downside of the Cube though is apparent when you compare it to what I ultimately want to get, a zPrinter - when I design for the Cube, I keep having to remind myself "no, you can't do that, this is extrusion." This is a completely unfair comparison but one that's on my mind - it's like driving a moped when what you really need is a Honda Gold Wing.

# Modeling Explosive Volcano

Previously, I showed you three projects to create using a primative pyramid. I already know the next question somebody reading this book will have - how would you make a 3D exploding volcano? I want to show you one of endless ways of accomplishing this.

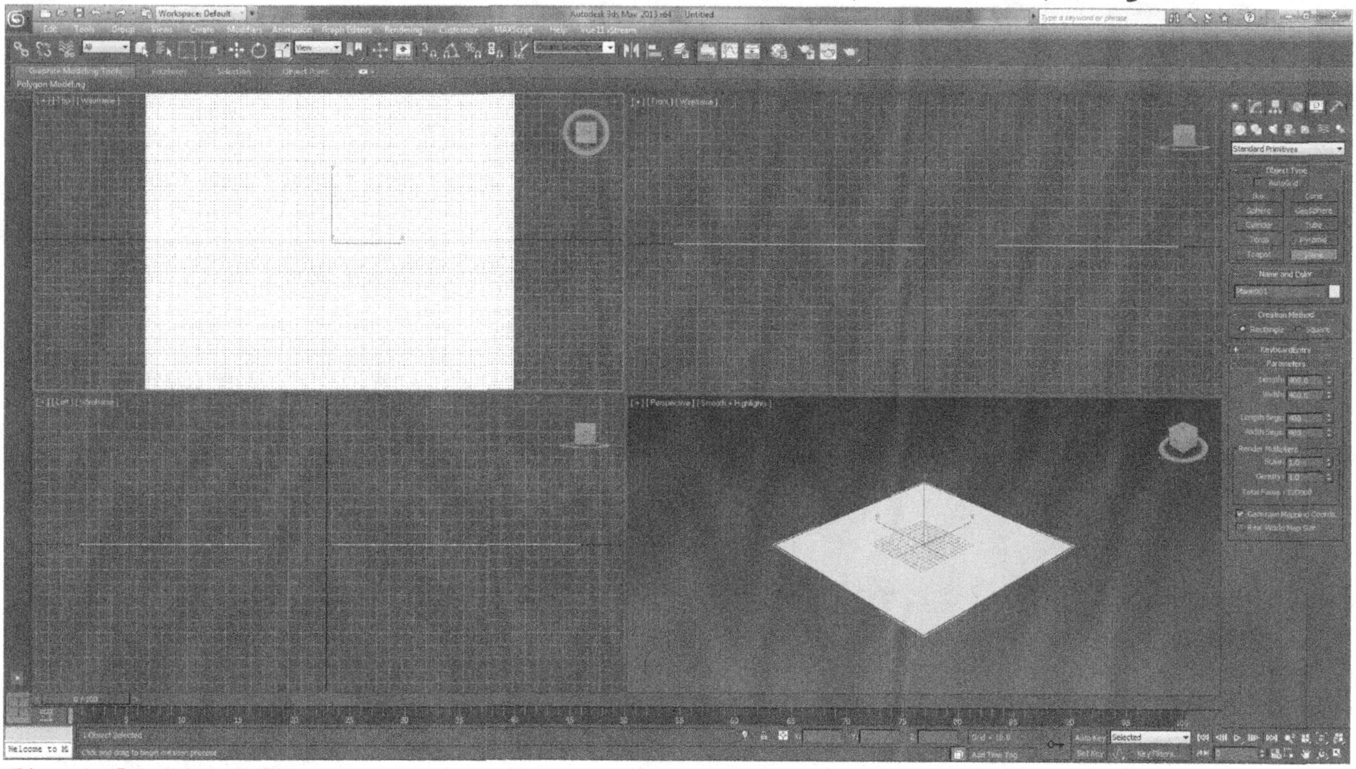

Step 1: Create a Plane

I clicked the "Geometry" (round ball) then selected a Plane. I want to make my volcano 4 inches cube so I made it 400 units square (length and width) with 400 segments across length and width. These squares will give me enough resolution to create a realistic rugged texture and, as unlike the exploding pyramid earlier, I want this to look realisitc.

Step 2: Vertical Displacement (next page top)

Next up, we'll give this flat plane a displacement map (under modifiers) with a strength of 100 and under "Maps" I selected a Gradient Ramp.

Step 3: Map Configuration (next page bottom)

What we have in Step 2 doesn't look like a mountain, so let's improve on the instanced gradient map settings. I want the volcano to be high in the middle and flat on the edges, so I chose "radial", then played with the settings till I was happy with the result.

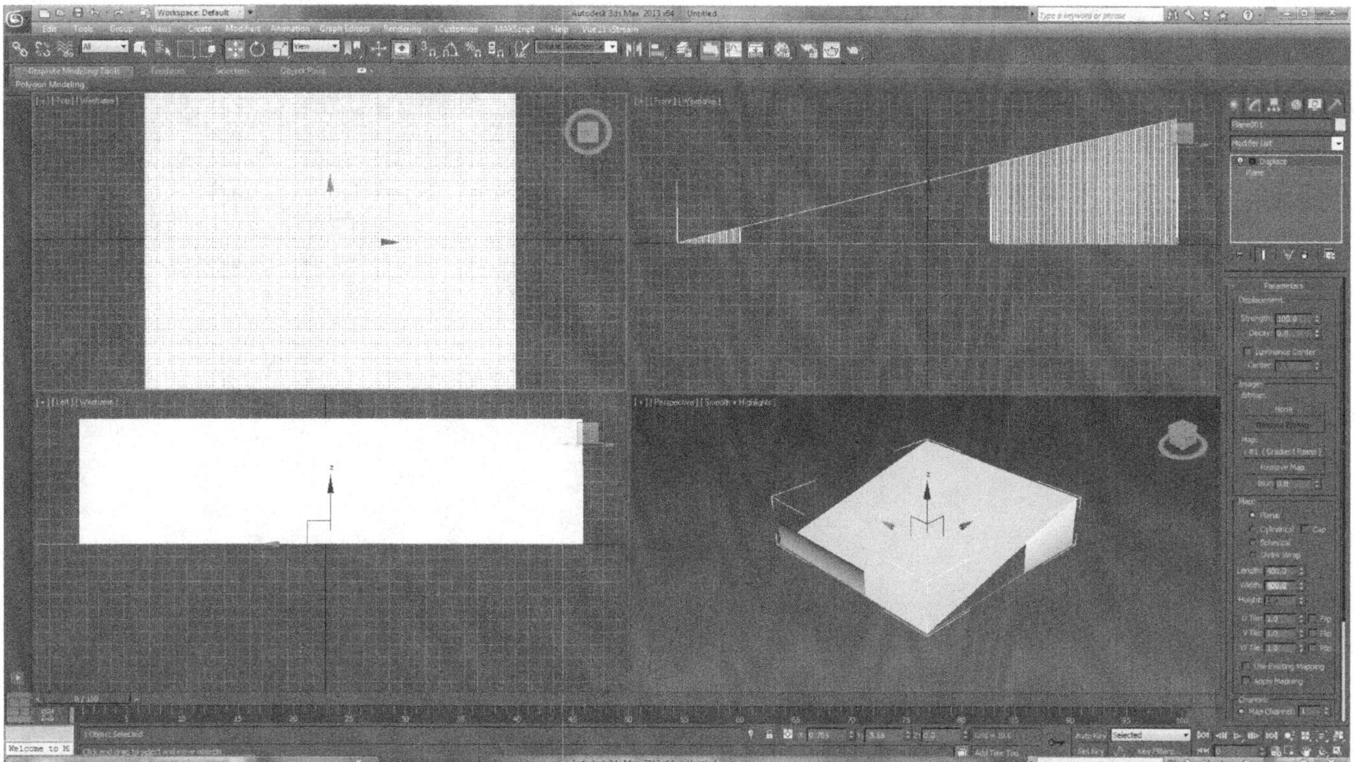
Workspace: Default
Autodesk 3ds Max 2013 x64 Untitled

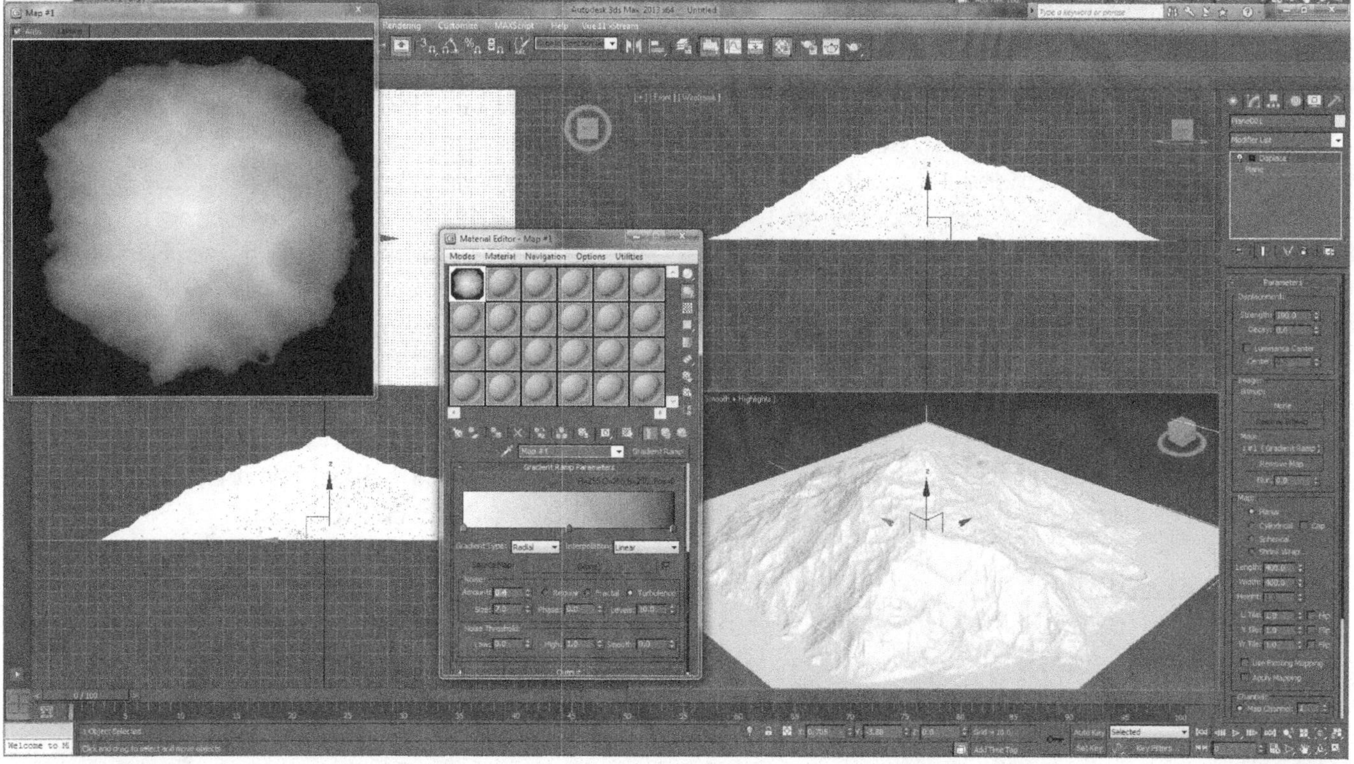
Map #1
Autodesk 3ds Max 2013 x64 Untitled
Material Editor - Map #1
Modes Material Navigation Options Utilities
Gradient Ramp Parameters

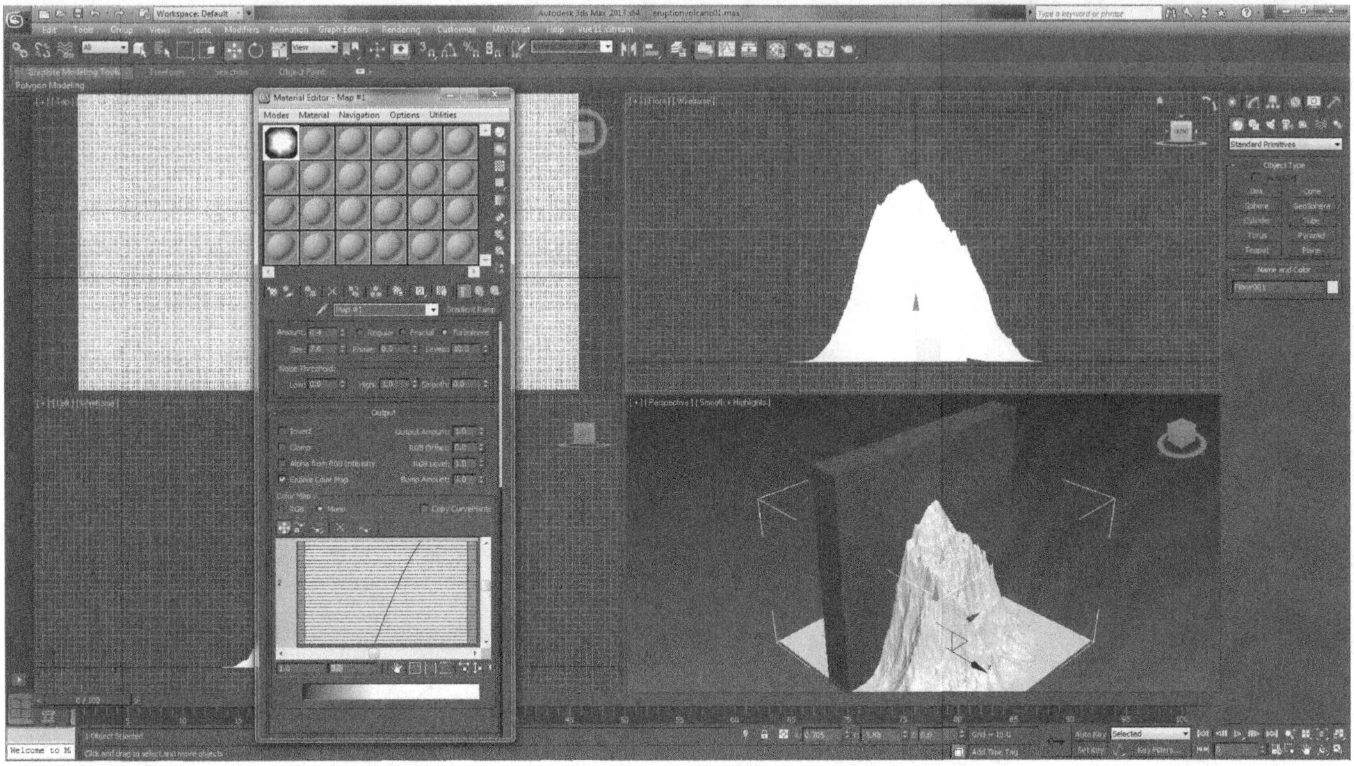

Step 4: Create some Height

Although the plane now looks like a mountain, on my 3D Systems Cube, I have 5 inches square to play with (I made my model 4 inches square for faster printing and to leave room for rafting). What we need to do now is make it proportionally taller so we have more room for the chemial reaciton to take place. What I did for easy reference is create a box that's 400 units cube so I know exactly how high I can comfortably make the mountain. I then enabled "Color Map" under the Gradient Ramp's "output" settings. I could also scale it manually but color mapping gives me more flexibility. I converted each end of the graph to a brezier corner.

Step 5: Hallow

As you can see, we have two problems with this mesh: the first is that the bottom is gone due to the vertical displacement, so although we could print this volcano, there is nothing to hold the chemical mix and the second is the top. We need a hole!

Step 6: Cap and Hole

The bottom is a quick fix, I just gave the mesh a "Cap Holes" modifier then added an "Edit Mesh" modifier, selecting then deleting the mountain cap to create the hole.

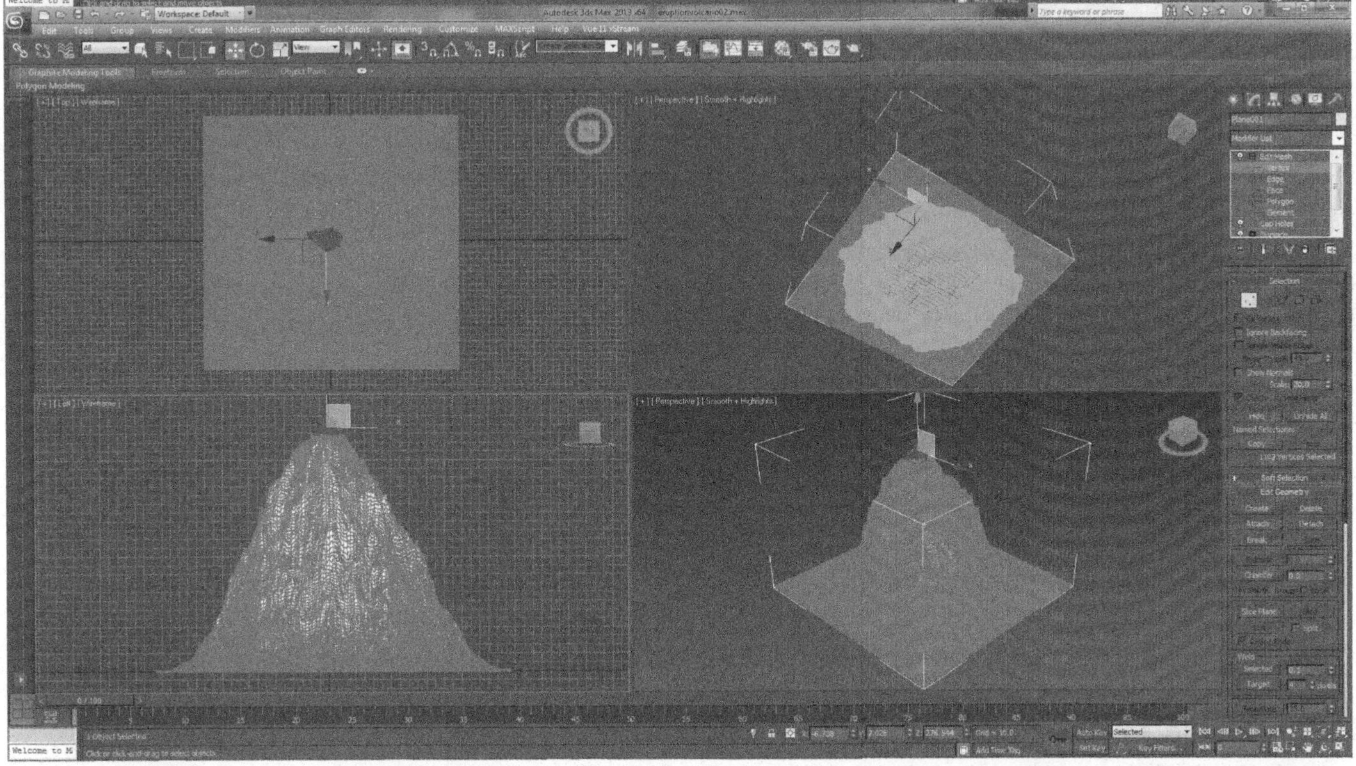

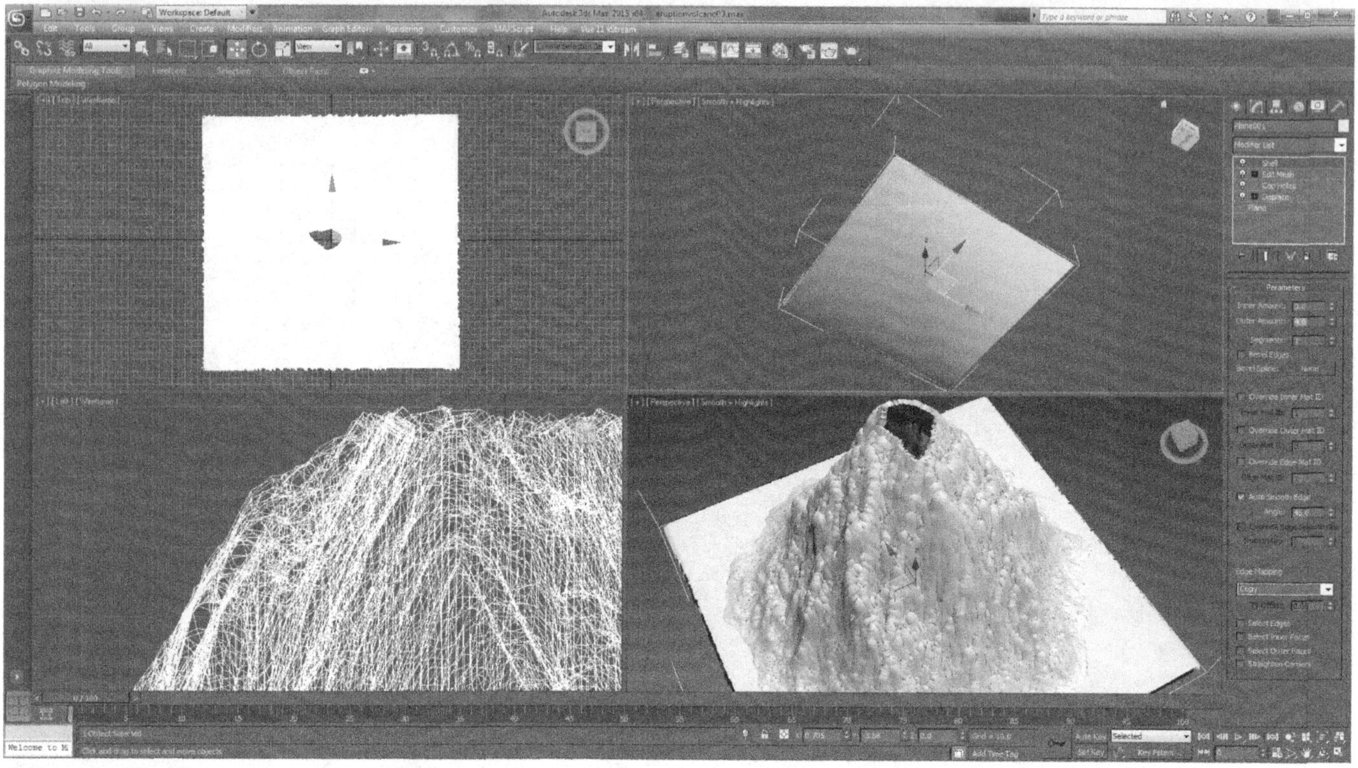

Step 7: Shell

Now that we have a hole at the top and a capped bottom, we have a new problem. The sides of this volcano are less than a 10th of a mm thick! We need to make it thicker - the easiest method for this is to add a "Shell" modifer. As this mesh is very dense, it's incredibly difficult to see thickness, so I zoomed-up on the edge to help guage thickness.

Step 8: Export into netfabb and Import

I'm happy with the results, let's export it into netfabb, change the scale to 10 cm square (4 inches), fix the errors and review the mesh more carefully. I then export the STL file.

Step 9: Import into Cube SoftwareExplosive Results

Yes, that's a 12 hour build time with supports and rafting! This is almost 3x longer than the erupting pyramid, that's due to a combination of supports and the roughness of the surface; a smooth surface has a smaller area than a rough one.

What can we do to solve this issue? The easiest method would be to scale the model down. I think making it 2/3 scale would cut the production time in half. Another option would be to get rid of some of the roughness and even removing the base entirely so you only have a vertical mountain without the horizontal beginnings. This would remove a lot of the require supports. 3D printing is fun huh?

Parts
Slices
Execute cut
Reset

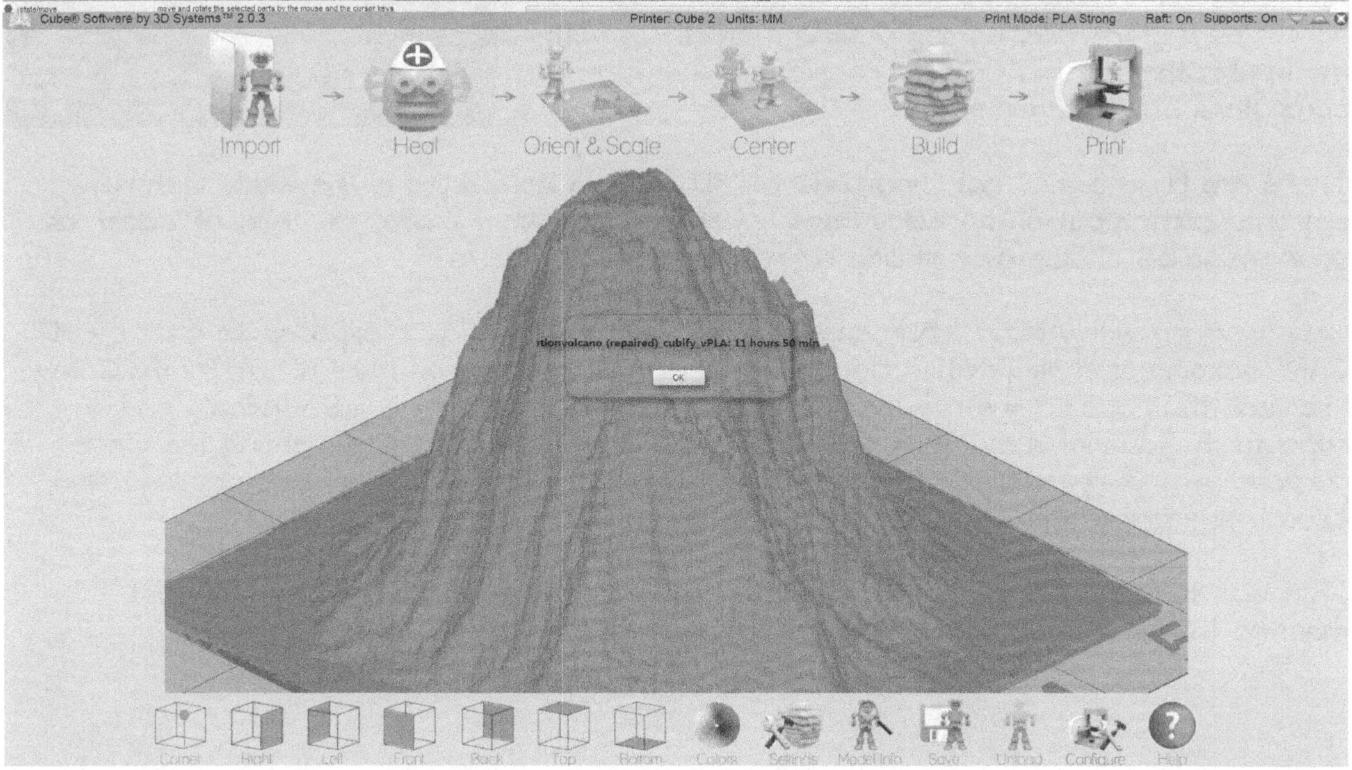
Cube® Software by 3D Systems™ 2.0.3
Printer: Cube 2 Units: MM
Print Mode: PLA Strong
Raft: On Supports: On
Import
Heal
Orient & Scale
Center
Build
Print
tionvolcano (repaired)_cubify_vPLA: 11 hours 50 min
OK

# Interview with Dr. Conor MacCormack
# CEO of Mcor Technologies

Dr. Conor MacCormack is the co-founder of the world's only 3D printing company that uses plain paper as their material, using a process that he invented.

**Can you explain from the beginning how you first got into 3D printing. When you were first exposed to this technology, and when did you realize the incredible potential it held?**

I saw 3D printing for the first time on a long-running BBC television series, showcasing new developments in the world of science called "Tomorrow's World in 1986."

I was just in secondary school at the time and I thought, "That's amazing!" It certainly appealed to the 'Space 1999' watching wannabe astronaut in me! But my interest in technology and engineering had its beginnings much earlier as a child tinkering and building motorbikes with my big brother Fintan. Engines, rockets, space, computers captivated the mind!

**There are hundreds, if not thousands, of 3D printing companies in the world with new entrants coming-out on a weekly basis -- realizing this, what made you think of paper as your media of choice over plastic or powder based systems?**

I was working with Airbus as an engineer on the A380, and I was exposed to a lot of 3D CAD packages and 3D printing in my job. Then, while working on my PhD at Trinity College, I realized that, due to the high cost of the materials, only one of the students had an opportunity to print a model at the end of the year -- just one -- defeating the whole purpose of a 3D printer. Fintan and I saw that while machine prices were coming down, consumable prices kept going in the opposite direction.

That was the seed of an idea that started with, "Could we make a super low cost machine that is also very eco-friendly for everyone to use?"

Given our complementary backgrounds in engineering, it had always been inevitable that Fintan and I would 'invent' a machine -- we both had the desire to 'upset the status quo' and we both really wanted to make a difference, and now we had found a mission -- to put multiple 3D printed parts into multiple hands!

We wanted to revolutionize how people design and communicate by making 3D printing accessible to the masses and make it as easy as printing on paper. Our mission was, and still is, to enable people to print in 3D without worrying about the cost, by creating a set of accessible tools that provide the freedom to innovate in an unlimited way.

Mcor was set up in 2005, and Fintan and I worked on developing the technology -- and from this the Mcor Matrix was born.

**How difficult of a decision was it to use such an easily sourced media over a proprietary blend that could give you another source of revenue after the 3D printer was sold?**

For us it was really about identifying a low cost sheet material that was eco-friendly and accessible -- this was the easy bit! Paper was the obvious choice. The inception of our technology was less about following in the footsteps of other LOM technologies and more about creating a low cost sheet fed 3D printer.

However, we meet a lot people who mistake our Paper 3D Printing technology for Helysis' old LOM technology. Many of them lament this old technology and the benefits it provided to many industries, but they also remember the issues with this printer. With this LOM printer, an adhesive was pre-applied to special paper on large rolls and then cut to shape with a laser.

Since the extra, non-prototype, support material in the old LOM process was glued with the same intensity as the part, this made weeding and post-processing prototypes extremely difficult and time consuming. And needless to say, machine part replacement costs were very high due to the use of lasers. We believe that we have improved on this process -- as Julie Reece, our Marketing Director, recently wrote in our blog -- 'It's not your Daddy's LOM'!

Mcor's Paper 3D Printing technology is different. In the same way that many other 3D printing technologies have greatly evolved from their early implementations, so too has Paper 3D Printing.

Our technology uses a Tungsten carbide blade (instead of an expensive laser) to cut each layer of regular business A4 or letter paper available from any office supply store, and the machine dispenses water-based adhesive only where it's needed -- more on the part and only the smallest amount needed on the paper surrounding the part. So post processing is very quick and easy, and the process is the lowest cost and most eco-friendly in the industry.

It was natural for us to move in this direction to deliver on our promise of making the process as easy and affordable -- and therefore, accessible -- as possible, while remaining friendly to the environment.

All of our technology decisions have been made in support of making professional-class 3D printing more accessible while remaining eco-friendly. So it was actually very easy for us to identify the most important capabilities of our technology.

The following are unique advantages of our technology:

Paper - Regular office A4 and letter paper is inexpensive, ubiquitous and eco-friendly. Pre-used paper can be used in the printers, and the parts and excess paper after printing can all be completely recycled. Paper is also clean to use and safe to handle. It doesn't require any special venting, disposal or chemical post-processing. When you think about it, ink was designed for paper, so it's the perfect medium for color ink with the IRIS.

Water-base adhesive - The method of dispensing the adhesive -- more in the area of the part and less in the area surrounding the part -- makes weeding parts after printing very quick and easy. And it's also safe and eco-friendly.

Full color capability - On the surface, the decision to develop a full color machine might not seem to clearly support our mission. But we wanted to make this once niche capability widely accessible with a low-cost, eco-friendly, high-quality color machine.

**3D printers haven't reached critical mass within the consumer marketplace yet. What do you think is the main reason for this? (price, model availability, creativity?) When will every home have a 3D printer, and what is Mcor doing to make this happen?**

Many people, both inside and outside the industry, talk about price as the key driver to mass adoption. I disagree. Prices will need to decrease (including ongoing materials costs,) but low price alone isn't enough. I think there are other key factors that will determine whether or not, and how quickly, 3D printing reaches broad consumer adoption, including purchase price, ongoing printing costs, part quality, 3D CAD software adoption, ease of use, printer/materials availability and capabilities, and new applications.

Personally, I believe that a low cost 3D printer isn't the path to broad adoption. Rather, I believe that a professional system capable of making high-quality parts and operated as a convenient, low-cost service to consumers, is the key to mass consumer adoption. After all, anyone can purchase a low priced 3D printer, but then what?

It's been proven that 3D printers provide tremendous value to commercial and educational organizations and that color 3D printers, like the Mcor IRIS, bring significant value to consumer applications. But those applications are quite different, and consumer applications can be handled more effectively by service providers. In the end, consumer 3D printing doesn't necessarily mean in-home 3D printing. The concept that a homeowner would need a 3D printer is predicated on the idea that they would print final products quickly, easily, and at a reasonable cost.

Let's explore that idea a bit further. For discussion's sake, let's consider a single part, a control knob for a home gas stove. The stove manufacturer would most likely design the knob and prototype it to make sure it looks, feels, and functions properly.

They might even print a tool and cast a few urethane or metal parts for further evaluation. Once they were happy with the design, they would order tooling and injection mould the knob using a high temperature, flame-rated material. If someone lost or broke the knob for their gas stove, they could, in this scenario, use their in-home 3D printer to make a new one. In order to do so, they would either have to find or create the 3D data before they could print the part. They would need some type of design software (that they learned how to use) in order to design a new knob. Or perhaps they could scan one of the remaining knobs and use that 3D data. For argument's sake, we'll assume there is a database of 3D product files to download free over the Internet.

The consumer would then need to ensure that the material properties are appropriate for the knob. In this case, the material must be suitable for high temperatures and have a high flame rating.

Even if this material existed for 3D printers, the homeowner would have to maintain a supply or locate and purchase the material. But, the next end product they want to print might require a completely different material.

Keeping a stack of document paper at home isn't difficult or costly, but stocking all the different materials that might be needed for real end products from a 3D printer, like the knob, would be nearly impossible for the average homeowner.

Ordering the right material might be the only option. But, in the time it takes to locate, purchase, set up and print the knob, it would be far easier, cheaper and faster to order the actual knob from a local stove repair shop or the manufacturer, or order a new 3D printed version from a print service bureau online or down the street that is already set up for this purpose.

For home 3D printing, I predict paper to be a much more suitable material than melted plastic or powder. And for broader adoption, this technology needs to be low cost, eco and full color, and through our partnership with Staples we can make this technology accessible to everyone. This will really turn the hype into reality!

**Do you have any advice for budding designers seriously looking into purchasing a 3D printer? What should they look for in a 3D printer and why?**

A product that is low cost, eco-friendly, full color -- should I say more!

**We are now in 2013 and it seems there are as many naysayers about 3D printing as there are enthusiasts -- it seems like half of the general public believe we are in the middle of a bubble while the other half thinks the market hasn't even started to gain steam. Where do you think we are in the life cycle of 3D printing as a method of fabrication?**

In general, I believe we are at the start of an exponential growth in 3D printing industry, and people will look back on this time in history and see it as the start of a new 3D age.

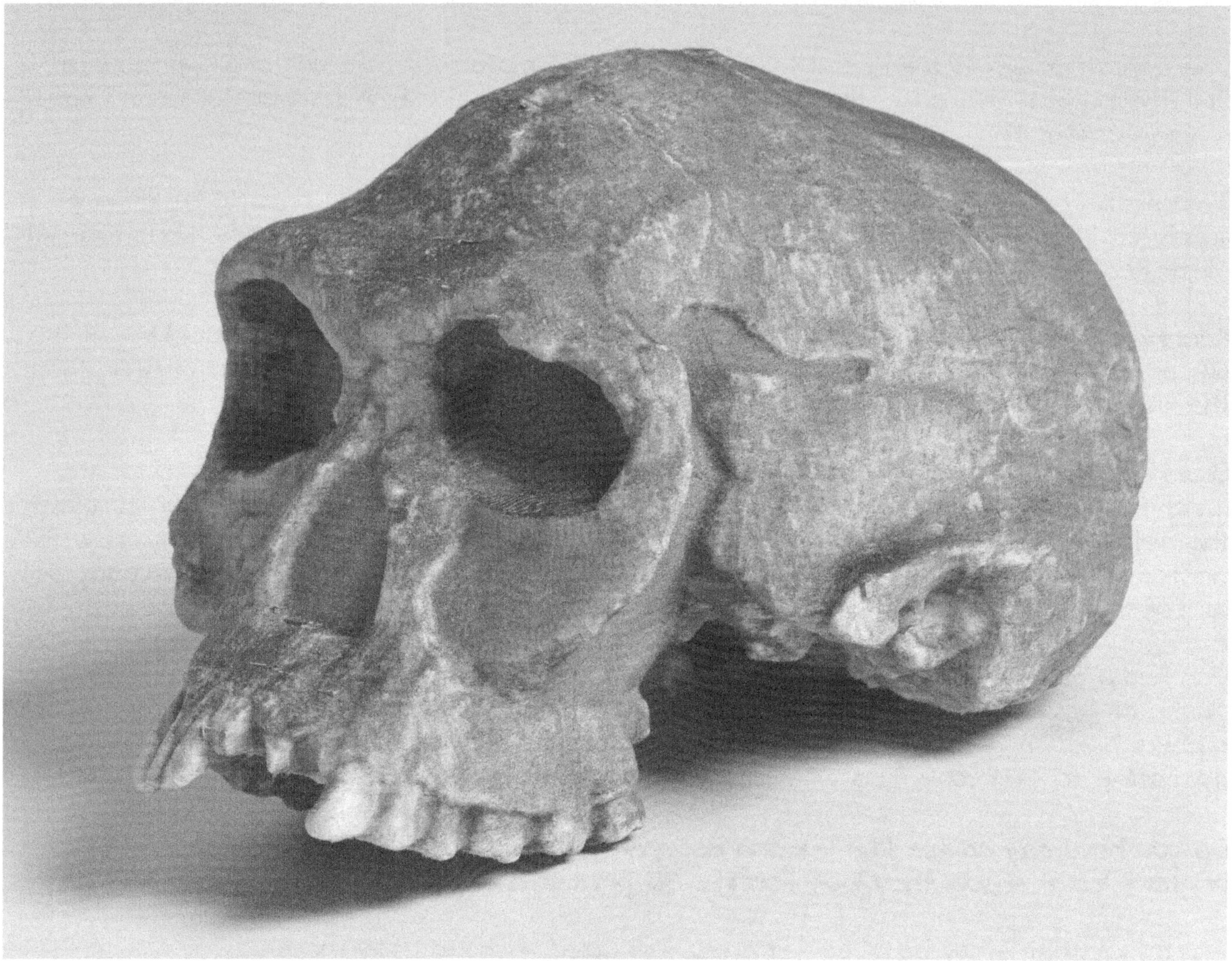

In the next five years we will see further blending of the open source and commercial equivalents, a plateau in the DIY community for building machines in the home, and increase in bureau services and walk-in off the street 3D print shops like Staples that will really bring 3D printing to the masses.

For Mcor, we believe we will change how people design, innovate and communicate -- transporting ideas into reality in schools, offices, universities, outer space. My vision is to develop the machine that will change the world we live in -- just watch this space!

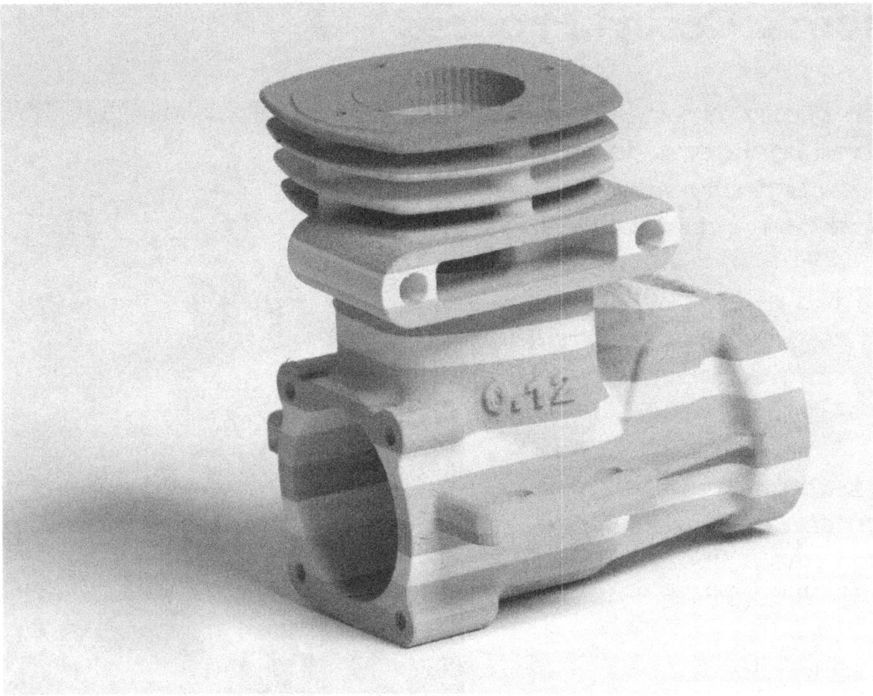
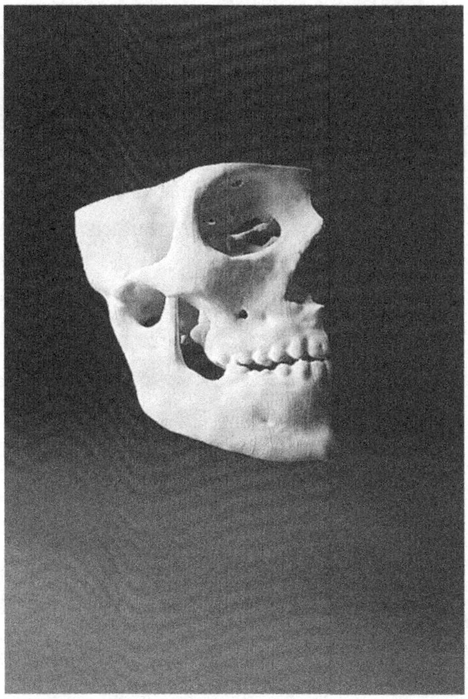

**Is there a reason why you chose a knife to cut out the paper in between layers of glue over a laser cutter? The speed and precision would be far higher and it would eliminate any consumables beyond paper and ink required for your printing process.**

True to our vision to provide accessible, low-cost 3D printing to everyone, we selected Tungsten carbide blades over lasers because they are much more affordable to install and replace. They are also much safer to use with paper than lasers!

We are able to achieve very high degree accuracy (0.1 mm [0.004 in]) with the adjustable blades.

**Do you think Mcor will differentiate into other types of environmentally friendly 3D printing methods in the future to augment your abilities, or are you squarely focused on using only paper well into the future?**

We are focused keeping to our ethos -- low cost and eco-friendly -- so whether paper or otherwise the material will have these core features.

# Kaetemi's Design Process

Jan Boon (Kaetemi) is a freelance programmer and artist from Belgium. He has a professional bachelor's degree in Digital Arts & Entertainment and is currently working towards an academic master's degree in Industrial Design.

I approached him when I first started offering 3D models on my website and was very impressed by his sculpting abilities -- he's incredibly talented and making amazing stuff for 3D printers.

As a programmer, he mostly works on computer game related projects, such as the open source Ryzom Core engine (http://www.ryzomcore.org/). As a 3D artist, he often works on projects related to 3D printing.

Recent works can be found at Kaetemi.be.

---

When I talk to people about 3D printing, they are quick to imagine a world where suddenly everything can be perfectly made using just a 3D printer. While that makes for some nice science fiction, the currently available technologies still do have their limits.

They are, however, already being used by many to create both decorative and functional products. Most types of products at the moment tend to end up quite different from traditionally produced items, though, sometimes not living up to expectations, other times also demonstrating new possibilities that were not possible with common mass production methods. Low volume production, with no upfront costs, is one of the major advantages.

At the moment I am mostly using 3D printing technologies for producing custom figures. The first figure I printed was a fairly simple design, printed entirely in one part using a single white material.

While the shape looked pretty good, there are some obvious issues with this approach. The surface is very rough and unreachable parts make it impossible to finish the surface by hand. It's similarly difficult to get it painted. While technologically a cool novelty item, it's far from a finished product.

Figure printed by Shapeways using an EOS selective laser sintering printer.

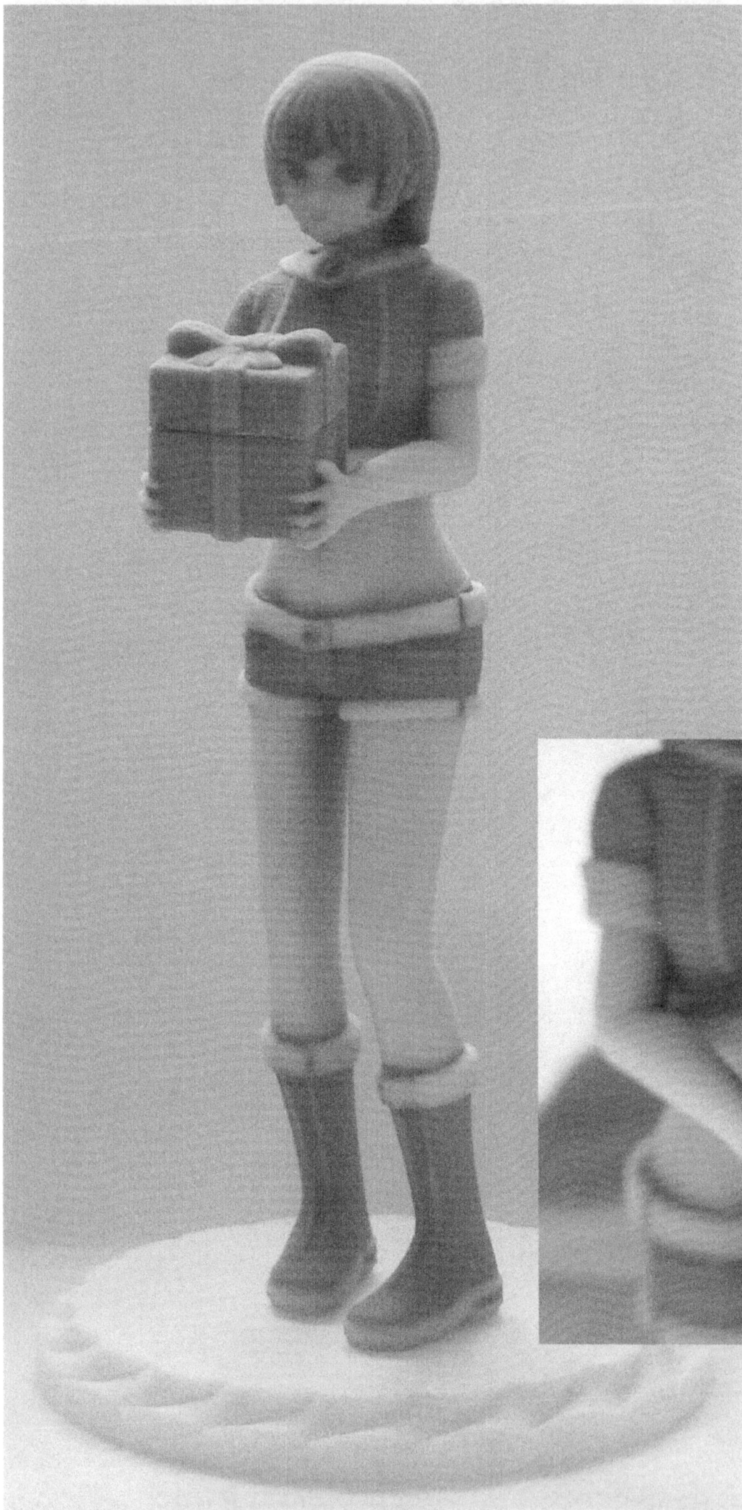

Fortunately, it's also possible to get fully colored prints. While this may seem to open up entirely new possibilities, with every product having a perfect color finish with incredible detail, in reality the control over the color accuracy is still really poor and the detail is far from being incredible.

This color print I made was also printed in one large part. It looks quite nice from a distance, but it's still very rough and still just a novelty item.

With the current technologies, just like a sculpture that's carved out of wood looks like a wooden sculpture, a sculpture that's 3d printed will look like a 3d printed sculpture.

(left) Figure printed by Shapeways using a ZCorp inkjet based printer. (top) Surface showing very rough printing layers.

Figure printed for the Tsunacon convention in 2013.

While it is often claimed that with 3D printing you can produce your product with no assembly required, in one part, directly to the consumer, in reality for most types of products we are still far away from that point. The larger a product part becomes, the more expensive it becomes to fix a design issue, and the higher the risk for a production issue. And when the entire product is printed as one part, a design or production error will require a reprint of the entire product, rather than a single part.

Hands holding on to an accessory.

For my most recent projects, I decided to print things in separate parts. In addition, this also allows me to make use of multiple materials, with different properties and different colors. Using different materials for some parts is not always just a cosmetic choice, though. In some cases there are limitations with one material that can be solved with another material that has other limitations. An example of this is the fingers of hands, which would break if they were printed using the color material without any support.

One solution is to make them hold on to something. The other is to use a stronger material that has no color.

Use a single color material and make the character wear gloves.

Of course, when using multiple parts, there needs to be a way to connect them together. Traditional press fit designs are unfortunately hard to pull off, since the precision needed is not sufficient.

printable connector model

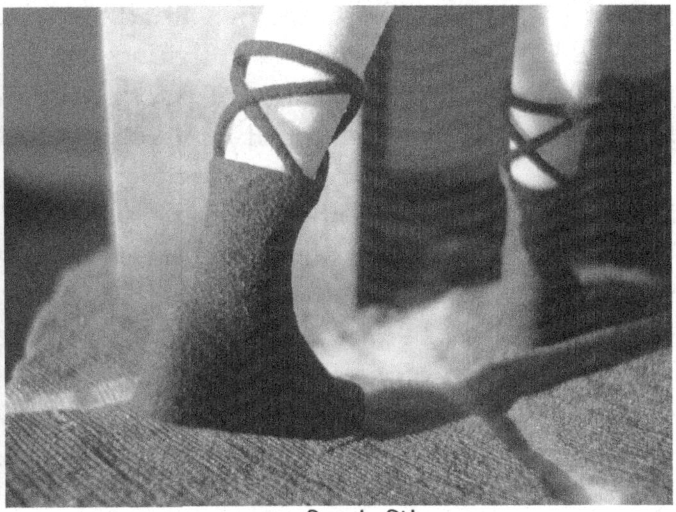
perfect fit

You're better off designing a special connector that depends on the properties of your materials, rather than just the shape. In this case, I'm using the flexibility of the material to create a hole that can slightly change its size so the connector pin magically fits, but that's still strong enough so it doesn't fall off easily. Not being afraid to destroy all of my prints in the process, I try to explore the limits of the materials, and try to find ways of improving them somehow. In a range of these experiments, I worked towards getting a better surface finish of the color material.

This turned into a rather complicated procedure which begins by first smoothing off the roughest texture mechanically, then spraying on a large number of gloss acrylic varnish layers, and finally finishing it with a matte alkyd varnish. While the end result of this does have a pretty good-looking surface, the downside is that some accuracy of the dimensions is lost. Even though the difference may be small, it may end up causing parts not to fit.

Measuring the difference in advance and taking it into account for the print data would be necessary in some cases, but unfortunately this smoothing process is not very reliable. It also takes a lot of manual work and time, that makes it very unpractical for working on a more commercial scale.

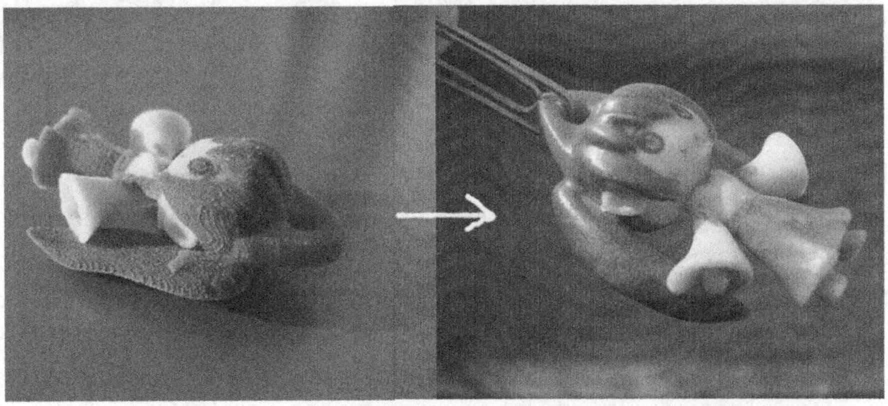
original print compared to finished version

Not all of the quality problems with 3D printed products are caused by technical limitations, though. It's actually kind of important to prepare your model correctly for 3D printing. Make sure the triangle density of your model is high enough.

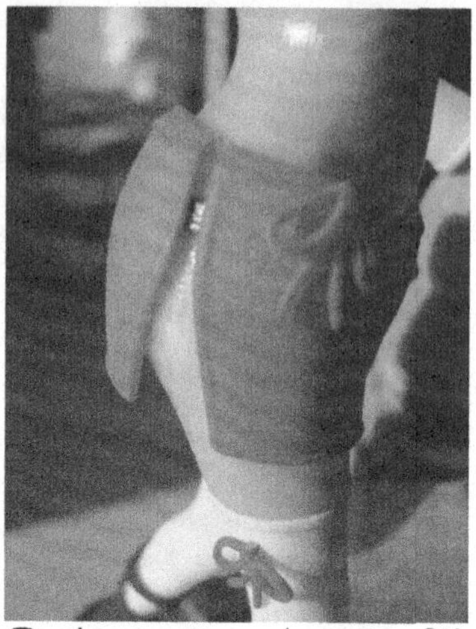

Parts may no longer fit together.

Look at your model without smooth polygon rendering enabled. That's what it will look like in real life. If it is possible to increase the polygon count through surface division, do so. Because surface division often causes clusters of needlessly small triangles, though, to give your model a much more evenly balanced spread of triangles, I recommend pushing it as high as your system can handle, and afterwards using a triangle optimizer to bring it within the limits that the 3D printing system allows.

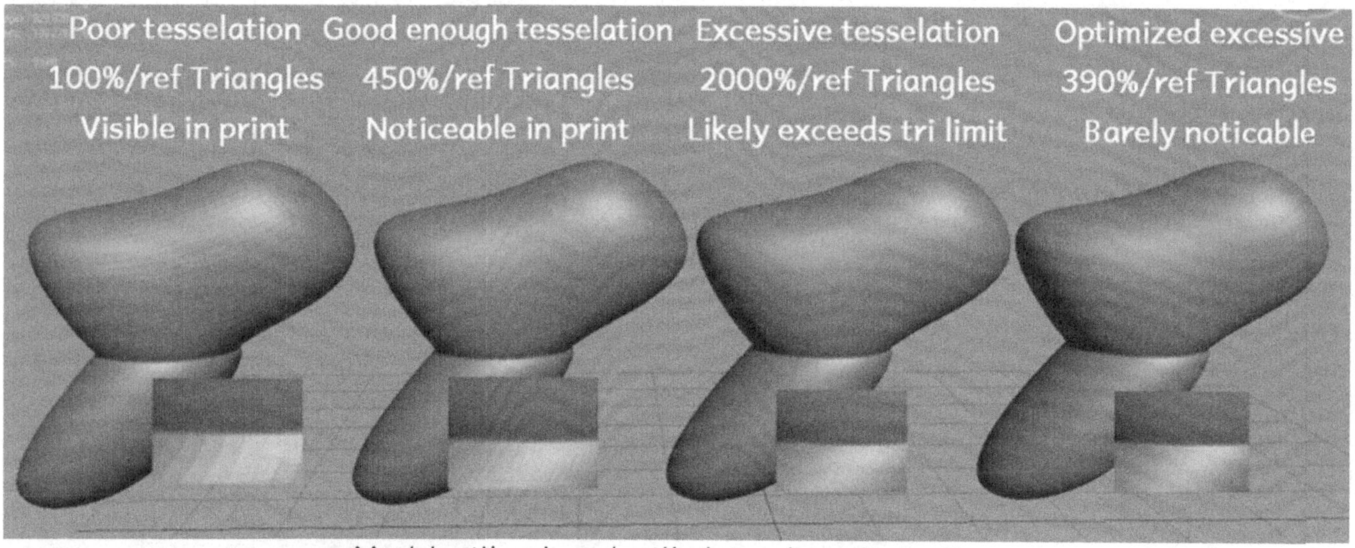


Model without and with triangle optimization.

By itself, 3D printing evidently isn't perfect yet. Rather than using 3D printing to directly produce consumer items, I believe at this moment it's still more practical as a powerful tool to use in combination with traditional building methods. For my next figures, I'm planning to use 3D printing just to create the prototype model, then bring it to a perfect finish by hand and create copies using simple casting methods. Using this method, which is already common practice in the industry, we can take advantage of the capabilities of 3D printing without sacrificing on quality.

Even though 3D printing may not be where we want it to be yet, we can already see where it can have unique commercial applications in the future. Mass customization is one of the greater strengths of 3D printing, as there is no setup cost per design besides creating the customized design, and even that can be reduced by using computer generated parametric designs. Mathematically generated models are some of the images that are indeed often associated with 3D printing. In the future, we might see more and more products with parts which are unique for every item, generated by random parameters and some cleverly programmed design, essentially making each one of them unique.

For figures, this may range from subtle variations in the pose and differences in the flow of the hairstyle to completely unique generated accessories. We can only hope that in the future 3D printing will reach a level of quality that can compete with commercially available products.

# Copyright & Liability in 3D Printing

So far, I've been very lucky in the CNC laser and table routing worlds -- the communities are small, and we generally respect each other's design skills and copyrights over our creations. The machines are somewhat expensive and mostly run by business owners who prefer to spend money or ask permission to avoid conflict and to help designers who try to make a living at their craft.

In the 3D printing world, thus far, this same tight-knit community also exists. The demographics are roughly the same with one major exception: the growing general public jumping into the lower tier of the industry. They do 3D printing for fun, and although there will no doubt be many small businesses opening up in this space, the bar for entry -- excluding design -- will be incredibly low. You don't need to attend a workshop or wait for the UPS truck to deliver a skid that's a few hundred pounds to your door -- you can just pick-up your 3D Systems Cube from the local Staples store!

This presents both incredible opportunity and challenges for designers. Like the music industry that has seen its sales plummet due to rampant piracy, the 3D designer will experience this as well, but at a whole new level. You see, a musician can always play live gigs and sell merchandise to make up for the fact that the music is now a commodity -- not true for a 3D designer. I don't see people lining up to see a designer spend several hours, days or weeks working on a model. In all honesty, the process of 3D modeling is a boring and repetitive task with endless amounts of rebuilding until something somewhat presentable is produced.

As mentioned earlier, 3D printing companies are very smart; they use proprietary blends that only work with their machines -- they both sell the machines and are the sole source of material for their machines. Like other printing companies, their job is machine sales. What that machine does is, completely and rightfully so, out of their area of responsibility. Sure you can 3D print a gun that "works" but nobody went back to the manufacturers and said "stop making printers, they can print deadly weapons." Instead, they go after the designer who made the file and run their feet over the hot coals.

One possibility is that a designer can patent his 3D design, but that's a waste of time because a 3D model can be easily modified enough to no longer be recognizable under the terms of a design patent, and most designers don't have the legal or fiscal resources to fight against a possible violation in the first place. Throw in consumer friendly 3D scanners and any hopes of branding or controling distribution of any 3D printable object gets completely thrown out the window.

A third option, one that I think will be inevitable, is like flat scanners that are unable to scan high resolution monetary notes due to an international industry standard treaty. In this case, it's easy: each country sends their notes to be scanned and integrated within an internal firmware DRM on a yearly basis, or there is something within the note that tells the scanner "no, you can't scan this." Although this sounds rather challenging at the moment as 3D models can be easily modified using more than a dozen different software packages, I think "big data" can create unique algorithms that would make this feasable for everybody involved. What needs to happen, like with the scanner, is for a print to be loaded. If it has a hidden watermark or matches an online algorithm, it will simply not print the design unless compensation is sent through.

I think we are several years away from anything like this coming into place, and I do see a very profitable business model in the making in this area where 3D models are uploaded into a centralized database and a small monthly fee is charged to protect the design regardless of the 3D printer manufacturer. Be sure to remember me if you develop this, I like royalty checks! This is all possible using today's technology; the challenge would be in signing up manufacturers and vendors along with marketing to make it a reality. Much like Apple's iTunes has convinced most people to pay for their digital music, a system like this, well executed, could also convince most people to pay for their 3D models.

Another issue yet to be worked-out is liability. If I buy a robot and due to some manufacturing defect it hurts or kills somebody - the company that made it - not importer, manufacturer of the machine that made the hammer or distributors - is responsible and would face litigation. They own and made the defective design and are ultimately responsible for the results. How does this work for 3D printed models?

Right now, if somebody got hurt with a model (purchased or free) due to a defect, who would be responsible? The 3D printer company wouldn't be as they just "made" it based on the defective design. The "internet" can't be held liable as it's just a distribution system so it falls back to the manufacturer or the designer? Up until this point, these two entities were usually one in the same. This is increasingly no longer the case.

The larger issue at hand is that of liability verse self reliance. If you have a design and got hurt as a result, well, YOU manufacturered it but you did use a designer's model. A case could be made either way and laws have not caught-up. The designer cannot make a 3D model that is 100% across all the different 3D printing machine types and processes. A 3D model made with a 3D Systems Cube will be different in material properties, tolerances and weight to the same model made out of glass or plaster. The consumer on the other hand, most likely got the file and printed it right away without any knowledge about 3D modelling or adapting a design to suit their needs or printer.

## Interview with Bathsheba Grossman

Bathsheba needs no introduction in the 3D printing world -- she's a world renowned artist who's had her work displayed in galleries and museums and whose influence in popularizing direct-metal 3D printing as an art medium is second to none.

Her work has appeared in the NY Times, London Times, Wired, and Discover to name a few. One of her lamps was even featured in TIME Magazine's 100 most influential designs of 2007. You may have also seen her work in hit TV shows, Second Life and even in a Japanese videogame commercial.

You can find her at Bathsheba.com

**Your beginnings in design came from doing manual casts and metal fabrication back in the 1990s, and you slowly evolved into metal 3D printing -- how has this transition affected your design capabilities and process? How much of your model development is virtual vs physical these days?**

On the design side, it's changed the process more than the product. Looking at my older work made by hand modeling and casting, you can see it was more difficult to work that way. The design space is more limited, but also that the ideas aren't too different.

I'm mostly virtual now, though I still make sketches and keep plasticene handy.

After the first ten years designing in CAD, I began to to be able to think about ideas directly in the computer rather than translating them from a physical 2D or 3D sketch.

But there's no reason to make a point of converting completely: it's always good to have more tools.

**Your Rygo 3D print is a massive 3600 lb project. How did it come about, and what were some of the design challenges, both virtual and physical, that you had to overcome to make it a reality? What are some other notable commissions you've taken part in?**

I prefer steady retail sales to a scatter of larger projects -- it's nice if my income is based on a big enough sample to obey statistical laws! -- and commissions are a market I haven't pursued. One that I did in 2006, a bronze in Urbana-Champaign fabricated by a hybrid 3DP/foundry process, is documented here:

http://www.bathsheba.com/gallery/beckman/

The Rygo was promoted by one JF Brandon, whose occupation seems to be "putting two and two together." He had the idea that one of my designs should be executed by D-Shape, Dr. Enrico Dini's architectural-scale sandstone printer, and being a man of action he made it happen.

I can't speak to the physical side -- the sculpture itself went from Italy to Vancouver, and I never saw it myself. I gather there were challenges at all phases: It had to be built in sections to make the topology cleanable, the original plan for a color-changing interior light had to be modified, people had to be kept from climbing on it ... it's a downside of being a digital designer: sometimes the party happens without me!

On the design side, there were two main considerations: first, making sure that the smallest features would print and clean well. Dr. Dini was forthcoming about how the machine works, its characteristic surface finish, wall thickness, and clearance constraints, and once I had that picture clear in my mind, I was able to tweak the design properly. And the second issue was simply it needed to be able to support its own weight. To address this I did some sneaky math to make the surface thicker lower down, as well as built on solid mount points and a plinth. Then Dr. Dini ran a stress analysis to check my guess.

Luckily everything worked out, since we only had one chance to get it right!

**With the rapid evolution of 3D printing, you've been outsourcing your 3D printing needs for quality and budgetary reasons. How has this approach affected your design testing? Do you test full scale or bring the model down to miniature size until a design is finalized?**

In the beginning it wasn't much different from working with a bronze lost-wax foundry: One expects to wait a month or six weeks to get metal back, and 3DP service bureaus were about the same. The turnaround I get now for direct metal prints is still 2-4 weeks, so still about the same. Patience.

Usually I just print what I hope will work, rather than scaling down to test. Occasional factory seconds are a cost of doing business, and I don't stress about them; also, I'm usually designing close to the available resolution of the printer anyway, so there isn't much scope for scaling down. What I might do with a design that I'm truly dubious about, or that needs functional testing, is print a full-size draft in a cheaper material such as sandstone or SLS, rather than going straight to steel.

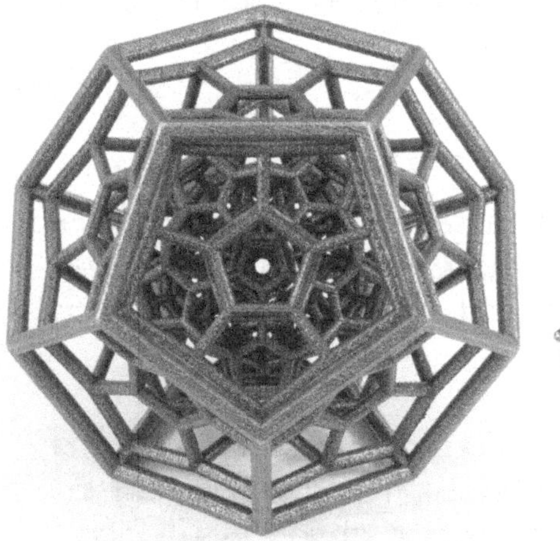

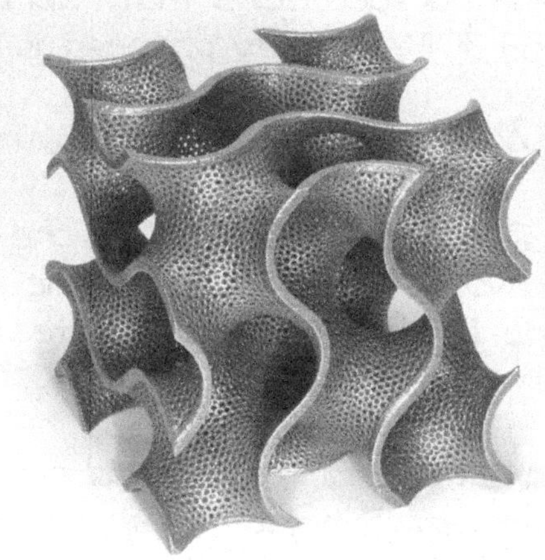

**I know you've been anxious for 3D printing to evolved to the point where it can print a mixture of metal and glass at the same time. What is it about these materials that interests you?**

Of course archival materials -- metals, precious metals, glass, ceramic, etc. -- are gratifying for their own sake: These are ancient and modern art media for very good reason. Closer to home, it's also true that art buyers will pay enough for these materials to cover the premium of 3D printing costs. It's more difficult to make that work with plastics:, however much one may value their inherent properties, because they're seen as lower in value. It's just harder to cover costs with those materials.

**Your 3D prints are surprisingly affordable, and you've also released some of your design files to the public at no cost. Is there a greater philosophy behind this?**

Maybe "I like customers and want them to be happy"?

Every digital medium -- and sculpture is now in that club -- comes with an open-source side as part of the territory. That will be there in any case, so it's better to see it as a friendly thing.

**You are especially well known for you 3D printed metal sculptures and jewelry. What is it about this material that attracts you over the typical plastic/resin/starch physical prints?**

Covered above, I think: it's pretty, and people will pay the freight for it.

**How did you get into designing in the first place?**

I went to art school and then simply refused to give up until something jelled ... it took a long time. I was over ten years from graduation to first dollar.

**How has your design process changed as you evolved from a purely mathematical approach to a more biomorphic slant?**

I can almost answer that with one brand name: T-Splines. This is a plug-in for Rhinoceros, formerly independent and now owned by Autodesk, which makes it easier to do organic modeling with that software. Before technology was invented, I spent a good deal of time making more lifelike forms, in parallel with the geometrical work, and it's been nice to come back to that with the new toolset.

**How long does it typically take you from design concept to finished physical proof? What are the biggest factors affecting this timeline?**

I think the time-trial record is something like six hours. Then again there are designs that were 15 years from sketch to finish, and of course I doubt the longest one has been finished yet. It's pretty arbitrary, really -- I haven't found a better way to decide what to put aside and what to work on next than "whim."

**Looking at your designs, you have obvious symmetrical tendencies. Where does this influence come from? Your original mathematical beginnings or your love of nature?**

Well, both: Nature and math aren't easily teased apart. People generally like symmetry, whether it is natural or artificial, so I think my feelings there are more the rule than the exception Perhaps I've indulged the taste more than most.

**What advice would you give to people interested in 3D printing that you wished you could have given yourself when you first began? What are some common issues that you experience on a regular basis with your 3D prints?**

It's such a different world now from when I started ... the problems I had were related to high costs and difficulty getting access, and those issues have largely faded out now. The central problem is still finding a design niche in which the unique attributes of 3D printing add enough value to justify the higher costs of production. Nothing makes that easier, and everyone has to make their own mistakes.

I will say that the main answer I find myself giving to people who write to ask how 3D printing can help with their project, is "not really." There's long been a mystique surrounding 3DP, more widely known now but not different in quality from what it was 15 years ago, that leads people to believe that it's a magic universal manufacturing method which will replace everything else. What the technology does do is wonderful, but inevitably when people come to grips with the reality after hearing the hype, they're disappointed. I can't wait for the media bubble to be over, so I can stop having to discourage people all the time.

The biggest recurring problem is, of course, that one gets the print back and it just doesn't seem as nice as it looked on the screen! Sometimes the printing process doesn't do quite what I hoped, and then I need to learn more about how the machine does what it does; but usually I just need to make better art. Back to the drawing board ...

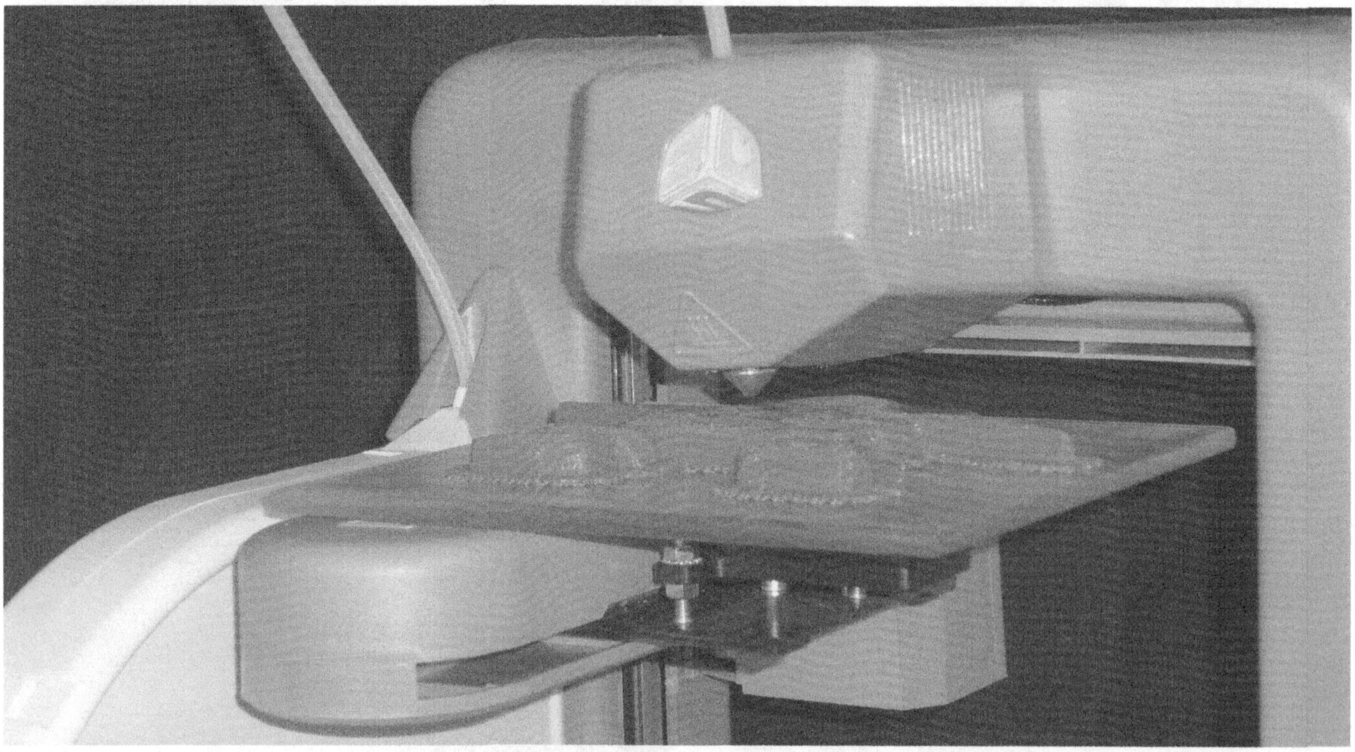

## CNC Plasma Cutting Technological Overview

Although you can technically cut steel using a high powered industrial laser cutter, a more affordable solution for small enterprises is to acquire a CNC plasma cutter. With this machine, you still benefit from the incredible precision and accuracy present with the other CNC machines with the added benefit of easily cutting through various thicknesses of steel and metals like a hot knife through butter. It's a beautiful process!

Although there are several methods of cutting with a plasma cutter, many use just compressed air and electricity, which means you don't have to deal with specialty and potentially volatile gases (with associated licensing and insurance overheads).

Before the advent of plasma cutting, people used oxyfuel to cut metal thicker than 5 mm with lasers being used for thinner cuts. The main advantage of plasma cutting over the use of oxyfuel is far quicker cutting times. It can slice though other metals (such as aluminium) by using other gas types such as nitrogen with the added benefit of using a less flammable cutting system -- meaning your plasma cutter's risks of exploding are far less than other traditional torch cutting systems.

What I really like about the possibilities of plasma cutting is the material you cut into, steel. It has incredible strength and unlike wood or acrylic, it can be left outside for extended periods of time without too much concern about it slowly breaking down. Steel is also a fantastic material safety-wise. You can drop it on the floor and it won't break. You can throw a model that's welded into the back of a pickup truck, and it's your truck that will get damaged, not your model! Shipping wise, it's rather heavy, even at the thinner gauges, but like acrylic, you can actually bend and curve it.

A unique feature with plasma cut models is that you can paint them using a host of processes that would melt or destroy most other materials. Do you want a gold-plated Zulu Main Battle Tank? That's not a problem with a plasma cut model, and if you built it right, you could probably build it to be full-scale and actually move around! Try that with MDF or particle board!

I've been interested in welding for years. I've never gotten much into it other than a high school course, but it is an area that I look forward to actively designing in ... of course, using a CNC plasma no less! As such, I wanted to include a small section about this technology in my book for those equally interested in this CNC technology.

I think, out of all the CNC machines types I've reviewed in this book, plasma seems to be the most challenging after 3D printers. They are loud, they are incredibly powerful, and if you don't pay attention, they are the quickest to get out of control safety-wise. That being said, they are easily shut down by simply hitting the emergency cut-off switch if needed.

Unlike CNC table routers, you never have to worry about a broken bit, only running out of fuel -- and unlike CNC lasers, they don't lose power over the extent of the platform. Getting a water table is also a quick fix regarding fumes and shard issues.

Where you'll be able to get away with running a laser, router or 3D printer in your garage, a plasma cutter is a whole other beast. But in the right hands, the results that it can produce are just as beautiful and astounding.

If you haven't already, I'd strongly suggest you give a call to a local manufacturer and get a shop tour -- if they are in heavy industry, they will have these amazing machines, and if you are kind enough, they'll show you first-hand the incredible cutting power they yield.

# Interview: Jeff Knoll, Marketing Manager for Lincoln Electric Cutting Systems

I wanted to get a perspective on the CNC industry from the world's largest welding solutions provider, Lincoln Electric Cutting Systems (which owns Torchmate,) so I reached out to Jeff to get some answers.

**Can you give me a little background about Torchmate? What is it's unique value proposition in the CNC plasma cutting field?**

Torchmate has been supplying metal cutting machine to many industries for three decades; in fact, we have been creating industries or movements -- like garage manufacturing..

We started from the ground up and before electronics and plasma cutters were developed. Our first machines were magnetic tracing machines that used OXY fuel, called pantograph machines. Pantograph machine "kits" were our bread and butter.

We sold the directions and where to supply the components needed to complete the machine. Kits and the "kit" mentality remain our mindset to this day. Yes we sell turnkey machines but our mantra is that you can save money by making some of the components yourself -- in fact, you can make the components with the CNC machine that you just bought from us.

The machine makes itself, and it does at our 30 000 sq ft manufacturing facility.

**What features should I look for in a quality CNC plasma cutter before purchasing one?**

Support, software, customer reference, design/construction, company history -- any one can make a CNC machine -- there are 40 competitors in North America that sell this stuff, ranging from "companies" operated out of their garage where one person is the salesperson, shipper, and the technician giving support.

Ranging to the other end of the spectrum: over-priced huge, margin behemoths that play on the "industrial look". Gantries so big that operators ride on them.

A contrived definition of "industrial" plays on the machismo look of their machinery. Let's remember that we are holding a 10 lb torch and if accuracy, low maintenance and tolerance are what the customer is really looking for, a rigid quick moving gantry is essential.

Torchmate is different from the rest of the integrators and CNC companies. We make everything from the software to the drive components to the plasma cutter and the support tables. There is no finger pointing and there are no excuses -- customer resolution is much faster and smoother. We own all the issues here in Reno, NV and we take it personally when a customer is down, realizing that the cutting system is the most important machine in anyone's fab shop or manufacturing line.

Who and how many people support your CNC machine should be a major deciding factor when you are putting your entire business and production line at risk. When the metal stops cutting, the business screeches to a halt. Software is often overlooked, too. TM CAD/CAM is ours -- we support it and make changes to it.

We have many competitors that outsource software and other components, and we have many competitors that use free or generic software that is supported by an independent forum and is worthless.

**How do plasma cutters compare safety-wise to other CNC machine types in the market?**

Compared to the rest of the tools in a shop, the Torchmate CNC plasma machines are very safe and controlled. The most important part of any electrical loop is always the ground. Having a good ground is also safe.

Any key on a Windows-based computer that controls our machine will automatically stop the machine from movement and cutting. Most plasma manufacturers, including Lincoln/Kaliburn, have what is referred to as a "parts in place" switch, assuring that the torch can not fire when changing the consumables.

**Are there any differences in cutting quality when you compare a plasma cutter to an industrial laser cutter that's also capable of cutting through metal?**

Sure, there are differences between laser and plasma, and each specific application or part has a specific need and tolerance. Plasma might not be a direct replacement for laser (yet) but plasma has a place in every shop.

Engineers love to use the most accurate materials specified for their components. The constant need to reduce costs will never subside, and in many situations a redesign of parts to allow for plasma tolerance can significantly reduce manufacturing costs.

I have always thought that if you are welding or bending metal, laser tolerance has been shot and you over-designed.

**Torchmate released its Growth Series that is geared specifically at smaller shops and individuals interested in plasma cutting -- do these machines function just as well as your larger machine range? Are they equally as powerful and capable?**

Is perception always reality? These smaller machines are designed to be just as accurate as their larger counterparts.

Plasma is plasma -- too many people try to make it out to be more sophisticated than it is. The smaller machines are just as accurate, in fact more accurate (over designed) than the plasma process can tolerate.

Routing, which the Growth Series is designed for, is a more accurate process than plasma; and our machine tolerance really shows up in the end result of routing different materials -- especially aluminum.

**What are some things that designers have to give special attention to when designing cutting files to be plasma cut?**

Tool pathing, software and balanced control/drive systems are what take a relatively "inaccurate" process like plasma and make it much more accurate than what is produced by other CNC machine manufacturers.

**What are some of the benefits that a plasma cutter has over a CNC waterjet or industrial laser cutter when it comes to cutting metal?**

Cost per inch is dirt cheap in comparison. As mentioned above, there are several ways to take advantage of this: redesigning your part for plasma tolerance, using the machine for prototyping/R&D, or just to alleviate the waterjet or laser machines.

**CNC table routers use bits, laser cutters use tubes, 3D printers powders/plastics -- what is the major consumable involved with plasma cutters? How does the price of this consumable compare to the others?**

Plasma uses copper electrodes that need to be replaced based upon many cutting variables. A Kaliburn High Density plasma cutter uses a variety of shield/cutting gases to achieve the best cut quality. Kaliburn has the longest life on consumables, and they use about half of the gas consumption compared to major competitors.

You won't hear these facts coming from gas supply companies, as they would be shooting themselves in the feet because the customer would be buying less from them.

**Two years ago, Torchmate was acquired by Lincoln Electric, a world leader in the welding field. What new innovations have been brought into the CNC plasma cutting marketplace as a result?**

Besides the fastest delivery of any CNC machine in the industry, 3 million of inventory on the floor, a full-time tech support department consisting of 15 employees on a toll-free number, and one-stop shopping, we make and support the entire machine because we make the entire machine.

**What are some technological limitations shared by all CNC plasma cutters that represent the next major hurdles for the industry to innovate over? What innovations do you see coming from the plasma cutting field over the next 10-20 years?**

I see truly integrated solutions as the answer and not just an empty and clever marketing term. Our major competitors are claiming to have this, but it's still just a catch phrase. Compatibility amongst all electronics, software, motors and plasma cutters and the elimination of 3rd party integrators -- there will be nothing left to ingrate. Remember how big and clumsy computers were in the '70s? Now they are in our pockets. Our customer is getting younger, and CNC plasma will be more intuitive shortly.

----------------

What I found most interesting about this interview is how, like Ted Hall, Cathy Lewis and Simon Moore, the future was about improving the usability of their relative CNC machines, not jumping to faster routers, more powerful lasers or larger beds. These machines have come a very long way and it's only a matter of time that they become so intuitive that just about anybody will be able to create optimal results with them with little to know knowledge or experience - much like digital cameras have made just about anybody into a professional photographer.

## Closing Thoughts

When you review this book with the others earlier in the series, a real comparative study can be made. I learned a lot about design building the initial model and even more when it had to be adapted to another scale or machine type. With the rise of 3D printing slowly gaining ground at the consumer level, this will push the creative opportunities available for designers around the world!

As for this book and my previous volumes... they will be translated into Chinese - after living in China and studying the language, I'm convinced that it's a huge market that I simply cannot ignore. This book will be an important vehicle to get me there.

Now that this book is done, and in my mind, the loop has been closed, I realize there is some repetition across my previous volumes but now that the slate is clear, time for all new designs, material choices and more variety across the board.

What would my dream shop have? That's easy -- a Trotec Speedy 400 Flexx, Torchmate X, 3D Systems Projet 860Pro (zPrinters have recently been rebranded) and a ShopBot PRSalpha -- WOW -- I'd be a one man designing and production powerhouse! These all cost serious coin so be sure to tell all your friends about this book so I can make projects featuring them in future volumes. Like the days designing away in my Korean shoebox apartment I called home years ago, I can already design for these machines (and more) -- I just don't have the machines in my hands YET. I have no doubt it will come and when it does, I look forward to sharing all I can about the world of wonder that CNC machines are capable of building. There are other machines I'd like to acquire too but that's later.

Although this book is expansive in scope, it represents but a sliver of the knowledge that one needs to know to be a well rounded one-man CNC powerhouse, but I feel I'm well on my way and writing books such as these are a fantastic way to gel it all together for others to follow and learn from.

I absolutely see myself getting into multi-CNC designs - each CNC machine has it's strengths - and augmenting a design to create ultimate projects sounds like a fun challenge to me. There are just so many possibilities that are open now that I intuitively know how to build optimally for each - I'm excited about what the future brings. The nice thing about this field is that there really are no limits - as soon as you break your perceived notion of what is and isn't possible, a whole new level of possibility opens up.

I want to get into capital equipment sales through lead generation as well, it's another market that I've had my eyes on for years - and another industry that I look forward to entering very soon. Designing is fun but the money that can be made in this field would go a very long way towards feeding my CNC machine and designing passions.

I really hope you enjoyed this book that was two years in the making! It's far larger than I originally had planned; the ideas and projects just kept on coming and I just had to get them in. Please feel free to contact me with your plans, ideas and thoughts. I'm always looking for new partnership opportunities and strategic mutually beneficial relationships to help spread the word about how much fun and capable CNC machines really are.

CNC machine may seem formidable at first, but they are INCREDIBLY EASY to work with and design for. I hope this book helped you along your path towards CNC design mastery.

## Your Turn: Design Your Own!

If you are interested in designing for a CNC machine, except for 3D printers, what you see on the next page is exactly what I start off with. A blank screen with lots of squares where each square represents the material thickness I'll be cutting the model out of. Each square can be 4 mm, 6 mm or whatever thickness you decide, but your first decision is just that -- what's the thickness of your material?

The next decision you need to make is a key one -- you can either make a free-style design not following the squares explicitly (such as having a part that's slightly off-square like you see in all my CNC table router designs) or dead-on like you see with my CNC laser designs. The reason MY table router designs are off is that I modified them in a graphics program and used physical trial-and-error proofs instead of a virtual environment for my testing.

Suffice it to say, you want things to be dead-on like my laser stuff. Why? Because if everything is according to material thickness, you can easily scale it to any size and all the interlocking slots WORK without having to make any additional design changes.

The next major decision you make is what will you design? If you are new to this, I'd suggest looking at the templates found in this book and my previous volumes and simply tracing them over BUT making slight changes -- adding more slots, removing or adding small details, and so forth. This may sound trivial, but it will teach you that every action on this grid affects another piece ON THIS GRID.

After you have a simple design figured out -- and make it only fit on the grid page you see, no multi-page or adding more squares along the x or y axis -- cut it out in some cheap material. My favorite is plywood -- you'll find MDF and cardboard break too easily and acrylic or steel is too unforgiving. Wood has just enough play that can be introduced with a few taps from your friendly shop hammer. I suggest getting a rubber hammer first as it dulls the impact; but if needed, feel free to use a carpenter's hammer.

When the design is done and assembled -- you've proven your design -- the next step you need to do is show it to your friends and family and get their opinion. Would they buy this from you? If so, how much would they be willing to give? If they have their own CNC machine, how about the digital files? Remember to keep tabs on your design, cutting and assembly times. Welcome to the wonderful world of CNC designing! Send me off a pic!

# Image Credits & Info

Unless otherwise mentioned below, all images and photography found within this book were taken or rendered by Jonathan Cantin and he holds all rights to them.

Page 33: Copyright David Newman of Blue Tongue Engraving - Coralie and Rudolph

CNC Laser
Page 38: Copyright Trotec Laser - engraving marble
Page 39: Copyright Trotec Laser - engraving wooden frames
Page 43: Copyright Trotec Laser - cutting paper
page 45: Copyright Michael Hughes (mjh_design@hotmail.com)
Page 47: Copyright Trotec Laser Australia - Simon Moore
Page 48: Copyright Trotec Laser Australia - tradeshow booth
Page 49: Copyright Trotec Laser - engraving a sign
Page 50: Copyright Simon Moore - his acrylic cutting table design
Page 51: Copyright Evright.com - Adelaide showroom
Page 52: Copyright Trotec Laser Australia - working QR code laser cut on acrylic
Page 53: Copyright David Newman (davidcncking.com) - Sydney VIIE 2013
Page 54: Designed by Jonathan Cantin, cut and assembled by David Newman
Page 61: Designed by Jonathan Cantin, cut by Evright.com in Adelaide, South Australia
Page 83: Photographed by Jonathan Cantin in the nearby Canadian forest
Page 91: Copyright Jonathan Engel - wooden laser cut track concept
Page 92, 94: Copyright Jonathan Engel
Page 151: Copyright David Newman - toys for his hospital program (top), Jon's prototypes
Page 153: Copyright Radecal Machine Sales & Service UK (radecalmachinesales.com)

CNC Table Router
Page 168: Copyright ShopBot Tools - founder Ted Hall using a Handibot
Page 169: Copyright ShopBot Tools - PRSAlpha
Page 170: Copyright ShopBot Tools - TJ showing off his skills!
Page 172-174: Copyright ShopBot Tools
Page 300-306: Copyright by Steve Glassel
Page 307-314: Copyright by Ralph Bagnall
Page 316: Copyright Edgar Castelo (http://faz-voce-mesmo.blogspot.pt/)
Page 317-323: Copyright by Brian Lawrence
Page 324: Copyright by David Newman

CNC 3D Printer

page 348 Copyright EOS: Dental: Building plattform I Material: EOS CobaltChrome SP2
Page 351: Copyright Stratasys Ltd.
Page 353: Copyright 3D Systems, Inc. - zPrinter 850
Page 354: Copyright 3D Systems, Inc. - zPrinter output
Page 356: Copyright 3D Systems, Inc. - SLA output
Page 360-365: Copyright 3D Systems, Inc.
Page 371: Copyright 3D Systems, Inc.
Page 406: Copyright Mcor Technologies - company founder Dr. Conor MacCormack
Page 408: Copyright Mcor Technologies - Orange part: 5.7in3 (93 cm3) $2.81 (€2.16)
Page 409: Copyright Mcor Technologies - White statue: 20.4in3 (334 cm3) $2.29 (€ 1.77)
Page 412: Copyright Mcor Technologies - Archeology skull: 18.6in3 (304 cm3) $13.06 (€ 10.05)
Page 413: Copyright Mcor Technologies Ltd. - Engine: 34.4in3 (564 cm3) $4.07 (€ 3.13)
Page 413: Copyright Mcor Technologies Ltd. - White skull: 6.6in3 (108 cm3) $1.12 (€ 0.86)
Page 414-421: Copyright Jan Boon (Kaetemi)
Page 424-431: Copyright Bathsheba Grossman

CNC Plasma

Page 434: Render by Jonathan Cantin, designed by Richard Lowe
Page 436-440: Copyright by Lincoln Electric Cutting Systems / Jeff Knoll

Thanks to all those who provided pictures and content to make this dream of a massive educational CNC design book possible. Your contributions are greatly appreciated!

# Other Books by Jonathan Cantin

WoodMarvels.com Volume 1: Top Sellers
102 pages, ISBN 978-1-896369-44-0

The book that started it all with woodworking projects that use round wooden dowels; our blueprints are completely measurement free. This book includes the following models: Abacus, Bird Feeder B, Bird House, Bread Box, Brontosaurus, Dolphin Future Car B, Parasaurolophus, Pen and Pencil Holder B, Portrait A: Sunshine, Portrait C: Waves of Hope, Ring Stacking Game, Showcase, Stegosaurus, Table Protector, Table Protector C, Tower of Hanoi, Wood Stacking Game, Abstract Bank, Chessboard, Crab, Cutting Board, Pterodactyl., Medieval Castle Walls, Platform Crane, Straddle Carrier.

WoodMarvels.com Volume 2: 50 Laser Cutting Plans
200 pages, ISBN 978-1-896369-46-4

This book includes woodworking projects from our site; our blueprints are completely measurement free. This book includes the following models: > Level 1: Centrosaurus, Dolphin, Elephant, Future Car A, Future Car B, Hammer and Stake, Pen and Pencil Holder D, Pet Bowl Holder, Rhinoceros, Submarine, Time Teaching Clock A and B, Whale. > Level 2: 1956 Porsche Speedster, Abstract Bank B, Bed Tray and Table, Crab, Crumb Tray, Desktop Organizer C and D, Elastic Marble Game, Halloween Serving Coffin, Halloween Sword Shelf, Mayan Temple Bank, Pen and Pencil Holder C, Pterodactyl, Pyramid Calendar, Serving Table, Small Table, Super Snail, Tracing Light Table, Trip Games. > Level 3: Amazing Train, Cantilever Side Table, Halloween Guillotine Tray. Level 4: Bus, Desktop Bowling Game, Ferris Wheel, Shuffle Board Game, Medieval Castle Walls B, Mobile Crane B, Rusk Fortress, Straddle Carrier B, Zulu Main Battle Tank. Dowel-Based: Full Armor Gear Series Bull, Tank and Transporter.

WoodMarvels.com Volume 3: Evolution of Design
200 pages, ISBN 978-1-896369-49-5

This book features woodworking projects at five different skill levels with 3D rendered step-by-step assembly instructions. What makes our philosophy unique is that all our blueprints are completely measurement free, which means you can build them to whatever scale you wish using only the thickness of the wood as your guide. This book includes the following models: 1911 Ford Model T Torpedo Runabout, Light Table, Medieval Tower Bank, Shopping Cart, Submarine B, TV Bank, CH47 Chinook Helicopter, Easter Island Secret Storage Head Bookend, Home Gym, Mangonel, Medieval Castle Bank, P38 Lightning, Physics Trophy, Platform Crane B, Spring Basket, Stereo Bank, Army Transport Truck, Baliste, Ferris Wheel B, Flying Car, Mangonel B, Mars Explorer, Sydney Harbour Bridge, Semi-trailer Truck, Mars Rover, Medieval Castle and Sydney ANZAC Monument among others.

HowToChinese.com: A Practical Mandarin Chinese Course
412 pages, ISBN 978-1-896369-50-1

Learning a new language is fun -- and having a book guiding you every step of the way, written from a student's point of view, is a great start!

Experience Lucky's passion for teaching Mandarin Chinese to students around the world through her comprehensive and professionally produced online video series to help you build fluency in record time!

This is a fantastic book for both beginners and advanced students with a step-by-step learning process with each chapter building onto the next and multiple in-chapter exercises to reinforce new learnings along with real-life conversation dialogs and pinyin and simplified Chinese characters throughout.